Philosophical Chemistry
in the Scottish Enlightenment

A.L. DONOVAN

Philosophical Chemistry in the Scottish Enlightenment

THE DOCTRINES
AND DISCOVERIES OF
WILLIAM CULLEN
AND
JOSEPH BLACK

AT THE UNIVERSITY PRESS
EDINBURGH

Edinburgh University Press
22 George Square, Edinburgh
ISBN 0 85224 281 6
Printed in Great Britain by
W & J Mackay Limited
Chatham
UK

Preface

The culture of the Scottish Enlightenment is remarkable both for its breadth and its integrity, a combination which we cannot help but admire even while finding it difficult to comprehend. The cultural assumptions of the twentieth century encourage us to draw a sharp distinction between reflective and artistic activities on the one hand and empirical and practical pursuits on the other. No one would deny the greatness of philosophers such as Hume, Smith and Ferguson or of writers such as Burns, Boswell and Scott; but one can ask if these are the only high peaks in this marvellously elevated landscape. What of scientists such as Maclaurin, Black and Hutton and engineers such as Watt, Telford and Rennie? Were these men any less influenced by the leading ideas of the Scottish Enlightenment and are their achievements any less representative of the high culture of their age? Scholars interested in studying Scottish literary and philosophical figures have by and large assumed that a mere glance at the scientific and practical questions of the age will suffice. Those who study physicians, scientists and engineers have tacitly accepted this characterization of the Scottish Enlightenment and have usually set their subjects against a highly generalized background of advancing science and technology. This truce-like separation may leave us at peace with the assumptions of our own time but it blinds us to the extensive range and inner unity of eighteenth and early nineteenth-century Scottish culture. One of my purposes in writing the joint intellectual biography that follows is to demonstrate that the advancement of scientific knowledge was a matter of both philosophical and practical concern in eighteenth-century Scotland. Perhaps readers who find this demonstration convincing will want to think again about the relations between science, society and culture in both the eighteenth and twentieth centuries.

The question that drew me to the study of chemistry in the Scottish Enlightenment was really quite simple: How did Joseph Black happen to discover fixed air? (chapter eight). A careful reading of the available evidence soon revealed that Black acquired his fundamental chemical ideas during his years of study with William Cullen, and I was therefore obliged to reconstruct Cullen's chemical doctrines by piecing together the surviving manuscripts of his lecture notes (chapter six). A more thorough investigation of Cullen's life, and of Black's education (chapter seven), revealed that both men were guided by the particular assumptions and expectations of their culture. Having recognized that Cullen and Black should be seen as figures in the Scottish Enlightenment, I then turned to the history of that movement and studied the ways it came to dominate the institutions in which they were educated and with which they were later associated (chapters one, three, four and five). Of course Cullen's innovations in chemistry were more than mere deductions from the general features of the new philosophy and I therefore looked elsewhere for the specific concepts he employed (chapter two). As my understanding of his doctrines grew, I was better able to identify the basic chemical questions that underlay all of Black's research. I therefore decided to study the development of his theory of heat (chapter nine), and finally, to survey the influence that Black's theory exerted upon his disciples and upon the development of physical science beyond the borders of his country and his century (chapter ten).

A great deal more could be said about many of the topics touched upon in this book, such as the relationship between Cullen's medical and chemical doctrines, and Black's involvement in industrial chemistry. Rather than attempting a complete biography of either man, I have tried to identify and characterize the ideas which guided them in their roles as professors and chemists. If Cullen and Black appear in these pages as eloquent and humane teachers endowed with a deep sense of intellectual integrity, then I have successfully conveyed the impression they made upon their contemporaries and, more recently, upon me.

Acknowledging those who have provided encouragement and assistance is surely the most pleasant task of authorship. My training an an historian began when, having been awarded a National Defense Education Act Fellowship at Princeton University, I came under the tutelage of Lawrence Stone, David Bien, William Coleman, Thomas Kuhn and Charles Gillispie. I emerged from this intense indoctrination with a heightened interest in the historical relations between science and culture. I am therefore particularly grateful to Professor Gillispie for the example of his work and for the patience and skill with which he supervised my graduate studies.

Research for this book was completed during a most pleasant year in Scotland. Dr Eric Forbes and Professors Denys Hay and George Shepperson welcomed me as a Research Student in the University of Edinburgh while Mr C. P. Finlayson, the Keeper of Manuscripts, helped me survey the Black MSS in the University Library. Malcolm

Macnicol opened the library of the Royal Medical Society to me and I was received with equal courtesy at the Royal College of Physicians, Edinburgh, the Glasgow University Library, the National Library of Scotland, the Scottish Records Office and the Public Records Office of Northern Ireland. In Scotland I was also most warmly encouraged to study Cullen and Black by Dr W. P. D. Wightman and Dr Andrew Kent.

While writing this book I discussed my ideas with many scholars. For their encouragement and criticism I would particularly like to thank Allen Debus, Arnold Thackray, Robert Schofield, Henry Guerlac, Robert Multhauf, Frederic L. Holmes, Robert Siegfried, Owen Hannaway, Lester King and John Christie. My good friend Perry Cartwright helped me keep my eye on the target, and Stanley Mellon, David Jordan, James Cracraft and George Bugliarello, my former colleagues in Chicago, provided crucial intellectual and institutional support. When I needed to consult rare scientific texts in Chicago I visited that distinguished public repository of science and technology, the John Crerar Library. When writing itself became a difficulty I joined a workshop organized by the Parents' School and staffed by members of the Columbia College (Chicago) writing department. Personally and pedagogically that was a most rewarding experience. The final typescript was skilfully prepared by Debra Qureshi and Nancy Heskett.

Curtis Benjamin took a likely interest in the publication of this book and provided deeply appreciated advice. The West Virginia University Foundation, acting on recommendations from William Doherty and Donovan Bond, showed extraordinary faith in a new faculty member by providing the generous subsidy which has made publication of this book possible. The reader who recommended my book to Edinburgh University Press kindly suggested two corrections, which have been made, and Mr Peter McIntyre prepared the index. Final thanks go to my wife, Carolyn, who provided vital secretarial and editorial assistance at every stage without ever letting production of 'the book' become an obsession.

Arthur Donovan, Morgantown, West Virginia
June 1975

Contents

PART ONE

William Cullen and the New Science of Chemistry

To Pretend to enumerate in detail how much we [the medical students in the University of Edinburgh] owe to him [William Cullen], how far he has contributed to diffuse an ardour for improvement and spirit of liberal inquiry in Medicine, were to waste your time and our own. We cannot, however, forbear to mention, that Chemistry, formerly little studied here, has been by him new-modelled, and raised to the highest reputation; so that his plan has been successfully followed in different parts of the British dominions, and that here there are now more Students engaged annually in the pursuit of this collateral branch of Physic, than in any other part of it, except Anatomy. [From an 'Address of the Students of Medicine to the Right Honourable the Lord Provost, Magistrates, and Town-Council of the City of Edinburgh, 1766.' Reprinted in Thomson, *Cullen,* I. 155–60.]

[1]

Scotland and the University of Glasgow during the Early Years of the Eighteenth Century

William Cullen completed his undergraduate studies in Glasgow in 1727, the same year in which a new set of statutes was issued for the governance of that city's university. The Parliamentary Commission that drafted the new statutes hoped they would bring to an end the disruptions that had marred university life since the turn of the century. But these disruptions could not in fact be isolated from the network of conflicts which produced such stress in Scottish society during the early decades of the eighteenth century and of which they formed a part. The struggle within the university was intense and important. The questions that occupied the university community during the years in which Cullen was a student were not resolved in genial debates between advocates of different points of view. They led instead to earnest struggles between parties of men having fundamentally opposed convictions about human nature and the way in which the youth of the nation should be educated. These men did not think the future of the university was controlled by an inexorable fate which would be revealed by history. They believed that their efforts would determine what kind of education would be offered in the Scottish universities during the remainder of the century. The issues that they divided over were specific and important, and their resolution was not foreordained. The men with whom Cullen later allied himself when he joined the faculty in Glasgow were thus not bland liberal educators who had adopted the most agreeable and least demanding principles suitable for men who wished to make a career in the university. And Cullen himself, like the best of his colleagues, was no mere conformist; he was a daring and innovative teacher and a lecturer whose instruction sparkled with a vitalizing purposiveness that looked beyond the confines of the specific subjects he taught. To feel, as his students did, the force of his arguments and the power of his example, we must realize that he advocated a significant break with the past, and to become aware of the magnitude and nature of that break, we must become familiar with the character and beliefs of his predecessors and opponents. Since Cullen began both his university education and his teaching career in the University of Glasgow, an investigation of the battles which were wracking that institution while Cullen was attending it will help us understand the social and intellectual issues which were to occupy him throughout his long life.

William Cullen was born in 1710 in Hamilton, a town that lies in the Clyde River valley midway between the falls at Lanark and the port of Glasgow.[1] The family that William was born into was well connected and comfortably situated in that pre-industrial centre of local commerce and agriculture. William's father was an attorney, a factor for the Duke of Hamilton, and the proprietor of a small estate which the Cullens had held for many generations. As professionals in a nation whose economy was still primarily agricultural, the Cullens, like the nobility, moved easily between city and country. William's older brother, the first son, received a classical education in the University of Glasgow and then went on to study law in Edinburgh. And William at various times in his extended career as a student and physician lived comfortably in Glasgow, Edinburgh and London, as well as in his home town of Hamilton.

Although we have almost no information about William's early years, what we know of his later life indicates that he enjoyed a happy childhood. In 1751, while writing to his brother about a farm that had belonged to the family, he volunteered that the 'place where I spent so much of my infancy, I have always had a particular affection for'.[2] This affection evidently extended to other aspects of his youth, for many features of Cullen's own life seem to duplicate the circumstances of his early years. Coming from a family of seven sons and three daughters, he fathered seven sons and four daughters of his own. Perhaps a more reliable indication of his attitude towards his family can be found in his decision, following his father's death, to return from London to oversee the education of his younger brothers and sisters. He also chose to travel down the avenues of opportunity opened by his family's connections when fashioning his own career. Like his father, who had been employed by the Duke of Hamilton, William served the Duke professionally, as a physician, and as a magistrate in the town of Hamilton. In the choices he made during his years of maturity and achievement, Cullen revealed his own favourable evaluation of the experiences of his youth.

The Cullen family's ability to combine a country easiness with an awareness of responsibility and place was no doubt the source of William's characteristic generosity and good sense. Although ambitious, he never sought personal honour alone, and while successful, he did not neglect the social responsibilities which went with the positions he obtained. He involved himself in a wide variety of contentious institutional and philosophic affairs, yet even when continuously attacked, as he was by John Brown, a former pupil and tutor of his children, he did not attempt to destroy his opponent or present his defence before the general public.[3] While it is tempting to say that this self-confidence and moderation are simply manifestations of an upper-class consciousness, in Cullen they were clearly not accompanied by the arch and distant sense of superiority usually associated with this term. Towards his students Cullen 'was so attentive, and the interest he took in the private concern of all those who applied to him for advice was so cordial and so warm,' reports former student James Anderson, that they found it 'impossible . . . not to be enraptured with a conduct so uncommon and so kind'.[4] Ever practical as well as concerned, Cullen inaugurated the custom whereby the medical professors in Edinburgh attended their ill students without taking fees.[5] Lacking any evidence to the contrary, we may assume that the adolescent Cullen who, around 1724 completed Grammar School in Hamilton, possessed, if only in latent form, the same characteristics that his colleagues and students found so noteworthy in his later years. Ambitious and self-confident, intelligent and interested in a broad range of subjects, generous yet shrewd, Cullen was ready to take on a challenge which was national in scope and philosophically profound. He found such a challenge and defined his own position in relation to it while living through the then turbulent experience of being a student in the University of Glasgow.

The troubles that disturbed the University of Glasgow in the early eighteenth century were a direct outgrowth of the circumstances by which the Protestant Reformation came to be established in Scotland. The 'College of Glasgow', like the University of Saint Andrews and King's College in Old Aberdeen—the other Scottish universities that date from the fifteenth century—was originally founded as a bishop's college. These other universities, however, as well as the Earl Marischal's College in New Aberdeen and the 'Tounis College' in Edinburgh, which date from the sixteenth century, are located on the eastern coast of Scotland, in or near seaport cities having easy and constant communication with the continent and with each other.[6] In these eastern cities the Reformation appeared as part of an international reconstruction of the Church and as a means of neutralizing the ever-threatening dynastic designs of the French. In western Scotland the independence of the national church and its relation to the more rural segments of the population were of greater importance. Lollardy, a popular religious movement drawing on the teachings of Wycliffe and Hus, had spread from England into western Scotland during the fifteenth century and had struck root in Ayrshire long before the doctrines of Luther and Calvin entered Scotland through her eastern, North Sea ports.[7] So long as the reformed national church was not under siege, its leading ministers, such as Andrew Melville, gravitated towards Saint Andrews, the ancient ecclesiastical capital of Scotland, and Edinburgh, the political capital. But when forced to defend themselves against the episcopal designs of the early Stuarts, the leaders of the kirk sought refuge among the more isolated fundamentalists of western Scotland. Although the General Assembly of the Church normally met in Edinburgh, the meeting of 1638, at which the Covenanters organized the resistance that touched off the English civil war, was held in Glasgow. Thus while all the Scottish universities were buffeted by the political and religious upheavals that disturbed the nation during the sixteenth and seventeenth centuries, the University of Glasgow, because of its geographic location, became more deeply involved than the others when the struggle for national independence was at its most desperate.[8]

The Scottish universities did not simply serve as forts or recruiting grounds for the captains of reformation, they were central to the new Scottish concept of the church. In 1560 the Scottish Parliament passed acts which destroyed the authority of the Pope and the practice of the Roman rite in Scotland. Unlike the English, who in the Elizabethan settlement of 1559 had transferred the papal privileges to the Crown and established a single form of public worship, the Scots did not place their reformed church under the direct control of the state. A Confession of Faith which followed the Swiss Confession was adopted as the sovereign authority for the new church while the structure of the kirk was not specified. The Scottish kirk therefore organized itself and its people according to its own *Book of Discipline*. During years of struggle, adaptation and compromise the Presbyterian form of governance developed and acquired its final shape. Ministers were to be elected by congregations and the members of the church, meaning the whole Scottish nation, therefore had to be adequately instructed. Having denied itself the central authority of an episcopal structure, the kirk placed its future in the hands of its members. The Scottish church therefore felt obliged to organize a comprehensive system of national education. 'Provision was made for primary, secondary, and university education; the importance of vocational training was recognized; the student's career was checked by examination and inspection; [and] the curricula were so arranged that the study of theology should be its very apex.'[9]

To carry out this educational programme the kirk needed to have at its disposal all the financial resources of the unreformed church. Needless to say, the government was more willing to recognize the institutional independence of the kirk than to endow it with the spoils of the medieval church. Although the great educational plan was never fully realized, its historical importance should not be underrated. The plan, formulated as a reformationary ideal, came to be recognized as one element of Scottish liberty, and the kirk, during its long struggle for independence, never abandoned its original educational programme. Thus throughout the period from 1560 until 1690, loyal Scots considered their universities the most advanced seats of learning in a system of education which was both national and religious. Evidence of the importance attached to education can be found throughout the era of struggle. Following the death of John Knox, Andrew Melville left Geneva and returned to Scotland to carry on the Reformation. In 1574 he became Principal of the University of Glasgow. Under his guidance the university was remodelled and committed to the militant defence of the kirk's independence, a commitment formalized in the *Nova Erectia* of 1577.[10] Although Melville later returned to his Alma Mater at Saint Andrews, the University of Glasgow continued to serve as a key institution for the kirk. During the English civil war, when the existence of the university was gravely threatened by a loss of endowment, the kirk found funds to sustain it. Clearly this faith in education, which was a distinctive feature of the Scottish Reformation, contributed mightily to the flowering of the Scottish universities during the eighteenth century.

When the firmly anti-papal and anti-French William of Orange claimed the throne vacated by James VII and II, the fate of the Episcopal church, which the restored Stuarts had imposed upon the Scots, was sealed. Confident of his ability to contain the rebellious Jacobites, William allowed the Scottish Parliament meeting in 1690 to re-establish Presbyterianism as the national religion. Many Episcopal ministers and professors were subsequently excluded from their positions and the kirk began the long job of reconstruction which awaited it.[11] By 1696 it had demonstrated both a victor's intolerance, by hanging Thomas Aikenhead for blasphemy, and, in passing an act compelling every parish to provide a schoolhouse and a salary for a schoolmaster, its traditional concern for education.[12] At the same time a Commission on universities, created by the Scottish Parliament and dominated by Presbyterian ministers, began planning the revitalization of higher education. Before the seventeenth century had ended, however, forces which the kirk, acting through the Commission, could not control had begun to transform the Scottish universities, and by the end of the first decade of the eighteenth century even the kirk's right to control the University of Glasgow had been brought into question.

The men who emerged as leaders in the kirk at the end of the seventeenth century had suffered great privation and abuse during the years of episcopacy and they assumed that once victory had been gained, they would be allowed to establish in Scotland a society which conformed to their sixteenth-century notions of justice and order. Among the older ministers, who had spent their years of harassment wandering in the fields and glens, learning and forgiveness had degenerated into fanaticism and bigotry. Among the younger men, except for the few who had studied in Holland, the Lord's word sufficed. These men knew little and thought less of those intellectual achievements of the seventeenth century which we now regard as the chief glory of that age. Having been cut off from the mainstream of European culture, they had nursed their prejudices and awaited their moment. When that moment arrived, it became increasingly evident to many Scots that the prevailing temper of the Presbyterian ministers was ill-suited to the demands of the new century.

During the first few decades in which the kirk enjoyed the dangerous pleasures of establishment, it appeared that the years in the wilderness would be amply rewarded. 'Civil and ecclesiastical authorities went hand in hand in disciplinary measures,' Graham tells us. 'Acts of Parliament, resolutions of Town Councils, and decisions of Sheriffs supported the Church.'[13] It also seemed that in the University of Glasgow the old pattern of collegiate life would be successfully carried forward into the new century. Before the eighteenth century, instruction in all the Scottish universities was in the hands of regents. In Glasgow there were four, each of whom taught their charges the four traditional subjects—Greek, logic, ethics, and natural philosophy—during their four successive years at college.[14] The only other member of the faculty in Glasgow was the professor of divinity, who offered advanced instruction to candidates for the ministry. In addition to teaching the arts candidates, the regents supervised their communal life with a prying and sanctimonious diligence which could only have been willingly tolerated by the Elect.[15] Although neither the method of instruction nor the mode of life was suitable to the new age, many years passed before they were displaced in Glasgow.

The years of wilful opposition to the world around them had also transformed the surviving Presbyterian ministers into uncompromising Providentialists. In so far as they recognized the existence of natural laws, they thought of them as conventional arrangements of Providence which could be lightly changed, stopped, or reversed. Prayer and the intervention of a saintly pastor were the means by which one influenced the natural world.[16] Not only was the power of nature held to be insignificant before the power of God, it was also believed that nature itself was infested with that evil which is the very opposite of God's goodness. Thus at the beginning of the century William Law, minister of Crimond, 'was deposed for saying in a Synod sermon that virtue was more natural to the human mind than vice'.[17]

The differences between such attitudes and the assumptions and procedures of seventeenth-century natural philosophy could not be blinked and the leaders of the kirk, ever fearless in their self-righteousness, threw themselves upon this new adversary with more gusto than wisdom. In 1696 the General Assembly of the Church passed an act warning 'her members against the "deists" who "make it their business to overturn and ridicule true and pure religion", and especially those who assert that "there must be mathematical evidence for each purpose, before we can be obliged to assent to any proposition" and that "natural light is sufficient to salvation". Clearly,' Cameron adds after summarizing the Assembly's warning, 'a rationalism, hitherto unknown, had begun to make itself felt and the church was in no doubt as to its own attitude.'[18]

Attempts to reform philosophy within the universities also met a similar response. The curriculum of studies in the University of Glasgow had not been significantly altered since the middle of the sixteenth century, when Andrew Melville had carried through the last general reform, and the Renaissance texts he introduced had continued to serve as the basis for instruction throughout the seventeenth century.[19] Although isolated individuals had lectured on aspects of seventeenth-century thought in the Scottish universities, the entire philosophy curriculum was clearly out of date and ripe for renovation. To remedy this situation, the Parliamentary Commission on Universities directed that a comprehensive philosophy course be drawn up and taught in all the universities. The representatives of the universities appointed to carry out this task diligently examined 'the severall courses of Philosophy which are extant,' but they found them all inappropriate. They could find nothing, they said, 'that wee can recomend or think fitt to be taught; for not only we cannot think on any one wholl course of Philosophy wryte by any one man, but wee cannot find any one entire pairt of Philosophy which we can recommend as sufficient to be taught, for many of them ar wryte by popish professors and therin they cunningly insinuate the heretical tenets mixing them with their philosophy which ar not so easyly discerneable by the youth.' The representatives also faulted the Cartesian systems of philosophy for being unhistorical and pedagogically useless. 'For Cartesius, Rohault and others of his gang beside what may be said against their doctrine they all labour under this inconveniency that they give not any sufficient account of the other hypotheses and of the old philosophy which must not be ejected, and weer never designed to be taught to students.'[20] Finding no text that would satisfy the needs of their universities, the Commissioners drew up plans to have a suitable text written. Parts of the proposed text were completed, but they aroused such a strong philosophical and theological reaction among the Scottish regents and professors that the reformed philosophy course was never completed. In philosophy, the Scots were evidently not yet prepared to practice the art of graceful compromise.

The forces which gradually enabled the University of Glasgow to turn its back on the doctrinal struggles of the previous century were in large part released by the desperate need for more ministers. Only about sixty of the Presbyterian ministers who had been among those ejected from their parishes in 1662 were still alive in 1689. These old men, having been returned to their former positions, provided the leadership for the kirk during the early period of restitution. Their antiquity and their years of campaigning gave them unchallenged sway over the other seventy-six ministers, who had been indulged to preach in 1687, and the forty-three elders with whom they met in the first General Assembly in 1690. It was these men who received from the Scottish Parliament the power 'to try and purge out all inefficient and scandalous and erroneous ministers by due course of ecclesiastical process and censure'.[21] The zeal with which these embittered warriors went about their task both angered more accommodating Scots and drastically reduced the number of active ministers. The Scottish Parliament was finally compelled to curb these inquisitors. In 1694 a policy of toleration was imposed upon the kirk and the General Assembly was told not to disturb Episcopalian ministers who agreed to take the oath of allegiance, subscribe to the Confession of Faith, and conform to Presbyterian rule. Nevertheless, the purge, though less extensive than the Presbyterians would have liked, created a serious shortage of ministers. Of the 900 clergy, 300 Episcopalian incumbents were left undisplaced and the remaining 600 were thrown out of their posts.[22] Thus even the moderated purge left several hundred livings vacant, for there were not nearly enough Presbyterian ministers available to replace all the dispossessed prelates. At the end of the seventeenth century, therefore, the kirk needed a massive infusion of well-trained ministers. Naturally it looked to the University of Glasgow.

To meet this need, the University of Glasgow increased its enrolment considerably. Before the English civil war, the university had been a small institution of about 120 to 150 students. By 1696 it had grown to 250 students and by 1702 the enrolment had reached 400.[23] This dramatic increase was accompanied by unplanned structural changes in the organization of the university; these made possible further developments which older members of the kirk must have considered unwelcome. As enrolment increased the collegial pattern of university life was not extended to include the new wave of students. Therefore the common table for meals had to be abandoned in 1694 and students soon began either lodging in town or they became boarders in the homes of their teachers. The breakdown of the scholastic community was accompanied by the collapse of the communitarian discipline that had held sway in the old college. Glasgow students, keenly aware of the profound changes transforming their own institution and their surrounding society, became notoriously factious and belligerent. Had their high spirits been opposed by a self-disciplined faculty and a purposive administration, their exuberance could have been contained, but the expanding faculty was itself divided and the administrators were torn by the conflicting demands being made upon the university. After the need for additional ministers eased during the early eighteenth century, the university continued to grow. As one astute Scot observed in 1733, the desire to gain social advancement through membership in the ministry was causing the number of ministers to increase far beyond the needs of the nation: 'At present,' he stated, 'we are over-stocked with young clergymen. The reason of that seems to be that many people out of vanity, because some of their relations are ministers, will educate a son in this way to push him onto a rank in the world above his birth and condition.'[24] An expansion undertaken to fulfil a traditional commitment to the kirk had irreversibly transformed the old college. The walls which once contained a community had been breached and students continued to pour in. By the 1790s there were over 600 students attending the university and the number continued to increase during the last decade of the century.[25]

There were, of course, other changes which weakened traditional restraints within Scottish institutions. A new generation was coming to maturity during the first third of the eighteenth century, a generation which was unimpressed with the heroics of their fathers. For most of these young men, at least for those who attended the universities, the opportunities opened by the Union of 1707 were more important than the defence of reformationary truths. They realized that the religious and political battles which had occupied their forebears had been brought to a close by the Act of Settlement, by the establishment of the Protestant Succession and by the Parliamentary Union of England and Scotland. As for the machinations of the increasingly feeble French monarchs and the trumpetings of the deposed Stuarts, the leading Scots of the new generation had no intention of being taken in by a conservative alliance of Catholics and Jacobites who wished to substitute a counter-revolutionary crusade for the advantages of union. Scotland had achieved political stability, and those not born to rule were encouraged to exercise their talents in other fields of endeavour.

Socially and economically, the Union brought the Scots into a burgeoning commercial empire. Ambitious young Scots could now leave North Britain and seek out the opportunities available in London without surrendering the rights and identity which were theirs by birth. Naturally the Scots continued to reveal their origins through their northern idioms and rustic pronunciation. But language and style, like manners and morals, could be polished to acceptability through careful training, and it was not long before the cosmopolitan Scots were importing tutors to teach them standard English. For those who remained at home or returned to Scotland to make their careers, the Union created opportunities for enrichment through the expansion of native production. Free to trade in all the markets controlled by the English, the Scots had to determine which of their products could be made competitive. This complex question involved considerations of natural resources, technical innovation, and social transformation. Scottish industry, at least those parts of it which could compete in the international market, had to be modernized and made responsive to demand. We now know that Scotland was being prepared for the Industrial Revolution, a transformation in the means of production overshadowing all the changes that took place during the first two-thirds of the century. But the driving desire for economic improvement and growth, the same emphasis which later encouraged the rapid exploitation of centralized sources of power and the factory system of production, was being championed in Scotland, as in England, from the beginning of the eighteenth century. Conscious of their past, confident of their future, those Scots who were in a position to profit from the political, economic and cultural implications of the Union welcomed the dawning century.

Although the conservative theology of the kirk set the tone for philosophic instruction in Glasgow well into the eighteenth century, by the end of the seventeenth century Gershom Carmichael was introducing ideas which spoke more directly to the interests of the new generation of students. In 1694 Carmichael left the University of

Saint Andrews to become a regent in Glasgow. Because he published an edition of Puffendorf's *De Officio Hominis et Civis*, which he used in his ethics class, Carmichael is usually remembered as a moral philosopher. Yet even in his moral philosophy he was anxious to acknowledge the new learning of the seventeenth century. He argued that 'no person liberally educated can be ignorant that within the recollection of ourselves and of our fathers philosophy has advanced to a state of progressive improvement hitherto unexampled; in consequence partly of the rejection of scholastic absurdities and partly by the accession of new discoveries.'[26] As a regent for over three decades, Carmichael was obliged to teach logic and natural philosophy, as well as moral philosophy. In all his classes he introduced his students to ideas which we now consider the commonplaces of seventeenth-century thought, but these ideas appeared as powerful new injunctions when he first taught them in Scotland. He spoke highly of 'ye two great Hinges of Natural Philosophy, or rather ye two constituent parts of it . . . Mathematical Demonstration and Experiment,'[27] and he introduced his pupils to the Cartesian-Newtonian debate by having them study Samuel Clarke's translation of Rohault's text on physics.[28] In the breadth of his philosophic interests, in the content of his doctrines, in his commitment to teaching and in the importance of his disciples Carmichael was so like the more famous Scottish philosophers who became prominent during the later years of the century that Sir William Hamilton believed he should be considered the founder of the Scottish School of Philosophy.[29]

The general expansion of the universities and the character of the education sought by the students who flooded into them during the eighteenth century made the appearance of new forms of instruction inevitable. One had to be a specialist to understand the philosophical achievements of the seventeenth century. Without a knowledge of mathematics and a familiarity with the experimental and observational data being discussed, one could not master the arguments of a Boyle or a Newton. The interests of the students also encouraged academic specialization, for they wanted useful knowledge, knowledge that was founded upon empirical investigation and that suggested practical innovations. The regents had been able to offer adequate instruction so long as the goal of education had been the cultivation of humanistic learning and consideration of what constitutes right conduct. But as the model of the Dutch universities suggested, competent specialized instruction could only be obtained by parcelling out the various fields of knowledge to teachers who were themselves specialists. Thus as the number of students increased the new teaching positions which had to be created were given to professors rather than to regents, and by the end of the eighteenth century all the Scottish universities had abandoned the regenting system of instruction.

The old system held on for many years in Glasgow. Regenting was abolished in the University of Edinburgh in 1708 but it survived in Glasgow until 1727, despite the growing number of professorships instituted there. In 1691 a professor of mathematics was appointed in Glasgow and four years later it was suggested that the teaching of Greek be 'fixed' as the responsibility of a single professor. The faculty welcomed this first step in the abolition of regenting, but they questioned the wisdom of piecemeal reform. 'With all submission,' they said, 'the pluralitic of our number doe think the fixation of all classes in everie Universitie verie necessar.'[30] Principal Stirling, who 'ruled' the university from 1701 to 1727, was particularly effective in appealing to the interests of the powerful individuals who provided the endowments for additional positions. Between 1704 and 1718 he established new chairs in Greek, humanity (Latin), oriental languages, ecclesiastical history, civil law, medicine, and anatomy.[31] In 1727 three additional chairs were added, but in name only, when each of the remaining regents became a professor of philosophy. A chair of astronomy, the only other professorship created in the University of Glasgow during the eighteenth century, was added in 1760. Thus by 1727 the shift from the regenting system of instruction to the professorial system had been completed and the composition of the faculty had been permanently altered. The new method of teaching transformed the university from a college serving primarily as a seminary into a comprehensive university. Naturally it took time to build a professionally distinguished faculty. Nonetheless, the opportunity and the promise were there, and Cullen was following the pattern laid down during the early decades of the eighteenth century when he returned to the university in the 1740s to occupy the newly created lectureship in chemistry.

The members of the Glasgow faculty were understandably concerned about the pedagogical implications of turning all instruction over to specialists. Even before the end of regenting enrolment in the philosophy courses began to fall as more and more students sought degrees in theology, law and medicine, for which graduation in arts was not a prerequisite.[32] Regents had also received graduation fees from their students, and according to Graham, when regenting was abolished 'and each professor no longer carried on his flock of pupils year by year to the end, he had no pecuniary interest in laureation, and the number of graduates sank in some colleges to zero, and scholars chose or attended what classes they or their fathers pleased.'[33] The 1727 statutes for the University of Glasgow reveal the anxiety of those who saw in this neglect of philosophy the decay of education. When the three philosophy classes were fixed, and the remaining regents retitled professors of philosophy, the hours for the philosophy lectures were staggered and the restrictions upon attendance were minimized so that as many as possible could receive this instruction. The statutes specified that 'the three Philosophy classes [shall] have each of them a different prelecting hour. . . . These hours may be attended by students without gowns [i.e., not MA candidates], provided they have passed their course, or are students of Divinity, Law, or Medicine, and that any person, not a student as is said may attend the lessons of Experimental Philosophy without a gown.'[34] The first set of lectures on experimental philosophy offered after the passage of these statutes was taught by Robert Dick, who became professor of natural philosophy when regenting ended.[35] The regents in Glasgow did their best to turn the shift to professorial instruction to the advantage of philosophy. The breadth of interest which regenting required of its teachers was forsaken so that faculty members could concentrate on mastering the new knowledge. Whether these new fields of knowledge could be linked together in a meaningful way and whether the implications of a philosophy which forged such links would be acceptable to all members of the university remained, in 1727, open questions. There can be little doubt, however, that the seriousness of purpose and intense moral concern characteristic of Scottish philosophy in the eighteenth century sprang largely from the conservatism of the university faculties. Regenting was abolished, but the didactic goals which had initially led to the institution of regenting were not forgotten until the last of the regent's disciples had retired from their professorial positions.[36]

In Glasgow during the first quarter of the eighteenth century the authorities were faced with a growing university and declining student discipline, a changing curriculum and an expanding faculty, the beginnings of a new outlook in philosophy and the reconstruction in theology it implied. Who was to bear final responsibility for resolving the divisions and tensions which accompanied these changes? Principal Stirling, who was also a minister, said it was his responsibility to protect the autonomy of the university while settling its internal disputes. The faculty, with the support of the students, said that they had the sole right to set policy for the university. The struggle between the two factions became public when they quarrelled over the procedure to be used in electing the rector of the university.[37] While this battle owed much to the conflicting ambitions of several colourful individuals, the legal question of who had final authority in the university was an important one. Since the contesting figures could not resolve the matter among themselves, the settlement had to be imposed by external authority.

A Parliamentary Commission was appointed in 1717 but it was dominated by churchmen, controlled by Stirling, and unable to force a settlement upon the faculty and students. The students responded to the commission's proposals by taking their case, which was founded on ancient privilege, to law. After failing to gain satisfaction in the Scottish Court of Session, they carried their appeal to the British Parliament. By that time the dispute had immobilized the university and was causing acute embarrassment to the Scottish members of the recently united Parliament. A new Parliamentary Commission was therefore formed, but this committee, unlike its church dominated predecessor, contained leading Scottish public figures and four lawyers. In 1727, after an extensive examination of university affairs, the committee issued a report denouncing Stirling's autocratic governance and turning primary control of the university over to the faculty. A new set of statutes was also issued. They codified the faculty victory, constituted the university as a closed corporation not directly responsible to the kirk, and remained the governing regulations of the university until 1858.

In 1727 the legal battle between the kirk and the university was resolved by separating the two belligerents, but this settlement clearly did not deal with the most basic problems raised by the ongoing transformation of the university. Although freed from the control of the kirk, the university had not redefined itself and its relation to Scottish society in a way which justified its independence. Claiming to be useful simply was not enough. The university could no more isolate itself from the profounder concerns of Scottish society than theology could be isolated from philosophy, and the kirk, realizing that the legal settlement had not clarified this central relationship, continued to press its traditional claims.

In 1714 John Simson, the professor of divinity, was accused of teaching Arminianism. His case was heard before the General Assembly of the Kirk where, after defending his orthodoxy, he was acquitted. In 1726 Simson was again accused, this time the charge being Arianism, and again he faced the church courts. In both cases Principal Stirling vigorously defended his professor, insisting that the university should be the sole judge of its members. Upon Stirling's death in 1727, however, Simson was found guilty and ordered not to teach. During the next few years the general issue raised by his case was ignored. The new Principal accepted the situation as he found it and both Simson and his successor died before the dispute was again forced into the open.

Cullen completed the arts curriculum at the University of Glasgow in 1727. Like the philosopher Francis Hutcheson and like his own pupil Joseph Black, Cullen avoided the regulations and expenses entailed by matriculation in arts and therefore did not receive this first degree. But, as we have seen, students who were not candidates could attend the arts classes and Cullen no doubt studied the three branches of philosophy—metaphysics, ethics, and natural philosophy—during the last years in which they were taught by regents. While Cullen evidently enjoyed the freedom which enabled him to avoid matriculating in arts, he later came to question the wisdom of allowing students to slight philosophy. In the 1760s, while expounding on the need to study logic, he told his students in Edinburgh that 'I cannot but lament that students of medicine in the University are not obliged to go thro certain preparatory branches of learning, for many of the gentlemen who come here are so deficient in that respect, that it is impossible for them to make tolerable progress in the several parts of Medicine.'[38] Clearly in the professionally oriented universities many students had done what the regents feared they would do, that is, neglect philosophy.

We know the names of two of Cullen's teachers in Glasgow. The only surviving official record bearing Cullen's name and dating from his early years at the university is a list of students enrolled in professor Robert Simson's mathematics class for 1727.[39] Mathematics was a separate subject, open to arts students but not one of their required courses. Professor Simson, one of the most learned yet engaging figures on the faculty, stressed the geometrical approach to his subject, an emphasis which helped make geometry the keystone of Scottish mathematics throughout the eighteenth century. Apparently Cullen did not study under either of the university's two professors of medicine. Neither John Johnstoun, appointed professor of medicine in 1714, nor Thomas Brisbane, appointed professor of anatomy and botany in 1720, lectured in the university. The neglect shown by these 'inert professors' did not, however, prevent eager students from securing a medical education in Glasgow, for one could enter the profession, as Cullen did, through apprenticeship. Cullen studied with a Mr John Paisley, a serious man known for his large collection of medical books, and the two doctors remained friends long after the apprenticeship had been completed. This happy and rewarding relationship, established during Cullen's early life, was, like the circumstances of his family life, duplicated in his later years. Cullen, a serious man with a large collection of chemistry books, accepted Joseph Black as an apprentice at a time when no advanced work in chemistry was offered in the University of Glasgow, and the two men remained close friends thereafter.

We know nothing else of Cullen's activities during his years as a student in Glasgow. His religious opinions, such as they were, were always marked by mildness and toleration rather than fervour or conviction and he no doubt avoided the sectarian battles which swirled around him. He would probably have agreed with the comment made by the Duke of Hamilton who, upon hearing that the surviving Presbyterian ministers had been empowered to examine all clergy holding Scottish livings, declared 'what is this but instead of fourteen bishops, to give unlimited power to fifty or sixty Presbyterian ones, from whom the Episcopal clergy can expect little justice or mercy?'[40] Cullen did not feel torn between the old truth and the new for he knew where he stood. The task he came to recognize as his, and which occupied him throughout his career, was to articulate and advance the best of modern natural philosophy and to make it the living centre of the institutions in which he taught.

[2]

English Natural Philosophy after Newton

William Cullen left Glasgow in 1729. Nineteen years old and backed by a prosperous and well-connected family, he decided to continue his education in the farther reaches of the British Empire. He travelled first to London, where he obtained an appointment as surgeon general aboard a merchant ship captained by one of his relatives. After nearly a year in the West Indies, where he botanized a good deal, he returned to London and apprenticed himself to an apothecary. Cullen's abiding interest in the classification of plants and the study of their medicinal uses was evidently well developed during this early period of his life. This interest no doubt suggested further questions to his philosophic mind: How do these medicines act and how can their action be compared to the newer chemical remedies finding increased favour among physicians? Perhaps questions such as these turned Cullen's attention towards chemistry while he was struggling to master the practical aspects of the apothecary's art.

Where and when did Cullen acquire his fundamental ideas about chemistry? As a student of natural philosophy and as an apprentice first to a physician and then to an apothecary, Cullen obviously had to gain some understanding of chemistry during the course of his education. That he knew something about chemistry is therefore hardly surprising, but the importance he attached to the subject and the way in which he explained its phenomena were unique in Scotland at that time. Although Cullen may have been entirely self-educated as a chemist, this seems unlikely. When he returned to Glasgow in the 1740s and began lecturing on chemistry, he presented a comprehensive and carefully considered set of doctrines which drew upon the whole range of chemistry and natural philosophy developed during the preceding century. From the very outset of his career as a teacher Cullen argued that chemists should study a certain set of phenomena, that their theories are related to those of natural philosophy in a certain way, and that chemistry has a particular role to play in the improvement of the practical arts. Such breadth of conception and clarity of purpose indicates, I believe, that Cullen took as his own the doctrines of one or two persuasive teachers, men who located the subject within a large context and provided the conceptual framework which guided his research and enabled him to present a coherent and comprehensive system of chemistry in his lectures. The conception of chemistry which Cullen advanced when he began teaching in Glasgow was, in the opinion of his colleagues and students, so novel that it is highly unlikely he was presenting the ideas of some Scottish master whose identity remains unknown to us. It seems far more likely that the teachers who instructed Cullen on the nature and importance of chemistry were those English chemists and natural philosophers who offered public lecture courses in London during the years Cullen studied there.

English natural philosophy was entering a period of reorientation when Cullen arrived in London. Isaac Newton had died in 1727, and the sudden removal of his authoritative presence had thrown his disciples back on their own resources. While he lived, Newton dominated London science. As president of the Royal Society he chaired the meetings, set the questions to be discussed and directed which experiments were to be performed. As the author of an exact and comprehensive system of natural philosophy he attracted brilliant minds anxious to extend and refine his ideas. And as a combative personality who relished polemics, he led a devoted band of disciples in doctrinal controversy.[1] So long as Newton lived, the pursuit of natural philosophy in London was dominated by those who were admitted to the Newtonian circle. With his death the spell was broken. After 1727 the tensions and ambiguities within Newtonianism, which had previously been treated as relatively unimportant, became starting points for diverging developments in natural philosophy. The era of Newtonian*isms* was beginning, an era in which the decreasing coherence of natural philosophy created an openness which made an independent science of chemistry conceivable.

Of course not all the London Newtonians saw the death of their master as the end of an era. Many of them continued to believe that it was only a matter of time before a truly adequate philosophy of nature would emerge from the patient extension of Newton's ideas. John Desaguliers was one of the disciples who kept faith in the future of the corpuscular philosophy.[2] As Curator of Experiments for the Royal Society during Newton's last years, he had loyally provided experimental demonstrations designed to confirm the theoretical interpretations judged orthodox by his master. Desaguliers was also the most successful private lecturer in London for over thirty years. In his course, which Cullen well may have attended, the interested public could learn of the latest research being carried out in the Royal Society. His lectures were also published in several editions, the earliest appearing in 1719, and Desaguliers described his experiments at greater length in the many papers which he wrote for the *Philosophical Transactions*. In Desaguliers' lectures and articles Cullen had at hand the best available introduction to the mainstream of Newtonian natural philosophy.

Desaguliers was a creative theorist as well as an ardent Newtonian and during the years following Newton's death he began a re-examination of the foundations of natural philosophy which must have infused his lectures with an appealing freshness. His new ideas were stimulated by his reading of the book titled *Vegetable Staticks*, which Stephen Hales published, with Newton's imprimatur, in 1727.[3] To determine the direction of Desaguliers' thought at the time that Cullen was studying in London, we must therefore first examine Hales' investigation of fixed air, which occupies nearly half of the *Vegetable Staticks*, and discover how his conception of air differed from that of the early Newtonians.[4]

During the seventeenth century natural philosophers used the recently invented barometer, thermometer and vacuum pump to examine the extraordinary 'spring of air'. Air's great elasticity convinced many experimenters, such as Robert Boyle and Francis Hauksbee, that elasticity is in fact a unique property of air, and being mechanical philosophers, they suggested that the particles of air are themselves springy, like tufts of wool or watch springs. This hypothesis obviously could not be incorporated into a system of mechanics which postulates the existence of forces acting at a distance, and Newton therefore asserted, in the thirty-first query, that the fixation and release of air, the most extreme examples of its elasticity, 'seems unintelligible, by feigning the Particles of Air to be springy and ramous, or rolled up like Hoops, or by any other means than a repulsive Power'.[5] He then distinguishes between two kinds of vaporization.

> The Particles of Fluids which do not cohere too strongly, and are of such a Smallness as renders them most susceptible of those Agitations which keeps Liquors in a Fluor, are most easily separated and rarified into Vapour, and in the Language of the Chymists, they are volatile, rarifying with an easy Heat, and condensing with Cold. But those which are grosser, and so less susceptible of Agitation, or cohere by a stronger Attraction, are not separated without a stronger Heat, or perhaps not without Fermentation. And these last are the Bodies which Chymists call fix'd, and being rarified by Fermentation, become true permanent Air; those Particles receding from one another with the greatest Force, and being most difficultly brought together, which upon Contact cohere most strongly. And because the Particles of permanent Air are grosser, and arise from denser

> Substances than those of Vapours, thence it is that true Air is more ponderous than Vapour, and that a moist Atmosphere is lighter than a dry one, quantity for quantity.[6]

Newton here argues that the particles in volatile substances, such as water, are relatively light and can be so easily agitated by a moderate heat that they fly off in the form of an easily condensed vapour. But when the particles of fixed substances, which are solid and gross, are so rarified by great heat or chemical action (fermentation) that they form a vapour, then these particles have acquired a repulsive force which keeps them apart and makes their vapour a 'true permanent Air'.

Hales' investigation of fixed air, as he stated in the introduction to his chapter on the Analysis of Air, was designed to confirm Newton's suggestion concerning the existence of repulsive forces in nature. 'The illustrious Sir *Isaac Newton* (query 31st of his Opticks) observes,' Hales said,

> that 'true permanent Air arises by fermentation or heat, from those bodies which the chymists call fixed, whose particles adhere by a strong attraction, and are not therefore separated and rarified without fermentation. Those particles receding from one another with the greatest repulsive force, and being most difficultly brought together, which upon contact were most strongly united. And query 30. dense bodies by fermentation rarify into several sorts of Air; and this Air by fermentation and sometimes without it, returns into dense bodies.' Of the truth of which we have evident proof from many of the following Experiments.[7]

Hales felt that new experiments were needed to establish the truth of Newton's conjecture because several prominent early disciples of Newton's, men such as John Freind and John Keill, had been unwilling to accept the notion that corpuscles of matter can acquire a repulsive force.[8] They, like the earlier mechanists, thought that elasticity was a peculiar attribute of air and that fixed air was merely trapped in a solid. They therefore ignored Newton's supposed repulsive force and based their systems of natural philosophy upon the action of the attractive force alone.[9]

Although Hales, like Newton, distinguished between two kinds of vapours, he was only interested in studying the behaviour of 'true permanent air'. He therefore developed a series of experimental procedures to prove that the air he obtained by strong heating (distillation) and chemical action (fermentation) was 'true air, and not a mere flatulent vapour'. According to one of his tests, the fixed air he collected had exactly the same specific gravity as atmospheric air. 'I filled,' he says,

> a large receiver which contained 540 cubick inches, with air of

tartar; and when it was cool, I suspended the receiver while its mouth was inverted in water. Then upon lifting the mouth of the receiver out of water, I immediately covered it by tying a piece of bladder over it. When I had found the exact weight, I blew out all the air of tartar with a pair of bellows which had a long additional nose that reached to the bottom of the receiver. And then tying the bladder on, I weighed it again, but could find no difference in the specifick gravity of the two airs, and it was the same with an air of tartar which was 10 days old.[10]

He also checked to see if the air he collected would condense when stored for a long time and whether it remained elastic when compressed. 'That this Air,' he reported,

which arises in such great quantities from fermenting and dissolving vegetables is true permanent Air, is certain, by its continuing in the same expanded elastic state for many weeks and months; which expanding watry vapours will not do, but soon condense when cool. And that this new generated air is elastical is plain, not only by its dilating and contracting with heat and cold, as common air does, but also by its being compressible, in proportion to the incumbent weight.[11]

Hales' experiments served his purpose admirably. By heating minerals, such as limestone, and by combining them with acid, he obtained such large quantities of air that one could no longer reasonably maintain that elasticity is an essential property of the air.[12] Clearly when fixed, air is no longer an elastic fluid. The weight and volume of the air Hales collected established this point beyond doubt and the chemical qualities of the various airs he collected were of no particular interest to him. What he wished to prove, and what his experiments had placed beyond question, was that 'elasticity is not an essential immutable property of air particles; but they are, we see, easily changed from an elastic to a fixt state'.[13] He then attempted to explain what his experiments revealed by saying that air is fixed when its repulsive force is transformed into an attractive force 'by the strong attraction of the acid, sulphureous and saline particles which abound in the air. Whence,' he continued, 'it is reasonable to conclude, that our atmosphere is a *Chaos*, consisting not only of elastick, but also of unelastick air particles, which in great plenty float in it, as well as the sulphureous, saline, watry and earthy particles, which are no ways capable of being thrown off into a permanently elastick state, like those particles which constitute true permanent air'.[14]

Hales, like the earlier mechanical philosophers, concluded that everything that has a repulsive force is air, but he differed from them in arguing that air is not always elastic. Thus for Hales elasticity is a state of a particular kind of matter. 'Since we find in fact from these Experiments,' he argued, 'that air arises from a great variety of dense bodies, both by fire and fermentation, it is probable that they may have very different degrees of elasticity, in proportion to the different size and density of its particles, and the different force with which they were thrown off into an elastick state.'[15] Hales did not claim he was being original: 'Those particles (as Sir Isaac Newton observes) receding from one another, with the greatest repulsive force, and being most difficultly brought together, . . . upon contact cohere most strongly.'[16] Therefore, he concluded, air itself is a heterogeneous mixture of particles. 'Those of the weakest elasticity, will be least able to resist a counter-acting power, and will therefore be soonest changed from an elastick to a fixt state. And tis consonant to reason to think, that the air may consist of infinite degrees of these, from the most elastick and repelling, till we come to the more sluggish, watry and other particles, which float in the air.'[17] Nevertheless, he insisted, all these particles have one property in common, for 'the repelling force of the least elastick particle, near the surface of the earth, while it continues in that elastick state, must be superior to the incumbent pressure of a column of air, whose height is equal to that of the atmosphere, and its base to the surface of the sphere of its elastick activity.'[18] Such a broad definition of the word air, while reasonable, would obviously create difficulties for anyone who wished to use the term to designate a substance having a more specific set of properties than those Hales assigned to it.

Hales' extension of his theory into chemistry, far from being an idle afterthought of a philosophical enthusiast, reflects an assumption shared by practically all English natural philosophers during the first third of the eighteenth century; they believed that the theoretical problems of chemistry would be essentially solved once the foundations of Newtonian corpuscular philosophy were adequately established.[19] Thus it is not surprising that having argued for a dualism of forces, Hales tried to explain the phenomena by attaching his theory to the great dualistic hypothesis of chemistry, the acid-alkali theory. His understanding of chemistry is most clearly stated in his account of the burning of sulphur. When sulphur is heated, he explains, it forms a vapour of highly agitated mutually attractive particles. When these particles meet atoms of elastic air, they transform them into an attracting matter and combine with them to form particles of fire. 'Fire,' Hales asserts, 'is chiefly invigorated by the action and re-action of the acid sulphureous particles of the fuel, and the elastick ones which arise and enter the Fire, either from the fuel in which they abound, or from the circumambient air.'[20]

It was this chemical theory which enabled Hales to apply his theory of air to observable phenomena. He argued that on physical grounds, we know that there are two kinds of particles in the universe, those which are attractive and those which are 'amphibious', and all natural change is caused by the interaction of these two classes of substances. 'There is,' Hales stated,

> diffused thro' all natural, mutually attracting bodies, a large proportion of particles, which as the first great Author of this important discovery, Sir Isaac Newton observes, are capable of being thrown off from dense bodies by heat or fermentation into a vigorously elastick and permanently repelling state: And also of returning by fermentation, and sometimes without it, into dense bodies: *It is by this amphibious property of the air that the main and principle operations of Nature are carried on.*[21]

Air is also, in chemical terms, the alkali principle, and the operations of the attractive and repulsive forces in nature may thus be correlated with the interaction of acidic and alkaline substances. 'May we not,' Hales asks, 'with good reason adopt this now fix't, now volatile *Proteus* among the chymical principles; notwithstanding it has hitherto been overlooked and rejected by Chymists, as no way intitled to that denomination?'[22] He argues, for example, that the caustic properties of quicklime are caused by the interaction of the acid-attractive and alkali-repulsive particles in the lime: 'From this manifest attraction, action and reaction, that there is between the acid, sulphureous and elastick aereal particles, we may not unreasonably conclude, that what we call the fire particles in Lime, and several other bodies, which have undergone the fire, are the sulphureous and elastick particles of the fire fixt in the Lime.'[23]

Although, as we shall see below, certain chemists among Hales' contemporaries were already beginning to question the reductionist programme of the corpuscularians, his book had a profound influence upon the development of natural philosophy in England. In reviving interest in the queries of Newton's *Opticks* and in drawing attention to the importance of repulsive forces, he changed the direction of the Newtonian tradition. These were the aspects of Hales' book which Desaguliers described in the long abstract-review of *Vegetable Staticks* which he published in the *Philosophical Transactions* for 1728 and he subsequently made repulsion the primary concern of the philosophical investigations which occupied the remaining years of his life.[24]

A year after proclaiming the importance of Hales' study of air, Desaguliers published a paper in which he attempted to apply the Newton-Hales notion of repulsion to the evaporation of water.[25]

Although Newton and Hales might have protested that water vapour is not true permanent air and that its particles cannot be made repulsive, Desaguliers argued that in all elastic fluids the repulsive force is present and exceeds the attractive force. Before demonstrating the truth of this conclusion, however, he examined several theories of evaporation. The first, advanced by 'Dr. Niewentyt and some others,' asserts that 'moleculae' of water vapour are composed of particles of fire and particles of water. These molecules are 'small Bodies specifically lighter than Air, which therefore, by hydrostatical Laws, must rise and form Clouds that remain suspended when they are risen up to such an height that the Air about them is of the same specifick Gravity with themselves.'[26] But, Desaguliers argued, Hales had performed experiments which render this hypothesis untenable. The hypothesis itself

> is built upon a Supposition that Fire is a particular Substance, or distinct Element, which has never yet been prov'd by convincing Experiments and sufficient Observations; and which the Reverend Mr. Hales has in his late excellent Book of *Vegetable Staticks* shewn to be an ill grounded Opinion, making it very plain, that in Chymical Operations those Bodies which had been thought to become heavier by Particles of Fire adhering to them, were only so by Adhesion of Particles of Air, &c. which he has shewn to be *absorbed* in great Quantities, by some Bodies, whilst it is *generated* (or reduced from a fixt to an elastick State) by others; nay, that it may be *absorbed* and *generated* successive by the same Body, under different Circumstances.[27]

After dismissing a second theory of evaporation as entirely geometrical and therefore incapable of explaining the phenomenon, Desaguliers described the hypothesis which he believed posed the greatest challenge to his own theory. 'The third Opinion' concerning evaporation, 'and which is most commonly received, is,' he says, 'that by the Action of the Sun on the Water, small Particles of Water are formed into hollow Spherules filled with an *Aura*, or finer Air highly rarefied, so as to become specifically lighter than common Air, and consequently that they must rise in it by hydrostatical Laws.'[28] But after listing the many problems which attend this theory, Desaguliers concluded that 'from all this it follows, that the Condensation and Rarefaction of the Vapours, which make Clouds, must depend upon another Principle than the Condensation and Rarefaction of the Air: And that there is such a Principle, I shall endeavour to shew.'[29]

Desaguliers then asserted as a lemma that '*The Particles of all Fluids have a repellent Force*'.[30] This statement is followed by a scholium in which he argues that 'by encreasing the repellent Force of the Particles, an unelastic or incompressible Fluid may become elastick, or a Solid (at least a great Part of it) may be changed into an elastick Fluid; and, *vice versa*, by diminishing the repellent Force, an elastick Fluid may be reduc'd to an unelastick Fluid, or to a Solid.'[31] Then, returning to his chosen subject, he says

> that Heat will add Elasticity to Fluids is evident from numberless Experiments, especially from Distilling and Chymistry: But what is needful to consider here is only, that it acts more powerfully on Water than common Air; for the same Heat which rarifies Air only 2/3 will rarefy Water very near 14000 times, changing it into Steam or Vapour as it boils it: And in Winter, that small Degree of Heat, which in Respect to our Bodies appears cold, will raise a Steam or Vapour from Water at the same Time that it condenses Air.[32]

This figure of 14000 was, he tells us, derived from

> a great many Observations made by Mr. Henry Beighton, F.R.S. and my self, upon the Engine to raise Water by Fire, according to Mr. Newcomen's Improvement of it; we found that the Water in boiling is expanded 14000 times to generate a Steam as strong (i.e. as elastick) as common Air, which therefore must be near $16\frac{1}{2}$ times specifically lighter. And that this Steam is not made of the Air extricated out of the Water is plain, because it is condens'd again into Water by a Jet of cold Water spouting in it.[33]

It is clear from these statements that Desaguliers, unlike Hales, did not believe that the repulsive force could only be impressed upon the particles of air. In his opinion the particles of all vapours are mutually repulsive. Yet by distinguishing sharply between true permanent 'common air' and the condensable vapour steam, he implies that air has specific properties not associated with its amphibiousness. In the paper being discussed, however, Desaguliers was only interested in proving that the elasticity of steam is not caused by the presence of trapped air released during boiling. He established this point by performing an elegant experiment,[34] thereby rendering his repulsion hypothesis at least plausible, and then proceeded to confirm its truth by using it to explain additional phenomena. 'Now to shew,' he says,

> that what has been said will account for the Rise of Vapours and Formation of Clouds, we must only consider;—whether that Degree of Heat, which is known to rarefy Water 14000 Times, being compared with several of those Degrees of Heat in Summer, Autumn and Winter, which are capable of raising Exhalations from Water or Ice; the Rarity of the Vapours (estimated by

> the Degree of Heat) will appear to be such, that the Vapour will rise high enough in Winter, and not too high in Summer, to agree with the known *Phaenomena. That the Effects are adequate to the Causes in this Case, I think I can make out in the following Manner. . . .*[35]

The 'manner' which Desaguliers employed was, as Cullen later pointed out to his chemistry students, most curious.[36] First he stated that 'the Heat of boiling Water, according to Sir Isaac Newton's Table (*Phil. Transact.* Num. 270) is 34, the mean Heat of Summer 5, the mean Heat of Spring or Autumn 3, and the least Degree of Heat, at which Vapours rise in Winter (*alias* the mean Heat of Winter) is 2'.[37] He then reasoned that since the density of steam at the boiling point is to the density of liquid water at zero degrees as 14000 is to 1, the density of water at the seasonal temperatures given by Newton must be directly proportional to this ratio. Or, as Desaguliers stated it, 'the Rarity of Vapour proportionable to these four Degrees of Heat, is 14000, 2058, 1235, and 823'.[38] He then proposed a third series of ratios, saying that 'the Rarity of Air [the density of water being taken as 1] is, in Summer 900, in Spring or Autumn 850, and in Winter 800'.[39] Given these figures he had little difficulty in explaining how water could evaporate and rise to form clouds in even the coldest weather.

Desaguliers' memoir on evaporation played an important role in the development of Cullen's understanding of the relationship between chemistry and natural philosophy. When Cullen began lecturing on chemistry nearly twenty years after the publication of Desaguliers' paper, he told his students that 'the evaporation of water & the ascent of vapours, as connected with the theory of meteors, has been often considered & deserves our notice as concerned in many chemical operations. Several hypotheses, the last [of which] is that of Desagulier in the *Transactions*, 407. He shews the fallacy of the former opinions, & which we shall suppose done, & consider only the Doctor's.'[40] Cullen probably became aware of this theory while studying natural philosophy in London. Later on, as his understanding of chemistry developed, he came to regard Desaguliers' theory as a paradigm of what happens when corpuscular philosophers attempt to explain phenomena they have not studied in detail. By the time he presented Desaguliers' arguments in his lectures, Cullen had a clear memory of only the last part of the paper. Although Desaguliers said, as Hales had, that fire is not a distinct substance and in spite of his clearly phrased statement that 'the Density and Rarity of the [water] Vapour is chiefly owing to its Degree of Heat,'[41] Cullen told his students Desaguliers had asserted that fire, rather than heat, 'rarifies bodies & water'.[42] When it came time to criticize Desaguliers' application of this theory, however, Cullen recalled his errors precisely. Those who had heard Cullen's lectures were well informed on the pitfalls awaiting anyone who, like Desaguliers, failed to notice that the expansion caused by heat is not uniform when there is a change of state involved.

Although thoroughly familiar with the theories developed by Newton's English disciples, Cullen was not himself a natural philosopher dedicated to the advancement of the mechanical philosophy of nature. Medicine and chemistry were his subjects and his approach to philosophy was more strongly conditioned by the character of the problems that arise in these subjects than by the Newtonian dream of constructing a comprehensive system of natural philosophy. Cullen praised without reservation the achievements of the corpuscular philosophers in the field of mechanics and he firmly believed that Newton and his followers had defined both the fundamental properties of matter and the proper approach to natural philosophy. Yet as a specialist he did not feel obliged to accept the brand of Newtonianism that asserted that once the foundations of physics had been firmly established, all the problems encountered in the other branches of natural philosophy would be solved by deduction. Thus Cullen's commitment to medicine and chemistry drove him away from the Newtonianism of Desaguliers and the mechanical philosophers and towards the methodological Newtonianism of those philosophers who, like the Scottish epistemologists and moral philosophers, dealt with aspects of philosophy for which Newton had not laid down the specific postulates upon which future theories were to be built.[43] Cullen was a Newtonian, but one who feared that disciples like Desaguliers might, by concentrating so narrowly on principles such as repulsion, in Newton's words, 'put a stop to the Improvement of natural Philosophy'.[44]

As a physician Cullen was primarily concerned with the treatment of patients and the explanation of the treatments used, two activities which could not be reduced to the principles of natural philosophy. How to combat illness effectively, the paramount concern of the practising physician, was a problem beyond the competence of eighteenth-century science. There were medical theories and there were medical schools, but physicians still gained their useful knowledge, not all of which was sound or beneficial to their patients, through apprenticeship and clinical experience.[45] Even in the realm of medical theory natural philosophy could provide few satisfactory answers to the questions asked by physicians. Why does illness cause fever? What is the relationship between diet and digestion? Why do the juices of certain plants relieve particular symptoms? If pathologists, physiologists and pharmacists could not find adequate explanations for their phenomena in the works of natural philosophers, then they would have to construct theories of their own. This of course they did, sometimes with an ignorant or resentful disregard for the mechanical philosophy, but more often with a conservative unwillingness to surrender valuable old concepts which had not been adequately replaced.

The Aristotelian concept of form is a particularly good example of an explanatory idea which was rejected by the mechanical philosophers of the seventeenth century but which continued to be used, in a sort of underground way, by eighteenth century medical theorists who were unwilling to ignore the specific qualities of different substances. The problems such theorists encountered were especially acute when they attempted to explain the properties of living bodies. We therefore should not criticize too quickly doctors like Hoffmann and Stahl who used the concept of an *anima*, for the *anima* they postulated was made necessary by the conceptual poverty of the corpuscular philosophy, not by their own unwillingness to take that philosophy seriously.[46] Like many other leading medical theorists of the eighteenth century, Cullen believed that the distance between the conceptual foundations established by modern natural philosophy and the specific phenomena studied by physicians would only be bridged if those who were attempting to explain these phenomena exercised a high degree of independence both in the formation of theories and in their application to medical subjects. In medicine as in chemistry Cullen was a methodological Newtonian and a devoted specialist.[47]

In the early eighteenth century chemists were trying to define a middle ground, similar to that staked out by medical theorists, for their subject. Chemistry could be easily and was traditionally distinguished in terms of its phenomena, such as the properties of acids and alkalis and the process of inflammation, yet it had less autonomy as a profession than medicine. Practical chemistry naturally flourished to the extent that it was useful, but theoretical chemistry had fallen into a dependent status. During the seventeenth century the natural philosophy of the Renaissance, which had been founded upon the chemical attributes of matter,[48] was discredited by the mechanical philosophers, who treated chemical qualities as secondary,[49] and the success of Newtonian force mechanics had completed this historic reversal. In the early eighteenth century it therefore seemed reasonable to assume that theoretical chemistry would develop as a distinct but subordinate subject within Newtonian natural philosophy. This expectation was, of course, held with particular strength by the London Newtonians working under the direction of the master himself. Not long after Newton's death, however, Peter Shaw, a Newtonian who may have been Cullen's chemistry teacher in London, began seriously to question this subordination of chemistry to natural philosophy.[50]

Shaw was a physician who, like so many other doctors, believed that it was chemistry's task to create the bridge which would connect the practical concerns of medicine to the theoretical foundations of natural philosophy. In 1725 he put together a three volume edition of Boyle's works and added an enthusiastically Newtonian commentary of his own. Two years later he translated the unauthorized Latin edition of Boerhaave's chemistry lectures, again adding an extensive commentary in which he surveyed the opinions of other chemists, such as Homberg and Geoffroy, and argued forcefully for the Newtonianism of the London circle. In 1730, three years after Newton's death, Shaw published a translation which made the chemistry system of G.E. Stahl available for the first time to the English speaking world.[51]

Upon first encounter chemists detected no contradiction between Stahl's chemistry and Newton's natural philosophy, and it took Shaw some time to realize that he could not champion the independent science of chemistry which Stahl advocated without admitting the need for a thorough re-examination of the Newtonian programme for chemistry.[52] Slowly, as the Newtonian grip on the content of Shaw's chemistry loosened, the change taking place in his approach to the subject began to emerge as a theme in his writings, even though in all his later works Shaw preserved as much Newtonian chemistry as Stahl's theories allowed. Cullen no doubt noticed these changes if he attended Shaw's public lectures in 1731 and if he studied the outline for a course in philosophical chemistry which Shaw published in the same year.[53] Once alerted to this shift in Shaw's thinking, Cullen probably also noticed that Stahl's ideas were even more strongly urged in Shaw's second translation of Boerhaave's chemistry lectures, published in 1741 and taken from the authorized Latin edition, and in his later works. Thus Cullen may have become aware of the need for an independent science of chemistry while studying with Peter Shaw at the very time that Shaw, under the influence of Stahl's writings, was publicly inaugurating the revival of chemistry in Great Britain.

Stahl argued that chemistry must be theoretically independent because the phenomena it seeks to explain are of a different order of complexity from those studied by natural philosophers.[54]

Explanations which begin with statements about the primary particles of matter may meet the needs of physicists, but chemistry deals with qualities of matter which are too various and too specific to yield to the general assumptions of natural philosophy. The successes of the mechanical philosophy were obviously impressive and a sufficient justification for this approach to physics. Nevertheless, natural philosophers should not claim that their branch of science provides a system of concepts which is adequate for all branches of science. Stahl argued that chemists must identify a set of principles having enduring chemical properties, even though they know that these principles are probably compounded out of smaller physical particles of matter. Acidity, alkalinity, inflammability and metallicity are real properties, and chemists are responsible for creating theories which can account for the distinctive and specific characteristics of these properties. While such theories should not contradict the established theories of natural philosophy, their acceptability depends primarily upon their ability to explain chemical phenomena.

Although chemistry was beginning to emerge as an independent branch of natural philosophy while Cullen was in London, it was many years before Newtonians like Desaguliers felt obliged to listen to what the chemists had to say. From roughly 1730 until 1770, mechanical philosophers continued to concentrate upon the problems raised by the repulsive force. They began, as we have seen, by arguing that repulsion was caused by heat or chemical action. Then, after Du Fay's discovery of the two kinds of electricity in 1733, they turned their attention to electricity and magnetism as the causes of the repulsive force.[55] It was only after the middle of the century, when the Newtonians' need to understand electrical phenomena led to the resurrection of the concept of a material aether, that the substantial fire and phlogiston of the chemists became important issues in British natural philosophy. Before then, the chemists had been quietly trying to establish suitable foundations for their subject. They were not at that point wedded to any of those material causes which the more rigorously mechanistic Newtonians later treated as such shocking offences against true philosophy. Shaw's translation of Stahl's text does not even contain the term phlogiston[56] and this concept did not gain currency in Britain until mid-century.[57] The idea that fire is a substance was well known, particularly through Boerhaave's long treatise on fire, yet even in this treatise, and more emphatically in the notes Shaw added to his translation of Boerhaave's chemistry lectures, Newton's dynamic theory of fire is acknowledged as equally probable. When Cullen was in London, chemists were less concerned with fire's substantiality than with determining its suitability as a fundamental starting point for a theory of chemical change.

Had Cullen followed Desaguliers, hc would have founded his chemical doctrines upon the universal forces of attraction and repulsion. But while he acknowledged that these forces are the two fundamental agents of change in nature, Cullen could not see how such highly generalized causes could be made to explain the specific phenomena which accompany chemical change. Cullen's opinion on this matter seems to have had an important effect upon the progress of natural philosophy in Scotland during the remainder of the eighteenth century, for while English natural philosophers were eagerly investigating electrical phenomena in hopes of formulating a general law of repulsion comparable to Newton's law of gravitation, Scottish natural philosophers, the best of whom were chemists, neglected electrical research because it seemed irrelevant to their primary concerns. They investigated instead the more specific forces of elective attraction and heat, the forces which Cullen considered the immediate causes of chemical change. In his lectures Cullen said that if chemists could explain their own phenomena in terms of the elective attraction between particles and in terms of the amount of fire or heat contained in bodies, they would have served their subject well. If chemists did their job well, then the task of accounting for their specific explanations in terms of the more general causes employed by mechanical philosophers would emerge as the central problem in the creation of a truly adequate understanding of nature. But, Cullen probably thought as he left London, it would be many decades before chemistry would stand as a mature theoretical science ready for incorporation into a comprehensive system of natural philosophy.

[3]

Medicine and Natural Philosophy in Edinburgh

In 1732 Cullen was again in Scotland. His father had died and, being the oldest surviving son, he had returned from London to supervise the education of his brothers and sisters. Twenty-two years old and fully qualified as a physician, he supported himself by practising medicine near Hamilton.[1] Cullen was an able physician and well known in Hamilton, and he could no doubt have settled into a comfortable and untroubled career as a practitioner in his home town. He was, however, too philosophical and too ambitious to be satisfied by the practice of medicine alone. Thus when he received a small legacy in 1734, he used the freedom it brought him to continue his education. During the following two years he spent his summers in England, living with a dissenting minister in Northumberland, and his winters studying in the medical school that had recently been established in the University of Edinburgh.

The University of Edinburgh did not become a centre of medical education until the second quarter of the eighteenth century.[2] Before the emergence of well organized schools of medicine within the Universities of Glasgow and Edinburgh, medical training in both these cities had been acquired primarily through apprenticeship, although it was often supplemented by lecture courses given in the municipal colleges of physicians and surgeons. During the seventeenth century several of the lecturers employed by these colleges had obtained the title of professor, but such appointments were accompanied by neither salary, responsibility, nor prerogatives, and the lectures, when offered, were seldom given in the universities. During the first decades of the eighteenth century the medical societies formalized the appointments of their lecturers and gradually positions were created for them within the universities. In 1705 the Edinburgh surgeons, who had recently separated their society from that of the physicians, gave Robert Elliot 'the sole and permanent charge of teaching anatomy'. Elliot then asked the Town Council to support his public lectures by granting him a salary. The council, which was the governing body of the 'Town's College', agreed to pay his salary on the condition that he accept the position of professor of anatomy and take care of the anatomical specimens owned by the university. Elliot accepted, and though his lectures were delivered in the College of Surgeons, the arrangements made for his appointment set the pattern by which extramural medical instruction was made a part of the university curriculum.

The formation of a fully fledged medical school in Edinburgh was the work of the Monros. John Monro, a distinguished army surgeon who settled in Edinburgh, became president of the College of Surgeons in 1712. Shortly thereafter he began urging that the physicians and surgeons take the lead in offering regular instruction in the various branches of medicine. The realization of John Monro's plan was the chief task of his son, Alexander. The first of three Monros to teach anatomy in Edinburgh, and therefore known as Monro *primus*, Alexander was an eager student who shared his father's enthusiasm for founding a medical school. Alexander began his training by serving as his father's apprentice and attending the medical lectures offered in Edinburgh. Then, when twenty years old, he began two years of travel and study, first in London, then in Paris, and finally in Leyden where 'he became the favourite and admiring pupil of the great Boerhaave'.[3] Upon his return to Edinburgh he so impressed his examiners at the College of Surgeons that the two members then sharing the appointment as professor both resigned in his favour. Shortly thereafter, in January 1720, the Edinburgh Town Council appointed Alexander Monro 'Professor of Anatomy in this city and college'.

After two years of lecturing in the College of Surgeons, Monro petitioned the Town Council for permanent status within the university. The council, which was eager to protect its right to govern, had previously refused such requests for unlimited tenure of appointment, but Monro's brilliance and success as a lecturer won their assent, thereby setting an important precedent for the establishment of a medical faculty within the university. The Town Council itself was eager to support the growth of a first rate medical school. The medical school also had a particularly powerful friend in George Drummond, 'the wisest and best disposed of all the long list of Town Councillors and Provosts, who during 275 years acted as patrons of the College or University'.[4] Drummond was Alexander Monro's most valuable ally. Having secured tenure, Monro looked forward to securing appropriate rooms in the university so that he could offer anatomy instruction within its walls. After this request was granted, in 1725, the need for a teaching hospital became paramount. Drummond, who became Lord Provost for the first time in 1725, took the lead in securing funds for this hospital, but the foundation stone of what was to become the Royal Infirmary was not laid until 1738.[5] When, in 1746, Drummond returned to the office of Lord Provost after an absence of two decades, one of the first acts of his administration was to institute clinical lectures in the infirmary. And ten years later, when several medical professors in Edinburgh opposed Cullen's appointment as professor of chemistry, it was Drummond and Monro *primus* who drew up the compromise which was accepted on all sides.

Naturally Alexander Monro could not teach all the medical subjects himself and by 1724 the Town Council was ready to create additional medical professorships in the university. A prominent physician, Dr William Porterfield, was appointed professor of the institutes and practice of medicine, an unsalaried position which entitled him to all the 'powers, privileges, and immunities' enjoyed by the other professors. Although Porterfield's contract required him to teach, he apparently never lectured. Shortly after this profitless appointment was made, four members of the College of Physicians received permission to manage the university garden, which had been neglected for several years. They proposed that they be allowed to grow pharmaceutical plants and process them into saleable drugs in a chemistry laboratory to be built adjacent to the garden. Apparently these four enterprising physicians had also 'undertaken the professing and teaching of Medicine in this City, and, by the encouragement which the Council had been pleased to grant them, had carried it on with some success'.[6] Thus in 1726 they were able to present a good case for converting their university connection and their teaching experience into regular university appointments. The Town Council, under the provostship of George Drummond, responded to their petition by promising to create a staff large enough to provide adequate instruction in all the main branches of medicine and of sufficient eminence to make the Edinburgh medical degree as distinguished as any. They then appointed Drs Andrew Sinclair and John Rutherford professors of the theory and practice of medicine and Drs Andrew Plummer and John Innes professors of medicine and chemistry.[7] These appointments, which were with tenure but without salary, completed the founding of the Medical Faculty in the University of Edinburgh.

These professors, who received their appointments just before Cullen began his medical education in Glasgow, had been teaching for almost a decade by the time he came to study in Edinburgh. The titles of their appointments did not, however, accurately indicate what each man actually taught. Innes chose to lecture solely on the practice of medicine while Plummer taught only chemistry. Although this degree of specialization suggests that Plummer presented chemistry as an autonomous discipline, he in fact taught nothing but pharmaceutical chemistry. Cullen, with his philosophic training and far ranging interests, must have found this treatment of a favourite subject discouragingly inadequate. In Edinburgh medicine had achieved an institutional position which clearly reflected its conceptual and professional independence. Within the medical curriculum, however, chemistry was treated as no more than a service course for aspiring physicians.

Andrew Plummer was by all accounts an able and erudite professor. As a pharmaceutical chemist he was known as the formulator of Plummer's Pill, a compound of antimony and mercury which was popular for nearly a century, and for his published analysis of Moffat spring water.[8] As a member of the medical faculty he distinguished himself as the co-editor, with Monro *primus*, of the *Medical Essays*, to which he contributed several papers. And as a learned colleague, such was 'his universal knowledge, that in any disputed point in science, the great Maclaurin always appealed to him as to a living library'.[9]

There is also ample evidence, especially in Plummer's two published chemistry papers, that 'he knew chemistry well'.[10] In the second of these papers,[11] Plummer made skilful use of Homberg's technique for determining equivalent weights. He first neutralized a standard weight of alkali with three different acids and then weighed the dry salt obtained. Since the dry salts all contained the same weight of alkali, the ratio of their total weights, which differed, was also the ratio of the relative weight of acid in each. These 'equivalent weights' of acid indicated, it was believed, the relative force with which the alkali attracted the different acids.

The general scholium with which Plummer closed this essay reveals his indebtedness to the reductionist view of chemistry championed by Boerhaave and by Newton's early English disciples.[12] The difference between the solid and the fluid states of matter is to be understood, he says, in terms of the powers of cohesion: 'We can have no other idea of a solid or consistent body in general, as distinguished from a fluid, than that the parts of a solid cohere together by some power, which retains the particles in their situations, with respect to one another, and makes them to resist their separation or division, till that power is overcome by a superior force.'[13] This power of cohesion can be overcome by the force of fire or by the force of elective attraction:

> The attractive force between the saline particles of the spirit of nitre and the silver, must be superior to the power by which the particles of silver attract one another. . . . That this power of attraction between the particles of spirit of nitre and those of

> silver must be very great, will appear, if we consider, first, that it requires a very great heat to melt silver; but this melting is nothing else, than, by the force of fire, to overcome the cohesion among the particles of silver, and bring the mass to a state of fluidity, so that the particles may easily move and slide one upon another while that degree of heat lasts; but spirit of nitre produces the same or rather a greater effect.[14]

And the force of elective attraction can be explained in terms of the mass and shape of the particles involved:

> The particles of copper have a stronger attractive power, with respect to the spirit of nitre, than the particles of silver have; which difference of attraction will probably arise from the different bulk, figure or density of the particles of the two metals, the acid being the same. The same thing may be said of iron, zinc, chalk, volatile alcaline salt and fixt salt, each of which attract spirit of nitre more strongly than silver does, and the last more than the preceeding.[15]

Despite his pharmaceutical skill, his erudition, and his Newtonianism, Plummer presented chemistry in a way which Cullen found old fashioned and unsuggestive. Having accepted the mechanistic-reductionist approach to chemical theory, Plummer had to decide whether to present explanatory hypotheses that could not be experimentally confirmed or simply to describe the facts and processes of chemistry. Given the choices available, he decided quite reasonably to do the latter. Furthermore, lecturing was a task Plummer found difficult and did badly. 'Laborious, attentive, and exact, had not a native diffidence veiled his talents as a praelector, he would have been among the foremost in his pupils' esteem. . . . So great was his modesty, that he spoke to young audiences upon a point he was perfectly master of, not without hesitation.'[16] But it was the narrowing of purpose, more than shyness, which lessened the impact of his course. As a teacher, Plummer took as his subject the preparation of medicines and he stuck to his subject with great persistence. All chemistry teachers in the middle decades of the eighteenth century were troubled by the gulf that separated chemical theory and chemical fact, but Plummer seems not to have made this problem a central concern of his course. As we have seen above, he accepted the current Newtonian ideas about the causes of chemical phenomena, but he isolated his discussion of these causes from the rest of his chemical work. In his second chemical essay, where the need for theory is at least recognized, he collected his explanations into a highly generalized scholium. This paper was first read after Cullen had completed his studies in Edinburgh and it did not appear in print until two years before Plummer's death. It therefore seems likely that Plummer did not privately encourage the medical students in Edinburgh to engage in the kind of theoretical investigations which he so assiduously avoided in his chemistry lectures.

Cullen, who probably had formed his own conception of chemistry before going to Edinburgh, could not, however, simply ignore the pharmaceutical approach which Plummer presented for nearly 40 years. Therefore, when he began teaching chemistry in Glasgow in the 1740s, Cullen compared the goals of his course to those of a course like Plummer's.

> Chemistry is an art that has furnished the world with a great number of useful facts, and has thereby contributed to the improvement of many arts; but these facts lie scattered in many different books, involved in obscure terms, mixed with many falsehoods, and joined to a great deal of false philosophy; so that it is no great wonder that chemistry has not been so much studied as might have been expected with regard to so useful a branch of knowledge, and that many professors are themselves but very superficially acquainted with it. But it was particularly to be expected, that, since it has been taught in universities, the difficulties in this study should have been in some measure removed, that the art should have been put into some form, and a system of it attempted—the scattered facts collected and arranged in a proper order. But this has not yet been done; chemistry has not yet been taught but upon a very narrow plan. The teachers of it have still confined themselves to the purposes of pharmacy and medicine, and that comprehends a small branch of chemistry; and even that, by being a single branch, could not by itself be tolerably explained.[17]

There is another reason for believing that Cullen had a low opinion of Plummer's chemistry course. Joseph Black had probably absorbed his master's attitude towards the chemistry taught in Edinburgh before going there to complete his medical studies. Therefore, when describing the reception that Cullen's essay on the classification of salts received, Black knew his teacher would be pleased to hear that 'it would require a course of chemistry almost to make them [the members of the Edinburgh Philosophical Society] fully sensible of the truths of your observations'.[18] When Cullen was preparing to succeed Plummer in Edinburgh, Black wrote that the available laboratory equipment 'is probably very imperfect for a course of Philosophic Chemistry, but you need not be anxious, provided your course be better than Plummer's, which it is impossible for it not to be'.[19] As Cullen's disciple and as a student who had not known Plummer in his middle years, when his conception of chemistry had not yet been seriously challenged, Black probably attributed the narrowness of Plummer's course to an individual constriction of vision. Cullen, while perhaps admitting that there was an element of truth in Black's harsh judgment, was in a better position to recognize the fundamental change in the conception of the subject which lay at the root of the difference between his chemistry lectures and those given by Plummer.

Chemistry was not the only field in which the gulf between empirical knowledge and theoretical explanation separated the practitioners from the philosophers. In medicine too the activities of curing and explaining had little actual impact upon each other, even within the Medical Faculty. While physiological hypotheses were occasionally presented in lectures, they were in general as irrelevant to the students' understanding of their subject and their world as is most of the history of medicine taught in medical schools today. What the students needed were theories that touched the medical realities of the day, theories that could fill the vacancy left by the discredited iatro-mechanical and iatro-mathematical hypotheses of the late seventeenth century. Therefore, to relieve their dissatisfaction with the dominant empiricism of their formal instruction, the medical students in Edinburgh established a philosophical society of their own. In doing so they were no doubt attempting to duplicate the success of the Rankenian Club which had been organized in 1716 or 1717 by a group of faculty and students of divinity and law in Edinburgh and which included among its members many prominent members of the university faculty. The philosophical discussions initiated by the Rankenians evidently had had a profound influence upon intellectual life in the capital city, for 'between 1723 and 1740', Ramsay of Ochtertyre recalled, 'nothing was in more request with the Edinburgh *literati*, clerical and laical, than metaphysical disquisitions. These they regarded as more pleasant themes than either theological or political controversies, of which, by that time, people were surfeited.'[20] The founders of the new medical society no doubt hoped that their society would stimulate a similar interest in discussions of medical theory and natural philosophy.

In 1735 Cullen and several other medical students began meeting by themselves to hear and criticize medical essays of their own composition. After two years of meeting informally, the students incorporated their group as the Medical Society.[21] This student organization, which later became the Royal Medical Society, has flourished down to the present day, and it still has among its most valued possessions the volumes of manuscript dissertations read by its members in the eighteenth century.[22] Although the oldest surviving manuscripts date from the 1750s, we know that the earliest essays read before the society, including several by Cullen, were collected and bound together. The titles of the surviving essays indicate that in the eighteenth century, as ever since, the society encouraged medical students to examine the content of their professional training in the broadest possible historical and philosophical context. In 1770 Thomas Blackburn addressed the question 'Is an acquaintance with Metaphysics, Mathematics and the different Branches of Natural Philosophy necessary to the study of Medicine? if so, how far are these sciences applicable?'[23] The question is distressingly general and Blackburn's answer is jejune, but the fact that the students asked such questions among themselves reveals a philosophic appetite which formal course work did not satisfy. Naturally many members of the society spoke on more strictly medical questions, and the disciplines that link medical theory to the other branches of philosophy were not neglected. Included among the dissertations read in the 1770s is a discussion of the original cause of respiration in animals and a response to the question 'Is the Doctrine of Latent heat well founded'.[24] Thus to compensate for the philosophic self-denial that characterized Plummer's chemistry lectures and much of the medical instruction, Cullen could turn to a circle of medical school colleagues eager to begin the philosophic examination of medicine and its allied sciences.

While the founders of the Medical Society were more openly philosophic than their medical professors, they had no reason to feel isolated within the university as a whole. Natural philosophy and logic were important subjects in the University of Edinburgh when Cullen studied there, and he was no doubt familiar with the doctrines taught by professors such as John Stevenson, Robert Stewart, and Colin Maclaurin.[25] Although none of these men were chemists, the way in which they each carefully defined the range of their own subject must have been encouraging to someone who, like Cullen, believed that chemistry should be treated as an independent science.

A few years before Cullen arrived in Edinburgh, John Stevenson, professor of philosophy from 1730 to 1775, introduced modern texts into his course on logic and metaphysics.[26] Before 1730 the lectures on these subjects had been drawn from Aristotle's *Organon* and the writings of Ramus. Stevenson, who apparently studied in the Netherlands, had his students read a textbook on logic by Heineccius and a text on ontology by De Vries, while in the history of philosophy course, which he gave as a separate series of lectures, he followed the book on that subject by Heineccius. In addition to the Dutch texts,

Stevenson had his philosophy students read Bishop Wynne's abridgement of John Locke's *Essay Concerning Human Understanding*. Although Locke's writings were far from unknown in the early eighteenth century, their inclusion in a philosophy curriculum strengthened their influence, and Stevenson can be justly credited with having 'introduced Locke's philosophy into Edinburgh'.[27] Even more novel was Stevenson's use of Bishop Berkeley's writings. He encouraged his students to read Berkeley's works and he made criticism of his doctrine of immaterialism a central issue in philosophical education. Stevenson was a popular and forceful teacher who impressed upon his students 'the effect which criticism and rational Logic have upon the opening mind'.[28] Although not directly involved in the lively debates generated by Stevenson's interest in Berkeley, Cullen was no doubt aware of the metaphysical issues raised in his class. For his own instructional purposes, however, the more moderate philosophy of Locke would suffice.

Cullen was not a Lockean in the sense that he focused on the particular epistemological questions that occupied the English philosopher, but he did recommend that students study Locke's doctrines as a means of preparing for his lectures on chemistry. At the beginning of a chemistry course given during the 1760s, Cullen said:

> I shall now proceed to give some advice with regard to your conduct in theoretical enquiries . . . but to enable you to follow me, and to make any advances yourselves in Chemical philosophy, much preparatory knowledge is necessary. Logic is a very necessary part of the introductory learning, by logic I mean the analysis of the human mind, such as may be found in Mr. Locke's excellent treatise of the understanding. This is not only necessary in chemistry, but also in every science where there is danger of error.[29]

A familiarity with Locke's *Essay* would enable students to grasp the philosophic meaning of the terms and categories Cullen used in his presentation of chemistry. When Cullen says chemists should have 'a clear and precise idea of our art'[30] and criticizes Boerhaave's definition of the subject because 'it communicates no distinct idea to me',[31] he assumes that his students have thought about the philosophic criteria being invoked. And when he says chemistry 'treats of the powers in nature by which the physical changes of bodies are produced,'[32] he is using the term power in the abstract sense which is made explicit in Locke's *Essay*: 'The mind . . . considers in one thing the possibility of having any of its simple *ideas* changed, and in another the possibility of making that change; and so comes by that *idea* which we call *power*. Thus we say fire has a *power* to melt gold.'[33] Again, in saying that the confused state of chemical terminology can be attributed to the use of the language of artisans in philosophic discussion, Cullen was providing a concrete illustration of the difficulties Locke said attend the use of everyday language in philosophic discourse.[34]

There is another, more profound, reason Cullen encouraged his students to master the conceptual system presented in Locke's *Essay*. Convincing philosophic arguments for treating chemistry as an independent branch of natural philosophy were always welcomed and Cullen must have therefore found Locke's three-fold account of the origin of qualities congenial. As is well known, Locke distinguished between the 'primary qualities of body', which 'produce simple ideas in us, viz. solidity, extension, figure, motion or rest, and number', and 'secondary qualities' which 'are also produced, viz. by the operation of insensible particles on our senses'.[35] Because the ideas they produce are simple, we know that the primary qualities we perceive and the real qualities in the body are identical. 'The particular *bulk, number, figure, and motion of the parts of fire and snow are really in them*, whether anyone's senses perceive them or no; and therefore they may be called *real qualities*, because they really exist in those bodies.'[36] But the secondary qualities raise more complicated problems. The real qualities of bodies also give rise to complex ideas, that is, ideas which cannot be described in terms of the primary qualities alone. These secondary qualities can be caused either by the operation of the primary qualities upon our senses, giving rise to specific colours, tastes, and sounds, or they can be caused by the interaction of the primary qualities of two bodies, thereby changing the qualities which impinge upon our senses. In other words, secondary qualities can be analyzed in terms of the characteristics of our senses, which leads to the psychology of perception, or in terms of the pattern of those properties and powers which we perceive as secondary qualities. Therefore, Locke concludes,

> The *qualities* that are in *bodies*, rightly considered, are of *three sorts*:
>
> *First,* The *bulk, figure, number, situation,* and *motion* or *rest* of their solid parts. Those are in them, whether we perceive them or no; and when they are of that size that we can discover them, we have by these an *idea* of the thing as it is in itself, as is plain in artificial things. These I call *primary qualities*.
>
> *Secondly*, The *power* that is in any body, by reason of *its* insensible *primary qualities*, to operate after a peculiar manner on any of our senses, and thereby *produce in us* the *different ideas* of several

colours, sounds, smells, tastes, etc. These are usually called sensible qualities.

Thirdly, The *power* that is in any body, *by* reason of the particular constitution of *its primary qualities, to* make such a *change* in the *bulk, figure, texture, and motion of another body*, as to make it operate on our senses differently from what it did before. Thus the sun has a power to make wax white, and fire to make lead fluid. These *are* usually called powers.[37]

In this third category, Locke provided a philosophic description of those properties which Cullen was to call the objects of chemistry.

Although Cullen may have been encouraged by reading Locke, he certainly would not have campaigned for the independence of chemistry with arguments drawn from Locke and Stevenson alone. With Plummer confining himself to medical chemistry, one would expect that the explanatory system taught by the professors of natural philosophy and mathematics would have been extended, like those developed by their English contemporaries, to cover all aspects of physical nature, but in fact they were not. Neither Robert Stewart, professor of natural philosophy, nor, more importantly, Colin Maclaurin, professor of mathematics, was eager to reduce other fields of knowledge to their own set of assumptions. Perhaps specialization within the faculty, which followed the abandonment of regenting, led them to recognize as many distinct areas of learning as there were professorships.[38] But certainly an important factor was the critical examination of knowledge itself which became the hallmark of Scottish Philosophy. In the face of the sceptical attacks of Bishop Berkeley and David Hume, Scottish natural philosophers became increasingly cautious. By emphasizing those aspects of natural philosophy that were well founded, rather than expanding upon suggestions that would carry natural philosophy into other areas, the Scottish professors were admitting that there were many phenomena that slipped through their theoretical net. If the chemists could explain these phenomena more successfully than the natural philosophers, all the better. Thus Cullen was arguing for an autonomous and philosophic science of chemistry at a time when the university philosophers were prepared to welcome new ways of explaining the physical world.

Students who attended Robert Stewart's lectures on natural philosophy received an uninspired introduction to the mechanical philosophy. Newton's own books and those of his disciples John Keill and David Gregory were the texts through which they learned about mechanics, pneumatics, hydrostatics, colours, vision, optics and some astronomy.[39] Stewart himself had come to Edinburgh in

1703 as a regent, and, it was said, was a Cartesian before becoming a Newtonian.[40] By the time Cullen began studying in Edinburgh Stewart was 'worn out with age'.[41] Having already been introduced to Newton's doctrines by the London Newtonians, Cullen probably found little that was noteworthy in Stewart's lectures. But by teaching Newtonian natural philosophy while directing his attention mainly to the mechanical phenomena of nature, Stewart in fact helped create the need for new theories of chemistry which could explain those phenomena of natural philosophy that were beyond the grasp of the Newtonians.

The mathematical features of Newton's philosophy were presented with authority and great clarity by Colin Maclaurin. From 1726 when, with Newton's recommendation, he was appointed joint professor of mathematics, until his death twenty years later, Maclaurin was 'the life and soul of the University of Edinburgh'.[42] Each year he gave three or four courses of lectures. In the first two, which were devoted to basic mathematics and its practical applications, he taught arithmetic, plane geometry, trigonometry, algebra, solid geometry, surveying, fortification, geography, gunnery and some astronomy and optics. In the third course he presented a more complete discussion of astronomy and optics, taught parts of Newton's *Principia*, and introduced his students to the method of fluxions. The fourth set of lectures, which began in December and ended in early April, was devoted to experimental philosophy and observational astronomy. For years Maclaurin tried to improve the opportunities available for astronomical observation, but the university did not acquire its own observatory until after his death.

This description of Maclaurin's courses indicates that in the lecture hall he restricted himself to subjects which clearly fell under his purview as professor of mathematics. But as the historian of the University of Edinburgh points out, Maclaurin took a keen interest in the many scientific activities of the city and university.

> M'laurin was a man of remarkable social qualities, and was a prominent figure in all the scientific circles of Edinburgh in those days. He was the friend of the great physicians, and accepted the post of Secretary to the 'Society for improving Medical Science'. After editing their *Transactions* for some time he got them to enlarge their scope so as to take in all the parts of Physics, and the antiquities of the country, and to change their title to that of the 'Philosophical Society,' which in a later generation was transformed into the 'Royal Society of Edinburgh'.[43]

Maclaurin's close affiliation with his medical colleagues should be borne in mind when reading his general texts on Newtonian philosophy, for he was well aware of the gap between medical knowledge and natural philosophy. His two volume *Treatise on Fluxions*[44] is an extension of his reply to Bishop Berkeley's attack upon the foundations of calculus, and on this subject Maclaurin forcefully defended the achievement of his mentor Newton. But in the *Account of Sir Isaac Newton's Philosophical Discoveries*, which he completed shortly before his death, Maclaurin carefully restricted himself to a discussion of those truths that had been firmly established within the Newtonian edifice. He did not question the validity of Newton's philosophical method, he simply advised caution before the complexity of nature. 'It is not the business of philosophy, in our present situation in the universe,' he said, 'to attempt to take in at once, in one view, the whole scheme of nature, but to extend, with great care and circumspection, our knowledge, by just steps, from sensible things, as far as our observations or reasonings from them will carry us, in our enquiries concerning either the greater motions and operations of nature, or her more subtile and hidden works. In this way Sir Isaac Newton proceeded in his discoveries.'[45] Maclaurin went on to say that the difference between established truth and suggestive hypothesis can be clearly illustrated from Newton's own writings by comparing the astronomical sections of the *Principia* with the microscopic speculations of the *Opticks*.[46] Maclaurin therefore stood for a rigorous kind of Newtonianism, one which considers mathematical formulation the test of certainty. But a philosophy that adopts such a demanding criterion has little to say on subjects that have not yet been mathematized. Maclaurin, being a practical and sociable man as well as a mathematician, recognized these limitations and admitted them in the company of his philosophic colleagues. Cullen's biographer, writing in the early nineteenth century, has correctly described the fundamental lesson which Maclaurin impressed upon his auditors in Edinburgh: 'The labours of Colin Maclaurin, . . . by explaining the true and legitimate application of the researches of Newton, ought to have satisfied all sound thinkers, that medicine was not the field for the application of geometry, and that, however useful when applied to mechanics and physics, the application of mathematical reasoning to elucidate the actions of animal bodies is barren and unprofitable.'[47] Cullen must have been pleased when he realized that with Scotland's foremost Newtonian restricting the range of his master's doctrines to the realm of mechanics and physics, his own efforts to develop more adequate chemical and medical theories would be judged on their own terms, not in terms of their Newtonian orthodoxy.

The complexity of medical and chemical phenomena was not the only reason neither Maclaurin nor Cullen espoused Newtonian reductionism. The critical examination of human knowledge itself, as advanced by Locke, Berkeley and Hume, bore directly upon the conception of Newtonianism which finally gained acceptance in Scotland. David Hume succeeded Maclaurin as secretary to the Edinburgh Philosophical Society and it was Hume and Alexander Monro *secundus* who wrote the preface to the first volume of the Society's memoirs.[48] Hume, of course, drew certain lessons from Newton's works, which few of his contemporaries could accept, but the difference between his opinions and those of men like Maclaurin and Cullen was one of degree only.[49] On questions of natural philosophy, there was little disagreement among the circle of thinkers who dominated intellectual life in the capital during the second half of the century. And Cullen, who supported Hume's candidacy for the chair of logic in Glasgow,[50] who became active in the Philosophical Society when he returned to Edinburgh, who as vice-president of the Society in 1782 played a leading part in its incorporation as the Royal Society of Edinburgh,[51] and who was Hume's friend and physician,[52] was an important member of this circle during his years of professional eminence. He was also, it seems safe to assume, already a member in spirit when he studied in Edinburgh during the 1730s.

The completion of his second academic year in Edinburgh marked the end of Cullen's formal education. He returned to Hamilton in the spring of 1736 and began an eight year period of professional activity and medical practice. He was soon caring for the leading families of Hamilton, including the duke and duchess, and his circle of acquaintances grew rapidly. He met Dr John Clerk of Edinburgh when this eminent physician was called in to consult on an illness afflicting the duke, and the two doctors soon became friends and correspondents. In 1751 Cullen asked Clerk to obtain for him a copy of the chemical lectures delivered by Gaubius, Boerhaave's successor in Leyden, and in 1753, when Cullen wished to have a paper on chemistry read before the Philosophical Society in Edinburgh, he transmitted it as a letter to Dr Clerk.[53] Cullen also befriended a theology student named William Hunter who had been studying at the University of Glasgow. Hunter was intended for the church, but he was beginning to have some doubts concerning the articles of faith to which he would be required to subscribe. 'In this state of mind,' his biographer reports, 'he happened to become acquainted with Dr. Cullen, the present celebrated professor at Edinburgh, who was then just established in practice at Hamilton, under the patronage of the Duke of Hamilton.

Dr. Cullen's conversation soon determined him to lay aside all thoughts of the church, and to devote himself to the profession of physic.'[54] In 1737 Hunter began a three year apprenticeship under Cullen, following which he continued his medical studies in Edinburgh and London. Cullen had hoped that Hunter would return and take on the surgical responsibilities of his practice, but opportunities in London were too tempting and Hunter settled there. As so often occurred in Cullen's life, the master – apprentice relationship was transformed into a friendship founded upon mutual respect and easy familiarity, and the two men kept up an extended and intimate, if infrequent, correspondence. Although Cullen had lost the company of his apprentice, he gained a valuable medical and philosophical 'intelligencer' in the British capital.[55]

As he passed into his thirties, Cullen took on more of the responsibilities of a settled life. In 1738 and 1739 he served as a magistrate in Hamilton and as one of his tasks he had to deal with a riotous protest against high grain prices. Although Cullen had not received any degrees for all his years of study, a situation which speaks well for the openness of university education and medical practice in Scotland at that time, in 1740 he took the degree of Doctor of Medicine at Glasgow.[56] The following year he and Anne Johnstone were joined in a marriage that lasted 46 years and resulted in the birth of eleven children. And throughout this period he was an active 'improver', attending meetings of the trustees appointed for the improvement of the high-roads, consulting on agricultural and rural affairs, and in all concerns acting the part of an enlightened Scotsman responsible for the agricultural and commercial improvement of his country. Here was an interest which, together with medical theory and philosophic understanding, made the advancement of chemistry trebly important to Cullen.

[4]

Moral Philosophy and Economic Improvement in Mid-Eighteenth-Century Scotland

In 1742 Cullen decided to move to Glasgow. A Doctor Brisbane, who had been caring for several of the more prosperous families in that city, had died and Cullen was invited to take over his practice. It was an attractive offer, but the Duke of Hamilton was reluctant to see Cullen go and countered it by promising him an annual salary, a chemical laboratory and a botanical garden if he would stay in Hamilton. Cullen did delay his departure so that he could care for the duke during his final illness, which began in 1742 and lasted until his death at the end of the following year, but in 1744 he took his family, his extensive library, and his 'philosophical apparatus' to Glasgow and a new career.[1]

There were at that time two medical chairs in the University of Glasgow. Doctor John Johnstoun, the professor of medicine, had been appointed in 1714, but long before his retirement in 1751 his enthusiasm for teaching had declined and he had ceased lecturing. Doctor Thomas Brisbane, the professor of anatomy, had occupied that chair since it was first established in 1720. Although obliged by the faculty to give a course on comparative anatomy, for which there was a great demand, Brisbane could not bring himself to do dissections. He therefore lectured on botany alone, leaving his primary assignment to a series of unsalaried surgeons who taught anatomy within the university in return for lecture fees. Professor Brisbane died in 1742 and it was no doubt his practice that Cullen was invited to take on. Certainly Cullen also hoped to obtain Brisbane's chair, for while he was no anatomist himself, he would have happily delivered lectures on botany. But Cullen's delay in leaving Hamilton ruined whatever chance he had of succeeding Brisbane, and in 1742 the chair was given to Doctor Robert Hamilton, who had begun lecturing on anatomy before Brisbane's death. The chair of anatomy and botany then became, like the chair of anatomy in Edinburgh, a family preserve. From 1742 until 1790 it was occupied by three Hamiltons successively, with the exception of the one year that Joseph Black served as professor of anatomy in Glasgow.

Before leaving the town of Hamilton, Cullen received a letter from a friend in Glasgow encouraging him to plan on lecturing as soon as he was settled in his new surroundings. Naturally in a profession whose members jealously guarded each privilege obtained, such plans had to be cloaked in secrecy. 'What you were so good as to hint to me in confidence,' Cullen's informant wrote,

> I was mindful enough to keep a secret; but I reheard it talked of in several companies, that you are to be here, and to teach medicine next winter. On this occasion, your friends cannot help expressing that esteem of you, which a man of distinguished abilities in science justly deserves. If you are determined to come, and make no secret of it, I should be content to know it. Nothing can give me greater joy than your success; and I think there is a probability you will meet with more, at least in teaching, than you could have had for some years past, as a good many have applied to anatomy this season, some of them foreigners, who, I know, design to study medicine.[2]

Towards the end of 1744 Cullen began delivering lectures extramurally, scheduling them, we may presume, so that students could attend both Hamilton's anatomy lectures and his own lectures on medicine. From the outset the demand for medical education in Glasgow and Cullen's natural ability as a lecturer combined to bring him success. In February 1745 William Hunter reported from London that 'the accounts Mr. Macauly gives me of your situation at Glasgow, afford me a deal of joy. Macauly talks much of your lectures, too; but that does not surprise me, for I knew you before.'[3] After just a few months in Glasgow Cullen's new career was well launched.

Cullen still hoped to obtain a position within the university. For two academic years he lectured extramurally. Then, in the summer of 1746, having demonstrated his competence, he approached the incumbent Dr Johnstoun, and the two doctors worked out an arrangement by which Cullen was allowed to give a course of lectures on the theory and practice of physic within the university during the following winter. Although Johnstoun continued to hold the chair of medicine, Cullen had successfully taken the first steps towards founding a modern medical school in Glasgow.

When Cullen left the University of Glasgow in 1729 the theologically conservative ministers who dominated the kirk were still struggling to maintain their influence over the university. Although the statutes of 1727 had given the faculty control of university affairs, the presence of conservative and liberal factions within the faculty invited manipulation from without. But by 1746, when Cullen returned to the university, the liberals had gained a majority in the faculty and control of the institution. This liberal victory owed a great deal to the efforts of Francis Hutcheson, professor of moral philosophy from 1730 until his death in 1746.[4]

Although Hutcheson preached liberal theology throughout his career in Glasgow, his party of like-minded professors failed to gain control of the faculty until shortly before his death. Their predominance was finally assured by Hutcheson's successful campaign on behalf of two liberal candidates for professorial appointments. The professorship of divinity was the first of these chairs to be vacated. The question of who was to judge the orthodoxy of this chair's occupant had lain dormant since the death of Principal Stirling in 1727, but in 1743, when William Leechman was nominated for the vacant chair, this highly charged question again became an issue. The election of the new professor was in the hands of the faculty, and Hutcheson, whose efforts to liberalize theology in both the kirk and the university had previously resulted in an accusation of heresy, became Leechman's leading partisan. But Leechman's opponents, who gave great weight to the kirk's criticisms of his theology, also had their allies within the faculty and the election resulted in a tie. This stalemate was broken in Leechman's favour when the rector voted for him but the church was loath to admit defeat. After his election, Leechman's enemies tried to keep him from being appointed by first refusing to free him from his rural parish and then, when this failed, by attacking him in the General Assembly of the Church. But this final onslaught was ineffective and in January 1744, a few months before Cullen moved to Glasgow, Leechman was admitted to the university as professor of divinity.

The battle for Leechman's election demonstrated that the university belonged neither to the ancients nor to the moderns. Leechman himself soon emerged as a leader among the liberal members of the university. Six months after joining the faculty he became its dean and many members wished to see him rise to a higher position. But the slim liberal advantage within the faculty was threatened when Alexander Dunlop, the professor of Greek and one of Hutcheson's oldest and firmest allies, announced his resignation.[5] Both factions put forward their candidates and the battle, which was to be Hutcheson's last, was joined. In 1746, the same year that Cullen began lecturing in the university, the Hutcheson group finally emerged victorious. 'From this event,' wrote Hutcheson's biographer, 'may be dated the triumph of the new party, which culminated in Leechman's election as Principal in 1761.'[6]

Until the beginning of the second third of the eighteenth century, the traditionalists in Scottish philosophy were primarily those Presbyterian ministers who favoured the old theology of salvation and reprobation and who thought in terms of the sixteenth-century philosophies which rendered these beliefs reasonable. But with the rapid decline of religious fervour among the upper class and the expansion and transformation of the universities, this kind of traditionalism lost its ability to command attention in the leading Scottish institutions and it gradually was reduced to being the distinctive creed of the dissenting sects.[7] In the 1740s, therefore, men like Hutcheson, who had struggled for the liberalization of the ministry and the universities, began to appear, when compared to many of their younger colleagues, as the new philosophical conservatives. They were well versed in seventeenth-century philosophy and they looked forward to a profound reconstruction of moral philosophy, but they were unwilling to accept conclusions which undermined their moderate religious principles or their ability to provide what they deemed to be proper moral instruction for their students. Thus when Cullen returned to the University of Glasgow, the rancorous struggles which had pitted the Elect against the profane had given way to the more genteel yet earnest debate between the philosophical sceptics and the common-sense moralists.

Hutcheson, of course, was much more than the leader of a faction within the faculty. He is chiefly remembered as a teacher whose pupils and disciples became the leading figures in the most notable generation of philosophers that Scotland has ever produced. In the study of moral philosophy his influence upon David Hume and Adam Smith was as determining as was Cullen's influence upon Joseph Black's study of chemistry. Hutcheson set the questions and defined the proper modes of investigation for his followers and yet, when compared to someone like Hume, he seems a faint-hearted philosopher, one who trusts reason only so long as it does not disturb his rather large set of preconceptions. But this judgment misconstrues Hutcheson's understanding of his own task as a philosopher. In his own mind he was above all else a teacher, a moralist, and in emphasizing the didactic aspects of his subject he was operating entirely within the traditional conception of philosophy as it had developed in the Scottish universities. A tension therefore emerged between the philosophies propounded by Hutcheson and his academic successors and the philosophies of his more notable disciples who operated largely outside the universities. This tension was created by the not easily reconciled desires to sustain the well-established educational function of philosophy while recasting the foundations of philosophy itself. Since philosophical modernization in eighteenth-century Scotland meant the reconstruction of philosophy according to the methods which had proven themselves so successful in the realm of natural philosophy, this tension troubled Scotland's moralists much more than her natural philosophers.

Although one would not expect this particular struggle between the traditionalists and the modernizers to have any significant effect upon natural philosophy, the setting in which the moral philosophers worked out their ideas encouraged several interesting developments that deeply influenced the Scottish understanding of Newton's achievement. Natural philosophers like Maclaurin, Cullen and Black understood and in a general sense shared the problems faced by their colleagues who taught moral philosophy: How can one organize and explain a body of knowledge in a way that is both philosophically defensible and pedagogically useful? This was a serious question that could only be answered by someone intimately familiar with the whole range of facts to be presented. Thus the Scottish natural philosophers who occupied university posts did not boldly build all-encompassing schemes of thought. The difficulties each professor encountered in teaching his own subject and his respect for his colleagues more than offset any desire he may have had to create a comprehensive system of philosophy by generalizing from his own area of competence. This restraint had a curiously tempering effect upon the search for new philosophical foundations. Since no single branch of philosophy was considered more fundamental than any other, all the philosophy professors stressed the method of Newton's science, which could be employed in all empirical investigations, rather than its content. Furthermore, they told their students that knowledge of the Newtonian method was not simply a means to truth, it also led, they claimed, to useful knowledge. It is therefore not surprising that gradually utility rather than the creation of a common ontology became the dominant characteristic of Scottish philosophy. Thus the philosophy professors in Scotland, because of their professional specialization and their shared commitment to education, came to agree upon a particular understanding of the Newtonian achievement. They valued Newton's methodological clarity more than his description of atomic matter and they saw natural philosophy as a distinct, but in no sense uniquely fundamental, branch of philosophy. In this setting it was relatively easy for Cullen and Black to be fully informed about current debates in the fields of logic and moral philosophy while devoting all their creative energy to the advancement of medicine and chemistry.[8]

A comparison of the philosophical programmes of Hutcheson and Hume will enable us to specify further the areas of agreement and disagreement in Scottish philosophy during the middle decades of the eighteenth century. Hutcheson did not consider himself an innovative philosopher. According to William Leechman, his protege and first biographer, he had 'no pretensions to new discoveries, but rather expressly disclaimed them'.[9] But in the early decades of the eighteenth century, Hutcheson's insistence that the empiricism of seventeenth-century natural philosophy should be accepted as the method for all branches of philosophy, including moral philosophy, still appeared a dangerously novel, if not entirely original, doctrine in the Scottish universities. Hutcheson's empirical point of view was supported by the deep attachment to Locke's philosophy that he acquired during his years as a member of the Shaftesbury circle in Dublin, yet he went beyond Locke in his distrust of metaphysics and in his rejection of ontological demonstration. As Scott, Hutcheson's modern biographer, has noted, he 'was painfully embarrassed by his indebtedness to Locke, . . . and he often speaks as empirically as Condillac'.[10]

A strong speculative urge saved Hutcheson's empiricism from degenerating into a mindless insistence upon the adequacy of experience. But whereas seventeenth-century natural philosophy assumed that mathematical formulation is the goal of all speculation and the guarantee of certitude, Hutcheson, in addressing problems of morality and taste, argued that mathematical certainty is not a proper goal in all subjects. According to Leechman,

> it was his opinion in this early part of his life, and he never saw cause to alter it, that as some subjects from their nature are capable of a demonstrative evidence, so others admit only a probable one, and that to seek demonstration where probability can only be obtained is almost as unreasonable as to demand to see sound or hear colours. Besides he was persuaded that attempts to demonstrate on such subjects as are incapable of it were of very dangerous consequence to the interests of truth and religion; because such attempts instead of conducting us to the absolute certainty proposed, leave the mind in such a state of doubt and incertainty as leads to absolute scepticism; for if once we refuse to rest in that kind of evidence, which the nature of the subject only admits of, and go on in pursuit of the highest kind, strict demonstration, we immediately conclude there is no evidence, because we do not meet with that kind of it which we expected; and thus the mind remains in a state of absolute uncertainty imagining there is no evidence, when all that the nature of the case admits of is laid before it, and enough to satisfy every one whose understanding is not disordered with unnatural thirst for scientific knowledge on all subjects alike.[11]

The historical importance of Hutcheson's argument can hardly be overstated. To avoid the intellectual demoralization so often bred by unreasonable expectations he asserted that there are realms of truth which can be empirically investigated but whose certainty cannot be judged by the logical criteria of geometry. By encouraging non-mathematical speculation, as well as reliance upon observable facts, Hutcheson made possible the development of different sciences, each having a degree of probability appropriate to its subject matter. Hutcheson's doctrine thus made possible the development of the human sciences, such as psychology, sociology, philosophical history and economics. It also fostered a view of science which justified Cullen's attempt to define an autonomous set of medical theories and a distinct science of chemistry.

In moral philosophy Hutcheson's methodology led him to postulate the existence of universal psychological properties. He opposed Samuel Clarke's attempt to deduce moral laws from logical necessity and argued instead that the moral sentiment or passion is the origin of all moral behaviour.[12] But this instinctive moral sense, while universal in man, manifests itself in diverse ways in different societies and circumstances. Thus Hutcheson's philosophy attaches great importance to the conditions of human existence, that is, to the subject matters of history, anthropology, and sociology, disciplines which his students avidly pursued.[13]

Adam Smith and David Hume were the most notable of Hutcheson's disciples. Smith, who studied under Hutcheson in Glasgow, later occupied the chair of moral philosophy himself, and he followed the plan of the 'never-to-be-forgotten Hutcheson' when he began lecturing on economic subjects.[14] Hume never studied with Hutcheson but, as Norman Kemp Smith has shown, his new view of philosophy developed under the influence of Hutcheson's earliest publications. 'As we can observe in Hume's *Treatise*,' Kemp Smith argues, 'it is with the eyes of Hutcheson that Hume reads and interprets both Locke and Shaftesbury; and, as we know from Hume's letters, the philosophy of his *Treatise* was already incubating in his mind as early as 1729, the year immediately subsequent to that of Hutcheson's second work. . . . This was no mere accident of temporal coincidence, and . . . Hutcheson's writings were in fact the immediate awakening influence that started Hume off on his independent and revolutionary ways of thinking.'[15] While Hume's statement concerning Hutcheson's contribution to the progress of his own thought is less emphatic than Kemp Smith's, he was completely open in his appreciation of the philosophy propounded by his friend and critic. In his *Philosophic Essays Concerning Human Understanding*, published in 1748, Hume commented that

> a late Philosopher ['Mr. Hutcheson,' added as a note] has taught us, by the most convincing Arguments, that Morality is nothing in the abstract Nature of Things, but is entirely relative to the Sentiment or mental Taste of each particular Being; in the same Manner as the Distinctions of sweet and bitter, hot and cold, arise from the particular feeling of each Sense or Organ. Moral Perceptions therefore, ought not to be class'd with the Operations of the Understanding, but with the Taste or Sentiments.[16]

But if Hume, in accepting the primacy of the passions, began as Hutcheson's disciple, he certainly pushed his thinking much further than the teacher cared or dared to. His unrelenting philosophic pressure revealed the unanticipated implications of Hutcheson's assumptions and created a new mode of reasoning, which has exerted a considerable influence of its own.[17]

Hume's *A Treatise of Human Nature: Being an attempt to introduce the experimental method of reasoning into moral subjects* was published in 1739 and, as Mossner has shown, its appearance was not neglected by the learned world.[18] True, Hume's argument was not immediately understood in all its depth nor was it acceded to, but his extension of Hutcheson's approach to philosophy gradually emerged as the statement against which Scottish philosophers had to define their own positions. Many scientists have read Hume's attacks upon the facile use of reason as attacks upon science itself and they in turn have often slandered the author of the doctrine they find so threatening. T. H. Huxley expressed the antagonism felt by many who have struggled with Hume's arguments when he said 'Hume exhibits no small share of the craving after mere notoriety and vulgar success as distinct from the pardonable, if not honourable, ambition for solid and enduring fame, which would have harmonized better with his philosophy'.[19] Yet it is clear from recent biographical studies that whatever the long-term effects of Hume's writings, his friends knew that he was not an ambitious misanthrope delighting in sophistry.[20] Far from mounting an attack upon man or reason, Hume sought a way, a philosophical method, by which the moral virtues described in the works of the ancient authors, which he loved to read, could be made a part of the realm of knowledge accepted and respected in his century. Having made this broadening of science his task, he used classical scepticism to beat back and hedge in the overarching claims of seventeenth-century rationalism. John Passmore, in his enlightening study of Hume's intentions, is careful to point out that although Hume used sceptical arguments, he was not himself a sceptic: 'The sceptical project which Bayle envisaged was one Hume could quite easily compass; and yet, at the same time, he had no intention of permanently persisting in a sceptical conclusion. Science, not merely peace of mind, was his objective; this fact sharply separated him from the classical sceptics.'[21] Hume also remained outside the schools of thought created by the French *philosophes* because, his biographer informs us, he could not accept either their rootless scepticism or their new faith.

> What must have struck Hume about the *a priori* character of the Holbachian atheism, the Helvetian materialism, and the Physiocratic economics, was that it showed a complete indifference to his own philosophy of mitigated scepticism. The old Aristotelian dogmatism of the schools was but replaced in France by a new dogmatism of inevitable progress. To some extent, therefore, Hume must have felt intellectually sequestered, intellectually lonely. The early conviction that he might have to wait for a distant posterity for the understanding of his philosophy may have been deepened by his experiences with the intelligentsia of France.[22]

Once the constructive thrust of Hume's philosophy is recognized, the function of his scepticism and his indebtedness to Hutcheson's programme for moral philosophy emerge as closely related aspects of his thought. A careful examination of reason and knowledge would lead, he believed, to the creation of a method which would serve the social sciences much as Newton's method had served the natural sciences. 'In his methodological moods,' Passmore declares, 'he thinks of himself as establishing the possibility of a *via media* between Cartesianism and scepticism by laying down rules for deciding what is probable; these rules are needed, because the social sciences cannot hope to achieve absolute certainty, and must work, therefore, with a logic of probability.'[23]

Hume's new methodology for the social sciences would have encountered much less resistance had he been willing to admit that there are fundamental differences between the three branches of philosophy, but unlike the more traditional professors of philosophy, he was unwilling to divide knowledge in this way. Truth about nature is knowledge of causes, he argued, and these causes are the creation of the mind reflecting on experience. Thus science becomes generalization; there is no knowable inner nature of things which science can discover. This limitation upon our knowledge is as true of Newton's own theories as it is of the social sciences. 'While Newton seemed to draw off the veil from some of the mysteries of nature,' Hume stated, 'he shewed at the same time the imperfections of the mechanical philosophy, and thereby restored her ultimate secrets to that obscurity in which they ever did and ever will remain.'[24] Before Hume could carry through the Newtonian revolution in the moral sciences, he had to purify Newton's theories of their rationalism. Unlike Newton, he did not even hold going beyond description as an ideal of science. As Passmore states, 'empirical science, to Hume, is the discovery that things in fact behave in certain ways; and it is not our incapacity which prevents us from discovering ultimate causes, but the fact that there are no such causes to discover. No connexion is any more rational than any other; all causes are equally "adequate", or, from the point of view of a rationalist, equally inadequate.'[25] General laws abstracted from experience were all that the scientific study of any subject could yield.

In the latter half of the eighteenth century Hume's devaluation of seventeenth-century rationalism became the common coin of Scottish natural philosophy. In 1780 Thomas Reid, the leading member of the Common-Sense school of philosophy, whose adherents opposed Hume's theological scepticism, wrote a letter to Lord Kames concerning causal explanation in natural philosophy. Like Hume, Reid argued that within natural philosophy we can do no more than determine laws. 'By the cause of a phenomenon, nothing is meant but the law of nature, of which that phenomenon is an instance, or a necessary consequence. The cause of a body's falling to the ground is its gravity. But gravity is not an efficient cause, but a general law, that obtains in nature, of which law the fall of this body is a particular instance. . . . A natural philosopher may search after the cause of a law of nature: But this means no more than searching for a more general law, which includes that particular law, and perhaps many others under it.'[26] But Reid did not follow Hume when it came to moral philosophy, for he believed that in this, the highest form of philosophy, truths of a kind inaccessible to the natural philosopher were discovered. 'Whether you call this branch of philosophy Natural Theology or Metaphysics, I care not; but I think it ought not to be confounded with Natural Philosophy; and neither of them with Mathematics. Let the mathematician demonstrate the relation of abstract quantity; the natural philosopher investigate the laws of the material system by induction; and the metaphysician, the final causes, and the efficient causes of what we see, and what natural philosophy discovers in the world we live in.'[27]

Whatever their stands on the relationship between natural and moral philosophy, all Scottish philosophers agreed that theoretical explanation means, in Adam Ferguson's words, 'referring particular operations to the principles or general laws, under which they are comprehended; or in referring particular effects to the causes from which they proceed'.[28] As causal explanation in natural philosophy was reduced to systematic description and as verification became identified with generalization mathematics and theories of matter lost the dominant positions they had occupied in the seventeenth century. Here then was a conception of causal explanation which Cullen could and did utilize in his lectures on chemistry. In a general statement on the nature of theory, Cullen told the students taking his chemistry course that 'theory consists of the doctrine of causes and of general rules for directing the operations of these causes. With respect to the doctrine of causes, all causes are known from experiment only. But when several phenomena are referred to one common cause, we are satisfied. Thus in explaining the effects of the mechanical powers, in finding the quantity of matter in the power and weight to be inversely as their velocities, we are satisfied in thus resolving it into a rule of the momenta.'[29]

In saying that theories succeed in so far as they bring psychic satisfaction, Cullen was merely conforming to the epistemological beliefs of his time, yet this conformity also had its effect upon his understanding of chemistry. Attraction, the first of the causes of chemical change, was thought of as a generalization, a law which accounts for certain regularities in nature, not as a physical entity caused by some hidden but knowable real property.

> Attraction is defined as a striving towards contact. An apology for this is unnecessary as it is liable to no objections besides such as are in common to other reasoning. Proofs of attraction: the spherical figure of fluids, the coalition of drops, the adhesion of fluids to the sides of vessels, the suspensions of fluids in capillary tubes. Others are not necessary as the doctrine is established and admitted. The cause of union and cohesion is not to be accounted for. Its laws depend upon the surfaces applied, the quantity of matter and the distance.[30]

And burning was considered to be the resolution of a compound composed of known constituents, not the manifestation of some phlogiston-like substance.

> Certain bodies are liable to and others incapable of inflammation. Therefore it must require a peculiar matter. Here a question arises whether inflammation depends upon bodies as mixts or whether it is essential to some elementary particle. I rather favour the former supposition because we know that in every inflammation an acid is present which is not inflammable as in oil, sulphur, alcohol, etc. There must be then some other matter requisite. This the chemists call phlogiston and imagine it to be an elementary matter of its self inflammable. Here, however, I differ from them, for the acid seems to have as great a share in

> the inflammation as the phlogiston. This phlogiston then of the chemists I suspect to be a mephitic air which we find arises from all inflamed bodies and enters sometimes into their composition, as in the vinous fermentation. Inflammation seems to consist in the resolution of this compound in the air.[31]

Hume of course not only argued against the rationalist assumptions, which made mathematics and matter theory the central concerns of traditional natural philosophy, he also attacked the notion that nature has a design and purpose which can be grasped by the human intellect. Here again his conception of causation provided him with his most telling arguments. 'All causes,' he asserted, 'are of the same kind, and there is no foundation for that distinction, which we sometimes make between efficient causes and causes *sine qua non*; or between efficient causes, and formal, and material, and exemplary, and final causes.'[32] By eliminating final causes and by denying that nature contains an essential structure waiting to be discovered, Hume implied that the only criterion by which scientific theories can be judged is utility. The mind, reflecting on experience, creates laws which give us a degree of control over our lives. 'The only immediate utility of all sciences,' Hume says, 'is to teach us how to control, and regulate future events by their causes.'[33] The distinction between the philosopher and the practical man remains, of course, even though Hume's approach to philosophy tends to blur it. The philosopher, seeking psychic satisfaction, creates laws of great generality, but even they must in some way anticipate the future if they are to be judged valid. Thus in Hume's reinterpretation of philosophy utility becomes the measure of philosophic as well as practical thought.

The fundamental utilitarianism of Hume's entire philosophy and of the natural philosophy of his contemporaries is also clearly evident in many of Cullen's statements about chemistry.

> Does the joiner want a particular glew? Does the mason want a cement? Does the dyer want the means of tinging a cloth of a particular colour? Or does the bleacher want the means of discharging all colours? It is the chemical philosopher who must supply these. In short, I need say but this one thing: Wherever any art requires a matter endued with any peculiar physical properties, it is the chemical philosophy which either informs us of the natural bodies possessed of these properties or induces such in bodies which had them not before, or lastly, produces new bodies endued with the necessary qualities.[34]

Had Cullen, in a different age, run practical and philosophic matters together in this way, he would have been quickly criticized for neglecting the true task of philosophy. But in Scotland during the eighteenth century the exclusion of final causes and intrinsic order from the realm of natural philosophy forced to the fore a conception of science that laid extraordinary emphasis upon utility, thereby providing yet another powerful argument for devoting great attention to philosophical chemistry.

Not all of Hume's colleagues were willing to accept the limitations he placed on the notion of cause. Henry Home, who became Lord Kames in 1752, although one of Hume's oldest friends and earliest patrons, used arguments based upon final cause throughout his life, and he maintained that 'he never did think as David does in that Treatise'.[35] Thomas Reid, as we saw above, accepted the need to dispense with arguments founded upon final cause in natural philosophy, but common sense alone was enough to convince him that the world was ordered according to a Providential design. Being a minister and a Christian, Francis Hutcheson could hardly hold the opinion that there is no discernible purpose to existence, and in his comments upon Hume's *Treatise*, Hutcheson used the term 'natural' in a way Hume found objectionable: 'Tis founded on final Causes; which is a Consideration,' Hume wrote, 'that appears to me pretty uncertain & unphilosophical. For pray, what is the End of Man? Is he created for Happiness or for Virtue? For this Life or for the next? For himself or for his Maker?'[36] Hutcheson, however, was more troubled by the tone of Hume's argument than by their disagreement over the appeal to final causes. Hume chose to argue abstractly while Hutcheson believed that the moral virtues were best served by fervour. In replying to Hutcheson, Hume said:

> what affected me most in your Remarks is your observing, that there wants a certain Warmth in the Cause of Virtue, which, you think, all good Men wou'd relish, & cou'd not displease amidst abstract Enquirys. I Must own, this has not happen'd by Chance, but is the Effect of a Reasoning either good or bad. There are different ways of examining the Mind as well as the Body. One may consider it either as an Anatomist or as a Painter; either to discover its most secret Springs & Principles or to describe the Grace & Beauty of its Actions. I imagine it impossible to conjoin these two Views. . . . Any warm Sentiment of Morals, I am afraid, wou'd have the Air of Declamation amidst abstract Reasonings, & wou'd be esteem'd contrary to good Taste.[37]

In carrying through his Copernican revolution in the moral sciences,[38] Hume chose to build upon the critical examination of human knowledge rather than upon the moral sentiments, final causes, common sense and warm persuasions favoured by many of his fellow philosophers.

Hutcheson, however, was above all a teacher, and while he admired Hume's brilliance he opposed his candidacy for the chair of moral philosophy in Edinburgh.[39] Hume's influence upon later generations has clearly established his stature as a philosopher, but it was Hutcheson who set the tone for philosophy in Glasgow during the middle decades of the eighteenth century. The most important medium for the expression of Hutcheson's philosophy was the lecture. He was a teacher who clearly loved his subject, his students and education itself. 'He was in no sense a system-builder, but rather a teacher who *preached* Philosophy, to whom a positive system was little more than a text.'[40] He refused to compartmentalize human understanding and he abandoned the old technique of delivering a Latin commentary upon a textbook. He lectured in English and took as his subject the whole field of 'Natural Religion, Morals, Jurisprudence and Government'.[41] Like many of the other professors, Hutcheson held private classes, largely for the benefit of young Glaswegians not enrolled in the university. Less typically, he devoted a good deal of time to counteracting the isolation fostered by the rank of professor and by membership in the faculty. He was particularly solicitous towards the lonely and often unruly Irish students, acting as their 'banker, friend and guardian'. Hutcheson's commitment to teaching and the extent of his influence serve as a reminder that a didactic tradition which is transmitted primarily through oral instruction can play an important role in a highly literate society.[42] Dugald Stewart, the son of Dr Matthew Stewart, who succeeded Maclaurin as professor of mathematics in Edinburgh, has described the particular influence which Hutcheson's lectures exercised over all university philosophy in Scotland:

> My father, who attended Dr. Hutcheson's Lectures, never spoke of them without much sensibility. . . . His great and deserved fame in this country rests now chiefly on the traditionary history of his academical lectures, which appear to have contributed very powerfully to diffuse in Scotland that taste for analytical discussion and that spirit of liberal enquiry, to which the world is indebted for some of the most valuable productions of the eighteenth century.[43]

By means of his personal example and his lectures, Hutcheson 'not only trained the new thinkers who were to follow him . . . but also prepared an audience for them'.[44] It was an audience that found in Cullen a teacher as committed to the advancement of his students and as stimulating in his lectures as Hutcheson himself.

During his first year as a university lecturer Cullen taught only the theory and practice of physic. The following year he repeated these lectures and added a second course on botany and materia medica. During his second year in the university Cullen was also able to arrange for a series of lectures on chemistry. Alexander Dunlop, the newly appointed professor of oriental languages, had been given permission to remain in Geneva as a private tutor during the first year of his appointment, and he suggested that the £30 saved from his salary be used to establish the teaching of 'chemie' in the university.[45] Cullen also asked that instruction in chemistry be offered, since in his opinion chemistry was an essential part of medical education. In January 1747 the faculty acceded to these requests and alloted £22 from general funds, in addition to Dunlop's £30, to equip a chemistry laboratory and classroom. These funds proved to be quite inadequate as £136 was spent on the laboratory during the first year alone, and the £20 which was alloted annually to sustain the lectureship in chemistry barely covered the cost of maintaining the laboratory.[46] The lectureship itself was a form of appointment that enabled the faculty to offer new courses within the university without incurring the additional expense or the dilution of powers which the creation of a new professorship would have involved. Those holding lectureships did not teach without remuneration, however, for student lecture fees, paid directly to the lecturer, were often greater than the stipend attached to a regular appointment.[47] Although Cullen was appointed as lecturer in chemistry, John Carrick, the assistant to Hamilton, the professor of anatomy, was to give the lectures. After delivering the first few lectures however, Carrick became ill and could not carry on. Cullen then had to take on the chemistry lectures while waiting for Carrick to recover and relieve him of this third course, but the illness persisted and he was obliged to deliver the rest of the chemistry lectures given that year.

During his second year of lecturing in the university Cullen demonstrated that he was prepared to teach the theory and practice of medicine, philosophical and practical chemistry, and systematic and pharmaceutical botany. Except for anatomy and surgery, which never captured his interest, Cullen, it seemed, could offer a complete medical education by himself. But he could not in fact continue to teach three subjects at the same time, since each course normally involved five lectures a week. He therefore decided to lecture on medicine and chemistry during the regular academic term while reserving botany for the summer. By the summer of 1748 Carrick had recovered enough to assist Cullen with the botany lectures, and after Carrick's death two years later, Cullen continued to offer the summer lectures by himself.

Joseph Black probably first encountered both Cullen and chemistry when he enrolled in the chemistry course which Cullen began teaching in November 1748. Black no doubt decided to continue studying under Cullen because the subject of chemistry captured his interest, but we may assume that in the early period of their relationship matters of personal style and attitude also carried considerable weight. Cullen was by all accounts a most attractive teacher and one can easily imagine an intelligent student, who wished to continue his studies finding the lecturer as attractive as any one of his subjects. At the outset, of course, students seldom distinguish between the characteristics of a teacher and the nature of the subject. For Cullen's students, the distinction never became an issue, for he impressed the order of his own mind and the concerns of his own life upon the organization and presentation of the subjects he taught. The integrity and purposiveness of Cullen's thinking, which complemented his evident erudition and skill, were the keys to his great success as a lecturer.

Of all the courses Cullen taught, chemistry proved to be the most popular.[48] As Doctor Wallace, who heard Cullen lecture during his first few years in the University of Glasgow, recalled, 'the number of students in the Physic Class and Materia Medica were very few, not above twenty. But in the Chemistry, the number was considerable; the lectures being calculated not only for the medical students, but for the general students of the University, and for gentlemen engaged in any business connected with chemistry.'[49] There were many reasons students flocked to Cullen's chemistry lectures. As Wallace mentioned, Cullen, unlike Plummer, directed his presentation towards all those students who happened to be interested in chemical problems. Like Hutcheson, Cullen lectured in English, an innovation that was clearly introduced to facilitate communication and not, as his detractors charged, to hide a faulty knowledge of Latin.[50] Another former student, perhaps punning on Cullen's contribution to pneumatic chemistry, said 'he was the first person in this country who made chemistry cease to be a chaos'.[51] And Lyon Playfair, in a precise appraisal of his predecessor's achievement, also praised this aspect of Cullen's lectures: 'Chemistry owes but little to Cullen as a discoverer, but much to him as a clear and philosophical expounder.'[52]

Cullen was much more than a professor. He never treated the subject being taught as more important than the well-being and development of his students and the lecture hall was only one of the many places in which he aided and encouraged them. Wallace remembered that 'his pupils were very much attached to him; indeed it could not have been otherwise, as every attention on his part was paid to their improvement.'[53] And Playfair described some of the techniques Cullen used to individualize and personalize his students' understanding of chemistry.

> His clearness of exposition alone entitled Cullen to the admiration which he received from his students. But this was not the only ground of attachment to him. He saw that a science like Chemistry was not to be taught by mere lectures, but that there must be a free and unreserved communication between the teacher and the taught. He cultivated the personal acquaintance of his pupils, and zealously aided them to overcome those difficulties which we all experience in ascending the first steps of the ladder of knowledge. He taught professors of Chemistry to act as tutors as well as prelectors, and to descend from the elevated rostrum of the lecture table to mix with the students in the freedom of an individual intercourse. To his example we owe much of the modern success in the diffusion of our science.[54]

Even in matters not directly related to academic subjects Cullen, like Hutcheson before him, generously assisted his students. He regularly visited sick students without taking a fee, a practice that later became customary in the medical faculty, and he often gave money to impecunious students.[55]

Such a comprehensive and genuine interest in the students' welfare no doubt drew many young scholars to Cullen's lectures, and although he never avoided the more demanding aspects of his subject, the attendance at his lectures continued to grow throughout his teaching career. During the decade in which he taught chemistry in the University of Edinburgh, the enrolment increased from 17 in his first course to 59 in his second and it reached a maximum of 145 in a single session.[56] By 1766, of all the courses offered by members of the medical faculty in Edinburgh, only the anatomy lectures attracted more students than Cullen's chemistry course.[57] Clearly it was not the subject matter alone that made these lectures popular. When, in 1763, Charles Alston, the famous botanist and professor of materia medica in Edinburgh, died in the middle of the academic term, the university authorities asked Cullen to continue his course of lectures. Cullen accepted the assignment and soon the attendance had increased from the eight or ten who had been attending Alston's lectures to over one hundred.[58]

Since Scottish students enjoyed considerable freedom in choosing their courses, we know that the large audience which Cullen attracted was not simply a flock of degree candidates fulfilling requirements. And since lecture fees were paid to the teacher, the popularity of a lecturer who presented ideas and information that the students considered worthwhile was rewarded directly. Cullen obviously flourished under these circumstances. It is therefore somewhat surprising to find him, in 1774, drafting a long memorandum arguing for legal regulation of Scottish medical degrees. As his friend Adam Smith gently reminded him, the institution of such regulation would probably disrupt the social arrangements that had kept the medical school in Edinburgh vital throughout the eighteenth century. Smith wrote that

> the monopoly of medical education which this regulation would establish in favour of Universities would, I apprehend, be hurtful to the lasting prosperity of such bodies-corporate. Monopolists very seldom make good work, and a lecture which a certain number of students must attend, whether they profit by it or no, is certainly not very likely to be a good one. I have thought a great deal upon this subject, and have inquired very carefully into the constitution and history of several of the principal Universities of Europe. I have satisfied myself, that the present state of degradation and contempt into which the greater part of those societies have fallen in almost every part of Europe, arises principally, first, from the large salaries which in some Universities are given to professors, and which render them altogether independent of their diligence and success in their professions; and, secondly, from the great number of students, who, in order to get degrees, or to be admitted to exercise certain professions, or who, for the sake of bursaries, exhibitions, scholarships, fellowships, &c., are obliged to resort to certain societies of this kind, whether the instructions which they are likely to receive there are or are not worth the receiving. All those different causes of negligence and corruption, no doubt take place in some degree in all our Scotch Universities. In the best of them, however, those causes take place in a much less degree than in the greater part of other considerable societies of the same kind; and I look upon this circumstance as the real cause of their present excellence. In the medical College of Edinburgh in particular, the salaries of the Professors are insignificant. There are few or no bursaries or exhibitions, and their monopoly of degrees is broken in upon by all other Universities, foreign and domestic. I require no other explication of its present acknowledged superiority over every other society of the same kind in Europe.[59]

Despite his ability to attract students and keep them interested, Cullen was not satisfied with a lectureship in Glasgow. The judgment of the marketplace, which Smith extolled, was gratifying, but Cullen's age, temperament and growing family sharpened his feeling of institutional and financial insecurity. To avoid being buffeted by the complex and often self-serving alliances which marked university politics, he needed more than student approbation. Unfortunately for Cullen, the death of the Duke of Hamilton in 1744 had deprived him of a patron who could have been of great assistance. Further unsettled by the Jacobite rising of 1745, Cullen experienced considerable anxiety during his early years in Glasgow. Writing to his brother he said 'it would take too much time to tell you what difficulties I have met with from the late Duke's death, the minority of the present, the Rebellion and other accidents.'[60] Even after seven years in Glasgow his medical practice remained disappointing. As he informed his friend William Hunter, he had many patients but little profit: 'I am quite tired of my present life; I have a good deal of country practice, which takes up a great deal of time, and hardly even allows me an hour's leisure. I get but little money for my labour; and indeed by country practice, with our payments, a man cannot make money, as he cannot overtake a great deal of business.'[61] In Henry Home, however, Cullen had one friend who would not stand by while he slipped into penury and institutional isolation. Home respected Cullen's knowledge of chemistry and shared his practical turn of mind, and he was in a position to provide all the support that Cullen required.[62]

Cullen and Home first became acquainted through a common interest in certain chemical problems. In 1740 George Martine published a collection of *Essays Medical and Philosophical*.[63] Martine was a philosophically minded doctor educated in Edinburgh and Leyden and he quite naturally devoted four of the six essays in his book to the experimental study of heat.[64] Although George Martine died in 1741, his brother Arthur was a friend of both Henry Home and Cullen and he doubtless knew of their common interest in the study of heat and its effects. He therefore, in 1748, 'transmitted to Dr Cullen the draught of an essay by Mr Home on Evaporation, with a request that he would state his opinion of the hypothesis which it maintained, and assist in performing such experiments as might serve for its elucidation'.[65] Cullen reponded by composing a long critique of Home's argument, thereby gaining a spirited but respectful correspondent and a valuable friend.

Henry Home was above all else an improver.[66] While he enjoyed theorizing, and promoted the formation of a philosophic society, he did so largely because, like his contemporaries, he believed philosophy was a useful study. In Cullen he found a man of great learning and considerable scientific imagination who also believed that theories should be useful. The theory of heat was therefore not the most congenial subject for their correspondence and they soon focused their attention on the topic of greatest interest to Home and of genuine interest to Cullen: the improvement of Scottish agriculture.

Cullen firmly believed that 'the lands of this country are mostly improvable, and are really improving',[67] and he further believed that as a chemist he could make a particular contribution to this general improvement. Since, as he told Home, his chemistry course 'was intended to teach the elements of a chemistry applicable to arts in general, agriculture claimed a place, and though I was not well prepared on that subject, yet I thought it was proper to make a beginning, and at least to open young gentlemen's views on the subject'.[68] While the students may have been surprised to hear agricultural improvement discussed in a chemistry class, Home and his circle of national leaders, all of whom were fully committed to economic development, were no doubt pleased to hear that a lecturer in one of their universities was investigating practical problems. Cullen, sensing an opportunity to advance both his subject and himself, responded warmly to Home's interest in his ideas and he told him in 1749 that he planned to undertake further agricultural research. 'I now take the liberty you were pleased to allow me,' Cullen wrote,

> of consulting you about my plan for prosecuting the study, or, if you please, the improvement of agriculture. Your recommendation of this subject has very much strengthened my own inclinations this way, and this would have appeared sooner, if my practice and the teaching a course of chemistry had not entirely occupied me during the winter; but now the last is over, I resolve, during the summer, to apply to some experiments that relate to agriculture; but lest I lose my labour in random experiments, I wish to have your advice, and to know from you what experiments are likely to prove decisive and fittest to ascertain the principles of agriculture.[69]

After having discovered Cullen's interest in agriculture, Home encouraged him with a flattering intensity which at times became overwhelming. In November 1749 he wrote to Cullen saying 'I live in the faith that you are going on briskly with your experiments. . . . I propose you as a member in our Philosophical Society, and indeed we greatly want members of activity and genius. I expect to hear from you, as you give me hopes in your letter, and hope for the favour of any discoveries you make from time to time, even those of the smallest consequence. Every thing from you will be acceptable.'[70] If Cullen were to focus his great scientific knowledge upon the problems of farming, Home seems to have believed, his investigations would result in a series of discoveries that would both enlarge their understanding and vastly improve their productive techniques. Thus in encouraging Cullen, Home believed he was contributing to a scientific revolution in agriculture that would benefit the whole nation. 'Do not forget,' he pleaded in 1752, 'your letter upon husbandry; having been entertained with no theory now for a long while, I am sinking into a mere practical farmer. I have not a single new thing at present, except one experiment I am making to convert moss into dung, by endeavouring to rot it in a dunghil [sic], by mixing it with fresh horse-dung. I shall let you know the result. If I succeed, I shall be able to multiply my manure greatly.'[71]

Cullen also expected his science to bear practical fruit, but as the person charged with formulating the useful laws, he was more modest in his expectations than Home. While teaching in Glasgow he began work on a manuscript entitled 'Reflections on the Principles of Agriculture' which, unlike Home's *Gentleman Farmer*, was never published, and he read an essay 'On the Construction and Operation of the Plough' at a public meeting.[72] But Cullen saw his position within the university as a more pressing issue than agriculture during these years, and in resolving this problem, which had nothing to do with theory, Home's friendship proved to be of immense value.

A year after first meeting Cullen, Home arranged that he be introduced to the Duke of Argyll. Arthur Martine, again acting as intermediary, told Cullen how he should present himself: 'Mr Lind, at Mr Home's desire, talked very particularly about you to the Duke of Argyll; and your friends here desire that you will wait on his Grace upon his arrival at Glasgow, which will be to-morrow evening. When you are introduced to him, it is probable, on your mentioning Mr Lind's name, that he will be more particular with you.'[73] Argyll's attention was well worth capturing for he was popularly known as 'the King of Scotland' because of his unparalleled political power and he was the leading patron of the University of Glasgow. The duke, who was also an ardent improver, was apparently impressed by Cullen, and after meeting him again a few months later he agreed to support his candidacy for the professorship of medicine. This candidacy seemed an impertinence to Cullen's university colleagues, for Johnstoun, not having died, had not vacated the chair. Therefore, despite the duke's support, Cullen had to wait over a year before he was admitted to his new post. During this time he fretted over the delay, asked Home to remind the duke of his promise, worried about the changing composition of the faculty, and thought about the effects his delay would have on the terms of his appointment. 'I have lived too long,' he complained, 'without a settled plan, which has always kept me distracted, and unsteady in my application to study.'[74] The duke was informed of Cullen's uneasiness and in March 1751 he wrote saying, 'I was very sorry that your affair was so long delayed, which I had good reason to believe would have been finished half a year sooner.'[75] Having expressed his continued concern for Cullen's advancement, the duke went on to mention a piece of parlour chemistry with which Cullen had amused him. 'People are greatly surprised at the metal you gave me that melts in hot water. Pray send me the proportions of the ingredients, for I have forgot them.' Cullen's chemistry served him well, and in 1751, after Johnstoun had finally retired, he was formally admitted into the university as professor of medicine.

By the time he had obtained the long coveted professorship in Glasgow, Cullen, with Home's assistance, had already begun planning his next move. Glasgow had provided him with opportunities to display his talents and he in turn had done great service for the university, but Edinburgh was the political and cultural capital of Scotland, and the finest medical school in Great Britain, if not the world, was located within her university. Five years after becoming a professor in Glasgow Cullen received a similar appointment in Edinburgh. This 'call' to the capital city was in fact the fruit of a long and carefully waged campaign.

In 1751 William Hunter received a complete description of Cullen's new goal and his plans for attaining it. 'I have some thoughts of acceding to a proposal that was lately made to me of removing to Edinburgh,' Cullen wrote.

> Dr Plummer, Professor of Chemistry, is a very rich man, has given up practice, and had proposed to give up teaching in favour of Dr Elliot; but this gentleman died about six weeks ago, and, upon this event, some friends of mine, and along with them some gentlemen concerned in the administration of the Town of Edinburgh, have proposed to use their influence with Dr Plummer to induce him to resign in my favour. As the income of that office cannot be very considerable, and my success in the way of practice is uncertain, I have hesitated about agreeing to their proposal; but, provided they can make the establishment such as will afford me a livelihood, the situation and manner of life there will be so much more agreeable than at present, that I resolve to hazard something, and have agreed to accept the invitation when made to me in a proper way. However, Plummer's consent, and some other circumstances, are still in doubt; and this, with other reasons, requires the affair to be kept as secret as possible.[76]

For three years the campaign to bring Cullen to Edinburgh was conducted with considerable enthusiasm. Success depended upon the availability of a chair, an adequate private practice, and the approval of the Town Council. Late in December 1751 Henry Home reported that the necessary vacancy had not yet been created: 'The project about Dr Plummer I find is not yet ripe, but I keep my eye close upon it.' [77] Over a year later he was encouraged by the failing health of two Edinburgh physicians: 'Poor Sinclair, I am afraid, will scarce be alive when this comes to your hand, and Dr Lowis, another physician of note, is in a very bad way. In short, your presence here, which I formerly thought would be useful to yourself, will shortly be necessary for the benefit of the town.' [78] But both Plummer and the Town Council remained uncooperative. 'I have got no return as yet from persons I employed upon Dr Plummer,' Home reported. 'Do not expect from the town any thing farther than countenance. Had I the leaders only to deal with, I believe I could operate more than you expect. But the town is poor, and such factions in the Council, about calling of ministers and candidates for Parliament, that there is no such thing as venturing any extraneous business.' [79] Early in 1754 the faithful Arthur Martine suggested that Cullen lay his latest scheme 'before Drs Clerk and Sinclair, as a proposition of your friends, which you think of acceding to with their good will. At the same time, we shall think of measures to treat with Plummer or Alston about resignation.' [80] In August William Hunter reported from London that the Duke of Argyll was favourably disposed to Cullen's move, [81] and the following month Henry Home, now Lord Kames, found in the death of Dr Dundas the occasion Cullen had been waiting for. He told Cullen that there remained in Edinburgh

> scarce one physician of any considerable character. It is my serious opinion you can never have a more favorable opportunity to exert yourself. It will be neglecting your own interest not to lay hold of it. In all appearance you cannot fail of coming soon into business here; and, in the interim, you may give what may be called a private college (Course of Lectures) of Chemistry, which will be successful, and will make you a favourite of the College of Physicians, because it will increase the number of students. In a word, the opportunity is so favourable that it is needless to bestow words upon it. Dr Dundas had a world of business, not only in the town, but in all the towns about it. You may naturally take his place till multiplicity of business confine you within the city. In the mean time, you may find leisure to publish a treatise upon husbandry. [82]

Cullen clearly encouraged these interventions on his behalf, but he could not be as lighthearted as Kames in managing his own affairs. As he had indicated in 1751, he would only surrender his position in Glasgow for a professorial appointment in Edinburgh, and as the prospect of an opening diminished, Cullen lost his enthusiasm for the whole venture. By the end of September 1754 Kames had begun to understand. He ceased pressuring Cullen to move and the campaign was suspended.[83]

In his public life Cullen gave the impression of being fully satisfied with his situation in Glasgow. He continued to lecture on chemistry and on the theory and practice of physic during the academic term, he pursued his own chemical investigations, and he tried to improve the techniques used in farming and industry. Upon becoming a professor he also exercised his rights as a member of the faculty, where he found a like-minded ally in Adam Smith. Smith had been appointed professor of logic a few days after Cullen was admitted as professor of medicine. Later in 1751, after professor Craigie died, Smith, who had begun teaching Craigie's course during his final illness, succeeded him as professor of moral philosophy. This move left the chair of logic vacant. Shortly thereafter, Smith sent Cullen a letter presenting his opinion concerning the candidates for this chair and expressing his preference for David Hume: 'I should prefer David Hume to any man for a colleague; but I am afraid the public would not be of my opinion; and the interest of the society will oblige us to have some regard to the opinion of the public.'[84] Cullen and Smith were evidently of one mind, and despite their reservations, both professors supported Hume in his unsuccessful candidacy. There was, however, one happy consequence of this affair; the friendship which Hume and Cullen came to share. The philosopher wrote to Cullen thanking him for his support and concluded by saying, 'whatever the reverend gentlemen may say of my religion, I hope I have as much morality as to retain a grateful sentiment of your favours, and as much sense as to know whose friendship will give greatest honour and advantage to me.'[85]

During the summer of 1755 Cullen's chances of gaining a professorship in Edinburgh suddenly improved. Professor Plummer was stricken with an illness which, it appeared, would force him to abandon lecturing permanently, and several candidates, including Doctor Francis Home, Cullen, and Black, were put forward for his position. Home, whose candidacy was strongly supported by his relative Lord Milton, had previously published an analysis of the water from a Scottish spring,[86] and it seems likely that Cullen and Black, whose papers on evaporation and magnesia alba both appeared in the 1756 volume of essays published by the Philosophical Society in Edinburgh, hurried into print to strengthen their own claims to Plummer's chair.[87] Black seemed a particularly attractive candidate to the other members of the medical faculty because, while clearly qualified as a chemist, he was not an experienced physician and was therefore not likely to cut into their profitable practices.[88] Black, then in Edinburgh, kept Cullen informed on all aspects of the situation.[89] Kames of course threw himself into the competition wholeheartedly. He 'wrote immediately to Provost Drummond, urging Dr Cullen's claims to be Dr Plummer's successor;—to Dr Whytt, recommending him for a colleague;—and to Lord Milton, pointing him out as the fittest person in Europe to fill the Chemical Chair, and as the candidate who would be most acceptable to the Duke of Argyll.'[90] While the appointment itself was in the hands of the Town Council, the opinion of the medical faculty and that of certain powerful political persons, particularly the Duke of Argyll, could not be neglected. Once again Arthur Martine extended himself on Cullen's behalf: 'Since parting with you,' he wrote,

> I have had conferences with Drs Sinclair, Porterfield, Macfarlane, and Monro, and said what I thought proper to each upon the subject in question. I find from Mr Russel, that Dr Black will be no sanguine antagonist. It is from Lord Milton's influence with the Town-Council that any danger is to be apprehended. I have had two full conferences with Mr Kincaid, who is very hearty and active. He has spoken to the Justice-Clerk, and is to speak to all his friends in Council, and to exhort them all earnestly not to give any promises but agreeably [sic] to the Duke of Argyll's pleasure.[91]

Plummer's being incapacitated, rather than dead, posed the chief difficulty from Cullen's point of view, for a substitute lecturer rather than a successor would apparently suffice. At the request of Plummer's relations and the professors of medicine, Black agreed to teach the chemistry course while Plummer was ill. This ploy did not, however, settle the problem of succession, for the Town Council was not happy about losing control of the appointment and Cullen still had a powerful ally in the Duke of Argyll. For several months circumstances and the conflicting claims held the question in suspense. It appeared that Plummer might recover, and neither the Town Council which favoured Home, nor the faculty, which favoured Black, was willing to yield. During this period of stalemate the Duke of Argyll visited Edinburgh and brought to bear the influence which finally gained Cullen the appointment. 'At length, on the 19th of November 1755, when the session was about to open, and when it was generally believed that Dr Plummer would not recover, the Town-Council appointed Dr Cullen joint Professor of Chemistry during the life of his colleague, with the succession in the event of his death, reserving at the same time to Dr Plummer all the rights and privileges of a Professor, and particularly that of teaching, whenever the state of his health would admit it.'[92]

Although the competition for Plummer's post never threatened the friendship between Cullen and Black,[93] the medical faculty was slow to forget the manner in which Cullen had obtained his appointment. In late November 1755 Black told Cullen what kind of a reception to expect from his new colleagues:

> It is indeed very proper you should come as soon as possible, in order to settle matters with respect to the teaching for this winter, as I am afraid you will receive but little help or encouragement from the Professors. They all seem to be very much out of humour at the Town-Council's having managed this affair with so little ceremony, and as if the College had no sort of concern in the matter. Plummer himself will certainly be highly incensed; and as the laboratory is entirely his property, you need not expect to obtain the use of it for this winter.[94]

The professors could not reverse the decision of the Town Council, but they hoped to deny Cullen admission to the Academic Senate. To gain acceptance in the university Cullen was obliged to obtain Plummer's consent and purchase his laboratory. Otherwise, the eight members of the university opposed to Cullen's admission planned to hold out until a court of law had defined the rights of the faculty and those of the Town Council. Cullen was obviously not anxious to force a confrontation on a question of principle and privilege. He began teaching without obtaining formal admission and waited for Plummer's failing health to resolve the dispute. However, a compromise was soon arranged by Alexander Monro, the professor of anatomy, and George Drummond, Provost of Edinburgh. After settling the question of the laboratory, Cullen and Plummer both resigned and, on 10 March 1756, were elected joint professors of chemistry. Plummer's death the following July left Cullen in sole and complete possession of the chair he had sought for so long.

The lateness of the Town Council's decision exacerbated the difficulties which naturally attended Cullen's transfer from Glasgow to Edinburgh. In his first letter to Black after hearing of his own appointment, Cullen anxiously asked whether he planned to teach chemistry in Edinburgh that winter, and he suggested that they combine their efforts. 'I am pressed from Edinburgh to come to teach this winter,' Cullen wrote, 'but this, particularly on account of my engagements here, I would avoid if possible; however, if you really dropt your former engagements, and would now in any shape concur with me, I think I could make the matter easy to us both. In short, if you can be concerned with me in teaching, I shall ease you of a great part of the labour,—I want no part of the profits.'[95] At the same time Cullen was doing all that he could to have Black succeed him in Glasgow. In December 1755 William Hunter reported from London that 'the Duke [of Argyll] is most sincerely your friend. He has not yet received yours and the Edinburgh recommendation of Mr. Black.'[96] During the following year, however, the recommendations were evidently received and approved, for in 1756 Black was appointed lecturer in chemistry and professor of anatomy and botany in the University of Glasgow. The next year he exchanged his chair for that in medicine, thereby fulfilling Cullen's plan to have his most accomplished student succeed him in Glasgow.[97]

Cullen was meanwhile experiencing a difficult first year of teaching in Edinburgh. A new appointment, the need to establish a practice, illness and injury all contributed to making the first half of 1756 hectic. But towards the end of the summer Cullen told his brother-in-law that the worst was past.

> I came in here to teach in the month of January, and was obliged to be very busy that I might overtake the session. My hurry was much increased by the frequent interruptions which my wife's illness gave me by calling me to the west country. She has had a long and tedious illness this spring, which, however, I hope is now well over. Just when I was leaving Glasgow, about four weeks ago, I got a broken shin, which was a small matter at first, but being therefore neglected and very ill used, it has since given me a great deal of trouble, and confined me to the house, and almost to my bed, ever since I came to Edinburgh. My family are still at Parkhead, and to continue there for the rest of the summer. My wife has been here to attend me; but now I begin to go abroad, she went west on Thursday to see the children, among whom there are some small ailments.
>
> As to the effects of my change of situation, I really as yet can say nothing. The hurry I was kept in by my teaching during the time I was here in the spring, did not allow me to take the measures for getting practice that may be necessary. The business of my removing from Glasgow, and other affairs in the west, have still hindered me from doing any thing here. My sore leg has now spent a great part of the summer session, and in the harvest vacation there is nothing to be done; so that, upon the whole, I shall not think myself settled here till winter, and even then I must lay my account with business coming slowly. Some

> time must pass before I can either know myself, or tell my friends how I am to fare in this place. Only I shall tell you that I live in good hopes.[98]

By the beginning of the next academic term Cullen was securely installed as a member of the faculty, he was preparing lectures on the history of chemistry, and he was looking forward to teaching in the finest, most cosmopolitan medical school in the world. A generous, learned, imaginative man at the height of his powers and at the centre of Scottish society, he was ready to lecture on medicine and natural philosophy before a student body and a nation eager to be enlightened. And Black, his first great contribution to chemistry just published, was at the same time beginning a decade of teaching and research in Glasgow, work which culminated in his doctrine of latent heat and earned him a call to Edinburgh where, as his master's successor, he enjoyed international fame as a teacher and chemist during the final decades of the eighteenth century.

[5]

Cullen's Chemical Investigations

There is no Cullen process in chemical technology, and theoretical chemists do not speak of Cullen's law or Cullen's constant. Scientists quite naturally record only the outstanding achievements of their predecessors, and while Cullen's success as a teacher has never been forgotten, the chemical research which occupied much of his time in Glasgow has understandably attracted little attention from historians of chemistry. If one searches through Cullen's chemical investigations hoping to find a research programme that bore fruit, the yield will be nil. He proposed no useful new techniques for farming; he failed to discover a way to make inexpensive alkali from natural substances available in Scotland; he was unable to find an economical method for purifying Scottish salt; and he never succeeded in correlating the heat flow in chemical reactions and the qualities of the reacting substances. But if one examines his investigations to learn more about the man and his influence upon the development of his pupils, then the search becomes rewarding, for in his research Cullen focused on problems of considerable social and intellectual importance. In his choice and handling of specific research projects, as in his lectures and in his conduct as a teacher, Cullen reveals the degree to which he shared the concerns that were dominant in his particular historical setting and the extent of his contribution to Joseph Black's success as a chemist.

Cullen was interested in the progress of practical chemistry for two reasons. He knew that the controlling figures in Scottish society were eager improvers, men who would welcome and reward technical developments which made Scottish products competitive in the burgeoning markets controlled by British merchants. But Cullen was not a merchant himself; he wanted the security of a university position more than great personal wealth and economic power. To gain what he wanted, however, Cullen had to win the support of these aristocratic and commercial improvers. Through Kames, whose acquaintance Cullen considered 'one of the most fortunate accidents' of his life,[1] he brought his suggestions for improvement to the attention of the men who could powerfully influence his own situation. If he could give them what they wanted, and what Cullen sincerely wanted too, since he shared their faith in improvement, he knew that they would take care of him. The constant support of Kames and Argyll clearly reveals that Cullen had not misjudged the extent of the benefits available to those who joined the campaign for improvement.

Cullen also had a more disinterested reason to take up improving research. As a philosophical chemist he believed he had the knowledge needed to investigate and rationalize the processes of production. It is difficult, upon first encountering this conviction, not to judge the chemist arrogant in his estimation of his own ability and ignorant of the difficulties which plague the artisan's craft. But Cullen was no dreamer peddling a Faustian promise of escape from the complexity of the real world. He addressed practical problems because all chemistry had to be founded upon the close examination of phenomena. Since chemistry, in its modern form, was a new science, Cullen did not expect to recast practice according to the dictates of theory; he simple wanted to use the methodological techniques of science to examine the rich store of phenomenal knowledge available in the practical arts. Perhaps the philosopher would be able to suggest improvements to the artisan, but to do so he would have to master the practical techniques of the craftsman and then add some useful knowledge he had gained as a scientist. Certainly the philosopher would find in the practice of the crafts facts and relationships which, when translated into philosophical language, would enable him to construct a more adequate theory of chemistry. Cullen expected his studies to advance both philosophy and the arts at the same time. His research can be classified as practical or theoretical according to its emphasis, but in his own mind the two goals could never be separated. Chemistry, whatever its purpose, remained the study of chemical phenomena.

During the second quarter of the eighteenth century the British Government established several commissions charged with encouraging the production of Scottish goods for export.

> In 1727 two important Acts became law: (a) an Act for the better Regulation of the Linen and Hempen Manufactures in Scotland, and (b) an Act for encouraging and promoting Fisheries and other Manufactures and Improvements in Scotland. The latter brought into being the Board of Trustees for Manufactures (1727–1832), which made it its business, together with the Society of Improvers in Agriculture (1723), and the British linen Company (1746), to foster the linen industry. This industry was thus fortunate in receiving State encouragement for the greater part of a century.[2]

The Board of Trustees, which soon established itself as the chief vehicle for the promotion of Scottish industry, immediately set about improving the quality and quantity of native linen.

> The Proceedings of the Board reveal the problems involved in the development of the linen industry. At the outset it was realized that success would depend on availability of skill and capital and the creation of adequate marketing arrangements. When the Board took office Scottish linen was poor in quality. . . . At their first meeting on 20 July 1727, they were determined to tackle the subject on a broad front. . . . Of the £6,000 income, £2,650 was allocated to the linen industry, a like sum to the fishing industry, and £700 to the coarse woolen industry.[3]

Because bleaching was the weak link in the Scottish linen industry, the Board decided to use nearly all the linen funds to stimulate improvements in this area of production. A few years of experience, however, forced the Board to reconsider the wisdom of its investment, for it became evident that if Scottish bleachers continued to use potash, sour milk and sunlight as bleaching agents, the quantity of linen produced in Scotland could not be significantly expanded. Thus well before the middle of the century the lack of a cheap and efficient method of bleaching had been identified as the bottleneck restricting the development of the entire linen industry.[4]

Potashes are made from wood, a substance that had become, by the eighteenth century, as scarce as intense sunshine in Scotland. But Scotland contained many natural resources, one of which, it was hoped, could be made to yield an alkali suitable for bleaching. Limestone seemed a likely candidate, for when roasted it yields quicklime, and quicklime, when mixed with water, forms a mild alkali solution called lime-water. In practice, however, it is difficult to separate lime-water from the undissolved slaked lime suspended in it, and lime-water therefore proved to be too caustic for bleaching when used by itself or mixed with potashes. 'It may have been this that caused the passing of a series of Acts, both in Scotland and Ireland, forbidding the use of lime in bleaching. Under later Acts, officers of the Board of Trade were given powers of search and seizure, and bleachers found using lime were prohibited from bleaching for several years.'[5]

Although banned in practice, bleaching with lime continued to attract the attention of research chemists throughout the century. In 1736 James Dunbar wrote that 'I know there are a great many who are enemies to bleaching of cloath with Lime (and so am I), because the Lime in the bucking penetrates and lurks in the cloath, which burns it, and makes it a yellow colour. . . . But Lime-Water is of a different Quality, for it can be taken inwardly by men and women, and it will whiten cloath sooner than the best Lee [i.e., alkali solution made from potashes] you can make.'[6] Sixteen years later Charles Alston remained hopeful: 'I am much persuaded,' he said, that 'Lime-water, yea and Lime too by itself, without Ashes, might be of great Use in bleaching Linens, though it is commonly believed to burn the cloth.'[7] In 1754 Black revealed his own interest in this problem when he wrote to Cullen from Edinburgh saying 'I long vastly to hear what improvement you have made on bleaching ashes. Do you make a cheaper ley from pearl-ashes and lime than that commonly employed; and have you calculated this matter exactly?'[8]

The possibility of using lime in bleaching continued to interest Joseph Black for at least another decade. In 1752 his brother Sam bought a bleaching green in northern Ireland, and John Black, their father, began to see some value in his son's study of alkalis and caustics. In 1763 Joseph informed his father that 'I have at last wrote a paper upon the use of lime in bleaching which I transmitted some posts ago to Mr. Ferguson, & shall do anything further that is in my power.'[9] A month later Black sent Ferguson a long letter containing further considerations on the implications of his research on lime for bleaching, and in 1771 his essay entitled 'An Explanation of the Effect of Lime upon Alkaline Salts; and a Method pointed out whereby it may be used with Safety and Advantage in Bleaching' was published as an appendix to the second edition of Francis Home's *Experiments on Bleaching*.[10]

Despite this continued interest in bleaching with lime, at mid-century it seemed unlikely that lime could ever be used to replace potashes, and Cullen, who was not seeking a theoretical solution to the bleaching problem, looked elsewhere for a cheap domestic alkali. To give direction to his research and to gain the support of leading improvers, he set down what he knew about bleaching and described the experiments he proposed to undertake in a memoir entitled 'Remarks on Bleaching'. 'In the following remarks,' Cullen begins,

> I have attempted to apply the little knowledge I have of chemistry to the improvement of Bleaching, but I am sensible that a person who proceeds on general principles without being minutely acquainted with the practice and detail of the art is in danger of proposing refinements that are often impracticable in the way of trade or of little use when practiced. I have indeed endeavoured to obviate this by conversing with bleachers and looking in to their practice, but I find still so many experiments wanting that I dare make no conclusions and therefore give the

> following reflexions as so many conjectures only that may lead to experiments which I hope will prove servicable to the Art.[11]

Kames evidently made sure this essay got into the right hands. In July 1752 he could only report that 'Lord Deskfoord has not yet got your lucubrations upon bleaching and upon ashes,'[12] but five months later he had more encouraging news for Cullen: 'I have got back your paper from Lord Deskford, and by his directions I am to send for John Chrystie (bleacher at Ormiston), and to see what he has to say upon the hints you suggest. This may be of some use: and, if any thing come out worth your notice, I shall be sure to acquaint you.— I think Bleaching a great object for this country; and now that a rational account is set on foot, I wish to have it carried to the utmost length.'[13]

Although he had obtained a hearing for his proposal, Cullen had not been promised support. Convincing the Trustees of the Board that his project deserved funding and that his own interests were primarily, if not entirely, scientific was not an easy task. Early in 1753 he conveyed his most recent thoughts on bleaching to Kames and asked him to push on in his efforts to secure support for his research. 'I am,' Cullen wrote,

> much obliged to Lord Deskford for the remarks he has made. From some of them I have received new information; by others I am properly corrected, and I am very sensible of the defects his Lordship has pointed out. One of them concerning the examination of Water I shall give my opinion of very soon. The other, the examination of Ashes, I cannot properly direct without many experiments, which I could not hitherto possibly find time to make. John Chrystie's notes give me great satisfaction. If he is not very complaisant, I must be better pleased with my own piece than I could be before. I am glad to find a man of judgment and experience agreeing with me in so many particulars. The most part of the instruction I had received on the subject, I had got from his directions; and from his notes on my remarks I have learned a good deal more. In some particulars he seems to differ from me, and I am ready to believe he is in the right. If there be any thing about which I am in doubt, it is not to be discussed by conjectures and reasoning, but must be referred to experiment; and on this footing I would let the whole affair rest for some time. But I must now own to your Lordship, that I have so far a good opinion of my first attempt on this subject, that I have a very strong inclination to prosecute it. It is a pretty wide field for experiments, and my fancy grasps at the whole; but I know well that my hands can only comprehend a

> small part of it. After the hints I have given, I would leave the experiments to be made in the bleaching-field by persons who have more skill and better opportunities for these; and I would choose to confine my views at present to the subject of ashes. The more I consider this subject, the more fully I am persuaded that all kinds of ashes, and all kinds of alkalies, may be made in Scotland; in what quantities and at what expense, I cannot judge but by trial. These trials I have an inclination to make but cannot afford the expense. If your Trustees have so far an apprehension of the utility of such an inquiry as to hazard a little money upon it, I shall hazard my pains, and, what may be a much more expensive article to me, my time, upon it. If the trustees are not likely to be very prompt in this affair, I would not have it mentioned. I shall not pretend to be disinterested. On the contrary, I should be very glad to find my interest connected with schemes of this kind; but at present I would be hurt by any suspicions of interested views. I have a great ambition to establish Arts,—I have not avarice enough to practise them. If you do any thing in this affair, some memorial or plan may be necessary, and you shall have it on demand.[14]

Kames found Cullen's presentation of his request quite compelling. In March 1753 he wrote Cullen again, saying that 'I could not think of doing better than to put your letter into Lord Deskford's hands. If the Trustees do not grasp at your offer, by defraying the expense of experiments, I shall not be at a loss to form a judgment of them.'[15] A week later he reported that there had been further progress coupled with some indication of delay:

> I have delivered to Lord Milton a letter from Deskford upon your proposal, and what I may call generous offer, recommending the matter to him warmly. But great bodies move slowly, and a certain habit of business leads some people into a slow and regular train, which no peculiarity of circumstances can provoke them out of. You will perhaps reckon it too sanguine an advice to go on with your experiments, with a reasonable, or rather certain, prospect of a retribution. But this you may safely venture, to set out with the cheapest experiments, to give notice to Lord Deskford that you have done so by my advice, and that you are ready to go on if he gives you encouragement.[16]

Taking Kames' advice to heart, Cullen continued investigating the various kinds of ashes that might yield an alkali for bleaching and, as Kames had promised, the Board did not forget him. 'Cullen had several contracts with the Board of Trustees, and on 13th June, 1755, he was awarded a premium in the form of three sets of table linen to the value of £21 for his ingenious experiments and observations on the art of bleaching.'[17] In the same year Black was able to report from Edinburgh that Cullen's research had made a lasting impression upon one of his patrons: 'I had lately the honour of dining with Lord Deskford, who spoke very favourably of you. He asked me if I had seen ——'s new book upon Bleaching, and intimated that the whole, or greater part of it, was stolen from a paper of yours.'[18] Clearly this sustained concentration on the chemical problems raised by the process of bleaching, while yielding no significant industrial innovations, was one source of Joseph Black's interest in the classification of alkalis and the nature of causticity.

During the 1730s the Scottish fishing industry appeared to be as open to improvement and expansion as the linen industry. At first the Board of Trustees tried to stimulate Scottish fishing by paying premiums to those who would equip themselves to catch herring, but there was little response.[19] By the 1740s subsidizing bleachers seemed a more promising investment and the Trustees, therefore, practically abandoned the fishing industry. The fishing grounds which the Scots neglected were, however, providing rich returns to the Dutch. The Government in London, not at all happy about the economic and naval implications of this foreign intrusion, decided to act aggressively. Since the Board had shown that subsidizing the existing industry was ineffective, the Government intervened directly through the Act for the Encouragement of the White-herring Fishery, passed in 1749. The passage of this Act indicates that despite the attitude of the Board, concern over the problems plaguing the fishing industry remained high at mid-century.

The price of salt was the most serious of these problems. For centuries the Scots had produced salt by using the coal which was readily available near the Firth of Forth to boil sea water, and this industry, thanks to a preferential tax, survived the test of English competition following the Union.[20] But all this Scottish salt was of no use to a fishing industry trying to expand its market. Scottish salt and, to a lesser extent, English salt contained so many impurities that only imported salt could be used to preserve fish for foreign markets. The dilemma was clearly stated by William Brownrigg in 1748:

> [Rock] salt is in several countries found so pure, that it serves for most domestic uses, without any previous preparation (triture excepted). But the English fossil salt is unfit for the uses of the kitchen, until, by solution and coction it is freed from several impurities, and reduced into white salt. The British white salt also is not so proper as several kinds of bay salt for curing

fish, and such flesh meats as are intended for sea provisions, or for exportation into hot countries. So that for these purposes we are obliged, either wholly or in part, to use bay salt, which we purchase in France, Spain, and other foreign countries.[21]

The British Government quite naturally tried to ease these difficulties by modifying the excise regulations on salt. Foreign salt used to cure herring for export was lightly taxed. English salt could be imported into Scotland only to cure herrings to be sold in the English market. However, the market for English salt in Scotland was constantly being undermined by the illegal use of foreign salt. This prompted an administrative reaction so complex that it effectively throttled the importation of foreign salt for the Scottish fishing industry. 'Because of the difficulty of preventing foreign salt, free of excise duty, being directed to other uses, such as the curing of herring for the home market, extremely intricate and costly restrictions were imposed.'[22] Clearly the impasse created by these regulations could be most easily removed by developing a technique for producing pure salt in Scotland.

A technical solution to this problem was precisely what Brownrigg offered in his book on *The Art of Making Common Salt*. He was particularly concerned with the impurities which contaminated British salt made from sea water. 'Although the English have considerable improved their methods of boiling salt,' he wrote, 'yet *this art is still practiced with greater skill and success by their allies the Dutch*, as the superior goodness of the fish, cured with their salt, doth sufficiently prove.'[23] There was, he argued, no intrinsic reason why British salt had to remain impure. The causes were accidental, the result of ignorance, and could be easily remedied. 'The following sheets were wrote,' Brownrigg said, 'with a design to remove these inconveniences, by shewing how the subjects of Great Britain may be supplied with salt of their own manufacture, fit and sufficient for all their occasions.'[24]

To isolate the offending contaminants and to determine at what point in the process they enter the salt, Brownrigg studied the constituents of sea water in the order in which they separate out during boiling.[25] Patiently he explored every step of the art. Having thoroughly studied the chemical literature and being familiar with the practice of salt boiling, he was able to identify the calcareous substances that made British salt bitter. His findings were presented in a treatise which exemplifies all the virtues eighteenth-century chemists believed would flow from the marriage of philosophy and practice. Although Cullen did not write about salt boiling, he studied Brownrigg's book and learned much from his analysis of sea water.

As Cullen said in his essay on the classification of salts, 'the appearances and effects of an alimentary salt that contains a perfect alkali and one that approaches to a calcarious salt are very different. The ingenious Dr. Brownrigg fully explained the difference. There was hardly any thing wanted to render the useful work of this author very perfect.'[26]

Cullen was studying the salt problem at about the same time that he set down his Remarks on Bleaching. He sent the Duke of Argyll a letter describing a method for purifying contaminated salt and, from his introductory remarks, it is clear that Cullen was well aware of the economic importance of this topic. 'You will not be surprised,' he told the duke,

> to find a chemist become a projector. I have hitherto kept pretty clear of this character; but, like many other chemical enthusiasts, having spent more time and money than I could well afford, I begin now to wish that my chemical labours would afford me some returns. If your Grace shall find that my proposed improvement is likely to be of any use, I shall consider it as a mark of the greatest goodness to have your opinion in what manner I may best avail myself of the discovery—a very simple one indeed, but if real, I think the simplicity very much increases the value of it.[27]

In July 1752 he learned of the duke's reaction to his project. 'I came yesterday from the Whim,' Henry Home reported,

> where I passed a day or two with his Grace of Argyle. As I have always your interest at heart, or rather that of the public, I took occasion to talk to him of your project for purifying salt. He told me of a letter he had from you on that subject, but said frankly, he wanted to have an account of your method to judge whether it would do. I gathered his reason to be, that if he were once satisfied of the practicability of the plan, he would be extremely keen to put you upon the most advantageous footing with regard to a premium. He said he hoped you would think yourself in no danger in communicating your secret to him, and that you would find him a trusty confident.[28]

Cullen replied promptly to Home's letter, protesting that 'I have never dealt in secrets, and I never shall have any with his Grace'.[29] He then went on to describe his method for purifying contaminated salt: 'This is the whole of my process; . . . that the lixivium [of potashes] serves to precipitate the calcareous salts, which render our boiled salt unfit for curing provisions.' This technique, in which alkali was added to a solution of salt causing the bitter contaminants, primarily magnesium salts, to precipitate out, utilized a scarce commodity, potashes, to obtain a substance of less value, common salt.

Cullen must have realized that unless his research into vegetable ashes led to the discovery of a cheap native alkali, his scheme for purifying salt could not be made economical. Late in 1752 Henry Home wrote Cullen a letter which indicates that Cullen had given up his project for the purification of salt. 'Pray,' Home asked, 'what are you doing about your project for refining Salt? If you are sure it can be made a manufactory, you must be entitled to a premium for your discovery in one shape or other. I am willing for my part to bustle for it; and therefore, if you think I can be of use, pray employ me.'[30] Although no commercial enterprise was feasible, Cullen had, through his practical investigation of the salt problem, become completely familiar with the complex subject of calcareous salts, a familiarity which is clearly in evidence in his 1753 'Essay towards ascertaining the Different Species of Salt'.[31] Surely Joseph Black's interest in the chemical properties of calcareous substances, one of which is magnesia alba, was, at the very least, reinforced by his teacher's interest in the practical problem of purifying the salt produced in Scotland.

There is a comic alternation in the agricultural speculations that passed between Cullen and Kames, each man taking in turn the role of theorist and that of the practical experimenter. In 1749 Cullen argued that accounting for specific properties is the fundamental chemical problem facing the plant physiologist. 'With respect to the nourishment of plants,' he told Home,

> the first question that arises seems to be this, Whether soils contain different juices fitted to nourish different plants, and that roots are endued with an elective attraction, whereby they take up only what is peculiarly fitted for their nourishment, and reject or exclude every thing else? Or, if the nourishment of plants, as lodged in the soil, is the same with respect to all different plants, and taken in by all different roots under one and the same form, but that by certain powers in the vessels of each respective plant, this common nourishment is converted into the juices peculiar to each? I am inclined to be of this latter opinion.[32]

The same year that Cullen advanced this hypothesis, Henry Home proposed experiments designed to test the effect of nitre upon the fertility of soil.[33] Early in 1750 Home elaborated upon Cullen's suggestion that plants which live mostly on air be used to enrich the soil. 'There is,' Home said,

> a point I am extremely curious about, and which came from yourself, and that is, to set about a course of experiments, in

order to determine what proportion of their food plants draw from the air. If this were found out, husbandry would be much advanced by it; for, all things else being equal, it would be best to chuse such plants as live most upon what they draw from the air, without exhausting the soil. By an experiment we all know, I should take this to be the case with cabbage, which, being hung up within a house in the winter time, will keep plump and green a very long while; and this can be owing to nothing else but to the nourishment which is imbibed from the air. I have often thought of planting an acre with cabbage purposely to make dung of it. It is a succulent plant, having a great tendency to putrefaction; and, if I were sure it lives much upon the air, and little upon the soil, I would certainly put my project in execution, and by that means augument the quantity of my dung considerably.[34]

Upon receiving Home's letter, Cullen realized it was his turn to call for new experiments.

> The experiments fit to determine which plants draw most from the air, and least from the soil, are such as Mr Hales and others have made on the perspiration and absorption of plants. Those already made are but a specimen. To answer our purpose, experiments must be made on a great variety of plants, and must be made on many of them during the whole time of their growth. A course of such experiments, therefore, would occupy a person for a summer season, require a garden in the country, assistance, and a pretty close attention. Such circumstances I cannot command at present.[35]

Finally, to complete the cycle, Cullen played the practical farmer to Home the cabbage theorist:

> Cabbages require a good mellow soil recently manured; though they take little, yet perhaps they take somewhat, and I doubt if what they take and what is lost by the perspiration of the soil, with the rent and labour, will be compensated by the crop recommitted to the soil again. I take it the best way of converting any crop into manure is by passing it through the body of an animal, as I believe the dung got this way, and the profit of the fed cattle, will exceed the value of the dung got from the vegetable putrified any other way; and I am told that cabbage, in particular, employed in the way I propose, turns to very good account; but I am perhaps out of my depth.[36]

It is hardly surprising that no practical theory of farming emerged from these discussions of agricultural chemistry. Nonetheless, in trying to develop new means of enriching the soil Cullen and Kames were engaging a problem of major importance. Although the chemical science available to them was not equal to the complexities of compost and manure, the widespread use of lime, a relatively simple compound, posed a theoretical challenge which seemed to be within their grasp. 'If one examines the chemical steps taken to initiate the agrarian revolution in Scotland, application of lime and marl figures largely in the early stages. It was said that larger quantities of lime per acre were used in Scotland than in any other European country, though for generations the way in which it operated was not understood.'[37] Cullen and Home evidently had well developed ideas about the way in which lime affects the soil. In 1750 Home sent Cullen 'a new theory of Lime, which I thought worth your attention; how far well founded you will judge: and, supposing it well founded, it does not in the least contradict your theory, but will be a proper addition to it.'[38] In his reply Cullen mentioned the chemical investigation of lime carried out by one of their contemporaries and emphasized how difficult it is to determine how lime affects plant growth. 'I have read your paper on Lime,' he wrote,

> but have not had time enough to consider it. I heard of Dr Alston's experiment when I was last in Edinburgh. I believe that lime may continue long to give out its virtue, and that is enough for your purpose; but that the lime-water continues so long as the Doctor alleges of the same degree of strength, I must have other proof than the Doctor's taste. Your reasoning from this experiment seems to me probable and just; but, before we rest upon it, we must have it determined whether lime directly by itself goes to the nourishment of plants, or if it operates only in preparing the nourishment, or if both are true: on these points you fluctuate a little. Upon any of the three suppositions, we may account for lime being in time exhausted; but it is only on the second that I can account for exhausted limed ground being worse than before.[39]

The experiment which Cullen is referring to was performed by Charles Alston, the professor of botany in Edinburgh, and it convinced Alston that a quantity of quicklime, when slaked, will 'afford more than six hundred times its own weight of good lime-water'.[40] Alston reported that he began his study of lime 'in June 1743, having only some Experiments relative to Vegetation in View,'[41] but his discovery was more chemical than agricultural. That liming sweetened soil for three years was, of course, a well established fact of agricultural practice. What Cullen and others doubted was Alston's claim that the last washing of a sample of slaked lime, after it has been used to infuse great quantities of water over several years, will produce lime-water just as strong as the first slaking. Alston, however, was adamant. 'When I communicated to not a few,' Alston wrote,

> what Quantity of Lime-water I had got from a Pound of Quick-lime, it was generally disbelieved, and looked on as impossible. . . . And no wonder, for had it been told me before, I would probably have suspended my Assent until I also had made the Trial. To convince Dr. Rutherford, who is very deservedly at the Head of our Faculty in the University, I sent him a Bottle of the Lime-water, which the Pound of Stone Quick-lime, mentioned above . . . afforded, after about five hundred Pounds of Water had been poured on it. He was not a little surprised at its Strength, as he told me afterwards, and said, there appeared to be something very wonderful in it, for, by the taste, he would have taken it for the first Water.[42]

Cullen was merely sceptical about Alston's conclusions but Robert Whytt, professor of the theory of medicine in Edinburgh, was quite sure that his colleague was wrong. On the basis of his own chemical experiments Whytt concluded that 'fresh calcined Shells afford a stronger Water than such as have been exposed for any considerable Time to the Air, and that the first Water poured on Quick-lime is stronger than the succeeding ones.'[43] Does quicklime, when slaked by repeated washings, yield lime-water of the same strength until totally exhausted, or does the strength of the lime-water decrease with each washing? This was the chemical question which lay at the root of the lime-water debate carried on by Alston and Whytt.[44]

The two Edinburgh professors continued challenging each others conclusions even after Joseph Black, in his inaugural dissertation, had resolved the essential chemical disagreement between them. Although Alston, in the second edition of his *Dissertation on Quick-Lime*, acknowledged the relevance of Black's experiments on magnesia alba, he failed to grasp all their implications for his own theory of lime-water.[45] Cullen of course recognized the importance of his pupil's work immediately and he graciously conceded, in his later lectures, that Alston had been fundamentally correct all along. 'Lime-water,' Cullen told his chemistry students, 'was always said to be strong in proportion to the quantity of Lime that was put into a given quantity of water, but Dr. Alston has shewn that one Pound of [calcined] Limestone Impregnates 40 of Water as strongly as it does 10.'[46] This chemical investigation of agricultural liming was thus yet another specific source of Black's interest in causticity. Although Home and Cullen were unable to determine how lime affects the growth of plants, their concentration on this problem helped direct Black's attention to some of the phenomena which he was finally able to explain with his theory of causticity and fixed air.

Cullen's excursions into practical chemistry naturally encouraged him to investigate related theoretical questions. He believed that to advance their field chemists had to meet the demands generated by the arts and by philosophy, and in his own research he attempted to improve both branches of the science. In 1752 he bought a farm for his brother, who was doing well as a merchant in the West Indies, and since the property lay only eight miles from Glasgow, Cullen agreed to manage and improve it.[47] In October of that same year, a Frenchman named Dangeuse visited Glasgow where he met Cullen, Black, and other local luminaries. Upon returning to Paris, Dangeuse told his friend Duhamel du Monceau about his visit, and when, a few months later, he acted as intermediary in setting up a correspondence between Duhamel and Cullen, he stressed their common interest in agriculture. Despite Kames's disparaging comment—'your friend, at least your correspondent, Du Hamel is a ninny'[48]—Cullen was pleased to be in touch with a member of the French Academy of Sciences and he eagerly encouraged the exchange of scientific information.[49]

Cullen also employed many other friends and correspondents to satisfy his vast appetite for scientific news. In 1748 William Hunter wrote from Germany saying that in the course of his travels he had obtained all the books on Cullen's list except those by Kunckel, Réaumur and Stahl.[50] In 1751 Dr John Clerk acted as Cullen's agent at the sale of the library of Dr Elliot, who was to have succeeded Plummer as professor of chemistry in Edinburgh.[51] Dr Archibald Davidson, who later became Principal of the University of Glasgow, wrote to Cullen from Goettingen in 1752, saying 'I remember that you gave me a commission to inquire for a critical Dictionary of the German language. I spoke to Dr. Haller about it.'[52] There were also many friends and former pupils who bought books for Cullen and sent him accounts of the chemical doctrines being taught in London and Paris.[53] As early as 1747, the year in which he began teaching in the University of Glasgow, Cullen was asking his brother-in-law in London to purchase the chemical writings of Becher, Stahl, Bohn and Pott, and a German dictionary and grammar.[54] While a shared interest in agriculture provided the common ground for Cullen's chemical correspondence with Kames and Duhamel, it is clear that Cullen's own interest in theoretical and philosophical chemistry never waned. Throughout his long career he remained interested in all aspects of the subject, and during his years in Glasgow he actively investigated theoretical as well as practical problems.

Within the realm of natural philosophy, Cullen was primarily interested in the chemical manifestations of heat and fire.[55] Thomas Melville, who studied in Glasgow in 1748–49, wrote to his former teacher in 1752 asking for further news about his investigation of the relationship between the density of a body and the rate at which it absorbs heat.[56] And when commenting on Kames' essay on evaporation, which he received in 1748, Cullen based his arguments upon a well-developed set of ideas on the relationship between heat and vaporization.[57] He also used his theory of heat when investigating the properties of salts. Being a professor of medicine, Cullen was familiar with the analysis of spa water and the use of lime-water to dissolve urinary stones.[58] Evidence of his knowledge of salts can also be found in his first letter to Duhamel, where he describes a method of obtaining 'an ammoniacal salt, made with the vegetable acid in a solid form,'[59] and in the essay on the classification of salts which he sent to the Philosophical Society of Edinburgh in 1753.[60] Although at first glance there seems to be no link between his heat theory and his studies of salts, in Cullen's mind they were connected by a general theory of chemical action.

Cullen believed that chemical change is caused by the powers of attraction and repulsion. The power of repulsion is, he assumed, connected in some way to the heat contained in chemical substances. It therefore seems likely, he argued, that the law governing the effects of the repulsive force in chemistry can be discovered by correlating the heat flow in chemical reactions with the other changes in properties which occur in these reactions. Since chemists at that time were studying the relative attracting powers of different substances by arranging them in tables of elective attraction, Cullen began his search for the chemical laws of repulsion by constructing tables of heating and cooling reactions.[61] Careful study of these tables convinced him that reactions that give off heat and those that absorb heat constitute two distinct kinds of chemical change. In two essays which he read before the Literary Society in the College of Glasgow, Cullen presented evidence supporting 'the general position, that *every mixture generates heat, and that every solution generates cold*'.[62] Furthermore, he argued, every combination of an acid and an alkali generates heat and should therefore be called a mixture. Salts, which are formed by acid-alkali neutralization, therefore constitute a class of chemical substances definable in terms of the heat flow which occurs during their formation. By means of further investigations Cullen hoped to elaborate upon this link between the properties of chemical substances and the change in their capacity for heat which occurs during their formation.

Kames, who seldom showed reserve when theorizing, had little patience with Cullen's cautious presentation of his ideas. In 1754, while preparing the second volume of *Essays and Observations* for the Philosophical Society in Edinburgh, he pressed Cullen hard for a contribution. 'A demand was made upon me for your paper upon Salts,' Kames wrote. 'I either dreamed it, or it was said to me, that you reserved that paper to make part of a greater work. If not, transmit it to me; and if you reserve it, I must insist upon some other paper from you in its stead. . . . I insist upon one paper or another; and I insist upon it as a point of right.'[63] Cullen planned to make the paper on salts part of a textbook on chemistry and he therefore refused to publish it in Kames' volume. 'I should be proud to have a share in it,' he told Kames, 'but I find I cannot. The paper I sent you last year is now destined to another work, which I hope will appear soon, and I have nothing else at present fit for you.' However, he continued, 'that this letter may not be quite without philosophy, which you always require, I shall give you somewhat that has occurred to me lately, and which is, as far as I know, entirely new.'[64] In these new experiments, which grew out of his study of heating and cooling reactions, Cullen produced cold by the rapid evaporation of liquids under reduced air pressure. Kames, evidently intrigued, asked Cullen to prepare a more complete description of these experiments, and the resulting paper, entitled 'Of the Cold produced by Evaporating Fluids, and of some other Means of producing Cold,' became the much demanded contribution to the second volume of *Essays*.[65] Cullen never completed his general text on chemistry and his essay on the classification of salts was not published until the twentieth century. His paper on the cold produced by evaporation therefore constitutes his only published contribution to the chemical literature of his age.[66]

[6]

Cullen's Chemical Doctrines

1. CHEMISTRY DEFINED

In November of 1748 Cullen began his second year as a lecturer in the University of Glasgow. John Carrick, who had taught the chemistry course at the beginning of the previous year, had not yet recovered from his illness and Cullen was therefore obliged to prepare and deliver the chemistry lectures by himself. Two extensive yet incomplete manuscripts of the notes Cullen used when delivering these lectures have survived. One set of notes, written in Cullen's own hand, contains only headings, key phrases, and the general outlines of his lectures.[1] The other manuscript of these early lectures is scattered through the first two boxes of the Cullen papers in the Glasgow University Library.[2] In these Glasgow manuscripts, which are written in a copyist's hand, Cullen's ideas and arguments are presented in the polished language he employed when lecturing. These two sets of notes therefore provide the materials needed for an accurate and fairly complete reconstruction of the theoretical portions of Cullen's first full series of chemistry lectures. His own notes, which are bound together, indicate what he thought and taught in 1748–49 and the smooth copies of these lectures contain some of the implications and interpretations he drew from his doctrines. Since Joseph Black began his chemical education by enrolling in Cullen's course in 1748, a knowledge of what Cullen taught that year will enable us to identify the particular theoretical concerns that guided both the master and his most famous pupil during their years of experimentation and discovery.

Cullen believed that a correct definition constitutes the first step towards a proper understanding of chemistry.

> In entering upon the study of any art or science, especially of one not commonly known or taught, it is proper to fix our notion of such art or science, that our pursuit may be animated by a view of the usefull knowledge to be acquired & that all unnecessary pains & labour may be saved in pursuing what either does not belong to the study or what is not to be expected or acquired by it. This consideration is especially applicable to chemistry, which presents to us indeed very engaging views

> both of pleasure & profit but has also had the misfortune to engage persons in pursuits that I believe could not be attained by it.[3]

A good definition will identify both the objects to be studied and the kinds of knowledge to be gained through such study. Chemistry has in the past been viewed as a study which takes all of nature as its field and which promises extraordinary power as its reward. Cullen's first task, therefore, was to redefine what chemists may examine and what hopes they may entertain. Since he considered this task important, the way in which he went about it casts a good deal of light upon his own conception of the subject.

Cullen found none of the published definitions of chemistry suitable. He did not hold a high opinion of his own skill as a chemist nor did he consider himself a profound theorist, but he felt competent to judge all definitions by criteria which, when applied to the formulations of other chemists, exposed their inadequacy. But Cullen did not simply assert that this inadequacy was self-evident, for he wanted his students to think through the reasons which lay behind his conclusions. He therefore devoted a considerable amount of time to his critique of several well-known definitions of chemistry. 'I find,' he told his students, 'many people have false & mistaken notions of chemistry & that the common notion is at best loose, undetermined, & very narrow. It would therefore be convenient in the first place to give a clear & precise idea of our Art, but as the artists themselves have been much embarrassed in this particular, a full discussion of this matter must engage me in a dry & perhaps tedious piece of criticism.'[4]

Before plunging into a critique of eighteenth-century definitions of chemistry, Cullen disposed of the older conceptions of the subject which attract those who believe that antiquity is a sign of truth. The content of what we call chemistry is not the same in all ages. Chemistry deals with a segment of the world perceived by the senses and ordered by the mind, and since that world and our mental representation of it are constantly being modified and reordered by the progress of the arts and our growing understanding of them, previous definitions of chemistry are necessarily inadequate for present purposes.

> It would be, therefore, tedious to consider the definitions of the elder chemists, which are often as erroneous as the common notion. As arts often obtain a proper form & are distinguished from each other by a natural sagacity long before an artificial logic, which only explains the operations of the mind as applied to them, so the chemists had a tolerable exact idea of their art before they could properly express that in their definitions. I shall

> now only consider two or three definitions given by some of the moderns of greatest name both as chemists & philosophers.[5]

Boerhaave's definition, as translated by Peter Shaw, was the first of the eighteenth-century formulations Cullen examined. 'Chemistry,' according to Boerhaave, 'is an art which teaches the manner of performing certain physical operations, whereby bodies cognizable to the senses, or capable of being rendered cognizable, & of being contained in vessels, are so changed, by means of proper instruments, as to produce certain determined effects; & at the same time discover the causes thereof; for the service of various arts.'[6]

Here, Cullen told his students, is a definition, which fails to identify even one of the three characteristics by which we can distinguish chemistry from the other sciences.

> Now gentlemen, I do not know what you may learn from this definition, but I must say it communicates no distinct idea to me. The operations of chemistry are not characterized. They are certain physical operations, but if or how distinguished from other physical operations is not said. The objects of chemistry are not specified, for bodies as merely cognizable to the senses & as capable of being contained in vessels are not the objects of this art. Lastly, in this definition the end of chemistry is not pointed out with the necessary exactness, for we shall find by & by that the effects of chemistry respect certain qualities of bodies only, by which it makes a particular branch of Natural Philosopy & is subservient to a set of arts that all belong to one class, which should have been defined.[7]

If chemists follow Boerhaave's lead, they will be forced to treat their subject as a subordinate branch of a natural philosophy founded upon a closed set of mechanical postulates. But if, with Cullen, they carefully specify the operations, objects and purposes of chemistry, they will see that their subject constitutes an independent branch of natural philosophy, a distinct area of knowledge which has not yet been unified by an adequate set of general explanatory laws. If properly defined, Cullen argues, chemistry can co-exist with the mechanical philosophy within the broad realm of natural philosophy.

Homberg's definition of chemistry as 'the art of resolving compound bodies into their principles by means of fire & of composing new bodies in the fire by the mixture of different matters'[8] was, in Cullen's opinion, little better than Boerhaave's.

> I think it has the advantage over Boerhaave's in giving a more determined idea, but we cannot, however, admit it as perfect for several reasons. We gave it in a literal translation, but it may be put thus simply: The art of resolving & compounding bodies

> by fire. The faults of it seem to be these: First, that it confines the operations & effects of chemistry to separation & combination only, tho this is liable to be disputed & it may be alledged that chemistry produces considerable changes in bodies without any sensible separation of their parts & without adding any new ones.[9]

Fermentation, Cullen argues, causes a chemical change when it turns juice into wine and wine into vinegar, but clearly this change is not caused by the addition or subtraction of chemical substances. Perhaps we should say that here we have an internal separation and combination, but there are other chemical changes where there is clearly no internal reorganization of the elements: 'Cases might be put, as for instance the changes of water on one hand into ice & on the other into an elastic vapour, where either a separation or combination cannot be properly alledged.'[10] Evidently any definition which Cullen found acceptable would have to include change of state as one of the phenomena to be studied by chemists.

'A second & greater fault in Monsr. Homberg's definition is,' Cullen asserts, 'his confining the means employed by chemistry to that of fire alone. The effects of fire & heat are certainly very extensive both in nature & art, & fire is indeed the principal means employed by chemistry, but there are several other means, [such] as menstruums, or dissolving liquors, & ferments, that may be employed & have always been employed by chemists.'[11] Cullen was eager to keep the definition of chemistry abstract. If, with Homberg, he identified chemistry in terms of the instruments used by chemists, his definition would become inadequate as soon as new techniques of analysis, such as the use of electricity, were developed. Cullen felt that 'it is better to say with Dr. Shaw that chemistry separates & combines by all the means in our power'.[12]

The last definition Cullen criticized was that given by Stahl, recorded 'by his disciple Juncker in his Conspectus Chemiae,' and translated by 'Dr. Shaw in his Chemical Lectures':

> Philosophical chemistry, says Dr. Shaw, is a rational art of dividing, or resolving, all the bodies within our power by all the instruments we can procure, as well into integrant as constituent parts, and of joining these together again, so as to discover the principles, relations, & changes of bodies, make various solutions, mixtures, & compositions, find out the causes of physical effects, and hence improve the state of natural knowledge & the arts thereon depending.[13]

Although pedantic, laborious, redundant, and incorrect—the adjectives are Cullen's—this definition at least points chemistry in the right direction. But before the worthwhile parts of this statement can be separated from the dross, the distinction between integrant and constituent parts must be clarified. Cullen begins this task by journeying with the corpuscularians into the world of atoms. 'Most bodies we examine are divisible to a great degree of subtility, so that art, in this respect, seems to have no bounds. But we have great reason to believe there are certain parts of matter not farther physically divisible, such we call atoms & suppose them the elements of which all bodies are composed.'[14] Then, when describing how elements combine to form sensible substances, Cullen follows Stahl. 'These elements we suppose too of different qualities & when two or more different elements are joined together such an union makes what we call a mixt, & again, if two or more mixts are joined together they form a compound, at least in the strick sense the terms of mixt & compound used by Dr. Shaw & his disciples.'[15]

But Cullen, true to the Scottish philosophy, will not found his doctrines upon differences attributed to the imperceptible qualities of elementary matter. He therefore abandons Stahl's distinction between mixts and compounds and talks only of the difference between constituent parts, which differ qualitatively from the compound formed by their union, and integrant parts, which have the same properties as the aggregate. The terms mixt and compound are, he said,

> used promiscuously for any substance that may be divided into parts of different qualities, & such parts are called constituent parts. But when any mass consists of a collection of parts of the same nature & qualities, such a mass is called an aggregate, & when a body is divided into parts each of which have the same qualities as every other part & as the whole of the mass had, such parts are called integrant parts, as may be understood by this example. A pound, that is, a muchkin, of brandy is an aggregate that may be divided into gills, half-gills, ounces, or even much smaller parts, each of which will have the same qualities as the others & as the whole muchkin had, & such parts therefore are called integrant parts. But the same brandy is a mixt or compound which by a proper operation may be divided into two substances, that is, water & spirit of wine, which have very different qualities, & such parts are called constituent parts. In the same manner a mass consisting of lead & silver may be considered as an aggregate which may be divided into shot or grains, each of which will have the same specific gravity, the same degree of ductility, & contain exactly the same proportion of lead & silver as every other grain & as the whole mass before division had. Such grains will be integrant parts.

> But if the same mass shall, as it may, be divided into the lead & silver, these will prove to be parts very different from each other, & are to be called constituent parts.[16]

After having clarified the difference between analysis into integrant parts and analysis into constituent parts, Cullen resumed his critique of Stahl's definition of chemistry. 'It is faulty,' he told his audience,

> in mentioning the division of bodies into integrant parts as the business of chemistry. The division of bodies into integrant parts is often subservient to the purposes of chemistry & indeed performed by the chemist himself, but surely the chipping of wood or the filing of iron, which are divisions into integrant parts, cannot however be considered as chemical operations without extending the art too far. No one art subsists by itself but [all] need the assistance of some other—the same tradesman often practices different arts. Different arts are not therefore to be confounded, but the proper & accessory are to be distinguished.[17]

In the realm of the practical arts we constantly distinguish between chemical and mechanical crafts, despite the fact that a single individual may be proficient in both fields. Similarly, Cullen argues, rather than trying to reduce chemistry to mechanics, or mechanics to chemistry, in the realm of natural philosophy, we should recognize them as separate subjects concerned with distinct phenomena. Thus, he argues, 'another fault of these definitions' proposed by Stahl, Juncker and Shaw 'is their putting the end of chemistry too generally, which Boerhaave also does. For it is not physical effects & their causes in general that are explained by chemistry, but certain effects & causes only, & these should be particularly specified so as to distinguish the chemical from the other parts of Natural Philosophy.'[18] A definition which characterizes chemistry in terms of the phenomena it studies will enable us to identify its proper task and thereby further the division of philosophic labour, a division which the professionalization of the Scottish universities particularly encouraged.

Cullen's own definition of chemistry depends entirely upon the distinction between qualitative and quantitative analysis. Chemistry, he said, 'is the art of separating mixt bodies into their constituent parts & of combining different bodies or the parts of bodies into new mixts for the purposes of philosophy & arts, that is, for the purposes of philosophy by explaining the composition of bodies, the nature of mixture & the properties of bodies thereon depending, & for the purposes of arts by producing several artificial substances more suitable to the intention of various arts than any natural productions are.'[19] A later statement of this definition lays particular emphasis upon chemistry's independent status within natural philosophy and upon its concentration on qualitative change: 'I think chemistry may be considered as that part of Natural Philosophy which explains the properties & effects of bodies depending on their mixture, or on their habits with respect to mixture.'[20]

To clarify the meaning of this definition, Cullen plunged more deeply into the distinction between the two modes of analysis utilized by natural philosophers. He said,

> I think the properties of bodies & their effects may be considered as of two kinds: First, these that they have as aggregates of a certain bulk, figure & quantity of matter without any respect to the different parts of which that aggregate may be composed. Thus the power of the wedge is demonstrated from its form without considering whither it is of wood or metal, & in the other mechanical powers the weight & power are considered only with respect to their quantities of matter without regard to quality of matter—the laws of hydraulics are demonstrated from the form, proportion & heights of several tubes without considering whither they are of tin or copper—& the consideration of these properties makes one part of Natural Philosophy which may be called the Mechanical. But there are another kind of properties of bodies which they have as mixts, such as: Specific gravity, solidity or fluidity, hardness or softness, brittleness or ductility, elasticity, volatility or fixedness, miscibility, solubility, inflammability, fusibility, opacity or transparency, tastes, smells, sounds, colour, form assumed in concreting &c &c; all which depend on the constituent parts or their peculiar arrangement. Such properties supposed but mechanical & it may be so, but it does not appear at present & the application of the Mechanical Philosophy unsuccessfull. Many of these properties known as facts. Some of them traced to general laws, but these laws not to be explained mechanically—for instance, in the case of elective attractions.[21]

Mechanicians study substances as aggregates, that is, as bodies composed of integrant particles having the same qualities as the whole. Chemists study substances as mixtures, that is, as bodies having qualities caused by the way in which their constituent particles combine. 'The properties which chemistry considers [are] of three kinds: The habits with respect to mixture, example in elective attraction, the properties of mixts from different ingredients, example in soap, the properties of mixts from the manner of concretion, example in lead.'[22] Thus the qualitative changes which accompany changes of state, such as the freezing or boiling of water, fall within the chemist's range of inquiry for exactly the same reasons that lead us to consider the changes which accompany displacement reactions and acid-alkali neutralizations as chemical phenomena. What Black was to call latent heat was, like specific heat and specific gravity, a chemical phenomenon.

The distinction between mechanical and chemical analysis was, in Cullen's opinion, the key to all the distinctive features of chemistry.

> I think you may now understand the nature, at least my notion of the nature, of chemistry & the chemical philosophy: That it is that part of philosophy which considers & explains all the properties & effects which are peculiar to certain parts of matter only & belong to them as elements or mixts & under which term of mixt I comprehend not only bodies as composed of different parts but bodies also composed in a certain manner. After this account of chemistry it will be needless to add the manner of operating, the means to be employed, or farther specify the end of chemistry, for whatever contributes to, explains these properties of elements & mixts, or serves to give these properties to bodies is a chemical operation, & you may perceive indeed that separation & combination will be principal operations for this purpose but that new properties also may be given to bodies merely by altering the manner of concretion or mixture. I said too it was needless to mention the means to be employed, for whatever can separate mixt bodies or combine different bodies or alter the manner of mixture or concretion of bodies is to be employed by the chemist. And lastly, the end or purpose may be understood from our account of chemistry: Whatever arts work on bodies as endued with peculiar properties, these arts depend on chemistry, which explains such properties, produces bodies which have the requisite properties, or produces such in bodies which had them not before.[23]

Cullen knew full well that his definition might well be considered naive. Nevertheless, he argued, he was only seeking to establish chemistry in those areas of natural philosophy where the weakness of the mechanical philosophy had been clearly demonstrated.

> It may be alledged I have done wrong in setting the chemical part of philosophy in opposition to the mechanical, since it is probable that even these peculiar properties of bodies considered by chemistry may depend on the form, bulk & quantity of matter of the elements & on the arrangement of elements of a certain form, bulk &c in mixts, & I shall allow that it may be so, but I must observe too that at present we are not certain of that nor know not if ever we shall be so. The elements of bodies are

vastly too small for us to discern their form & size, we have nothing yet with respect to these but conjectures, & the application of such conjectures & of the mechanical philosophy has hitherto been very unsuccessful in explaining the properties of bodies which chemistry considers. To give an easy instance, the fluidity of water & other bodies has been constantly supposed to depend on the parts of fluids being perfectly round & spherical, so that, having little contact & cohaesion, they must be easily moved, which is fluidity. This is very specious but by no means satisfying, for [as] all bodies by a proper degree of heat may be rendered fluid, neither can we suppose that the fire, in rendering consistent bodies fluid, changes the parts formerly not spherical into spherical. I say that we can hardly suppose that the difference of heat which raises the thermometer from 31 to 33 degrees & which changes solid ice into fluid water, changes the angular particles of ice into spherical ones, but we must either suppose this or that the parts of all bodies are spherical if this is necessary to fluidity, & if we do suppose the elements to be all spherical, how shall we account for the consistency of one body or the fluidity of another. We must, I think, in either case have recourse to some other cause, & indeed, we can neither from the supposed spherical figure of the particle deduce fluidity nor from the fluidity conclude the spherical figure. This is one instance of the unsuccessful application of the mechanical philosophy to chemistry, & the truth is, a great many properties of bodies which chemistry considers we know only as single facts. Some [of] them, as common to several bodies, we can trace to general laws or to certain circumstances on which they depend or are constantly connected with, but still we are no nearer to a mechanical explication of these. Thus with respect to the elective or superior force of attraction in some bodies than others, there are so many specialities, sometimes of opposite kinds, with respect to these that I believe no body will think of accounting for them from form, bulk or solidity of elements. Who will not rather suppose it a property of matter distinct from all these & upon the whole that chemistry not only considers a set of properties different from these which the mechanical philosophy considers, but a set of properties also that may depend on different circumstances.[24]

For Cullen, chemistry must aspire to be more than 'the art of making gold, the art of preparing medicines and the art of distillations'.[25] It must identify its proper phenomena and then, like all sciences, use the empirical method of investigation to formulate and test its own laws and general explanations. The enthusiastic reception that Cullen's chemistry received in both Glasgow and Edinburgh indicates that his conception of the subject was at very least compatible with the new philosophy being taught in the Scottish universities.

2. THE HISTORY AND PURPOSE OF CHEMISTRY

Since Cullen believed that all definitions of chemistry should contain a statement of purpose, he was careful to include such a statement in his own. Of course he could have defined chemistry as a branch of natural theology, but unlike the deists, Cullen did not pursue chemistry as a means of finding out about God. Like many of his philosophical colleagues, he believed that through science one could attain the dual goals of practical knowledge and philosophic truth. Had Cullen devoted himself primarily to the task of uncovering more of the cosmic order postulated by the philosophers of the seventeenth century, he would have been more respectful of the Newtonian attempts to reduce the phenomena of chemistry to the axioms of the mechanical philosophy. And had he been no more than an improving chemist, he would have followed Plummer's example and taught only pharmaceutical and practical chemistry. Cullen, however, believed that truth and utility are the twin goals of philosophy. If the philosophical chemist hopes to advance his subject in a responsible and thoroughly modern way, he has to keep these two criteria in view at all times.

To master practical techniques and gain theoretical understanding simultaneously was, Cullen realized, a difficult feat, and as he told his students, the two activities had actually been treated independently by all but a few of the most recent chemists. True, some earlier chemists asserted that knowledge is power, but they sought knowledge by means of occult communication with an animistic nature and their dreams of power were therefore more chimerical than chemical. Only since the time of Bacon and Boyle has the search for useful knowledge been founded upon the correct empirical method. Chemistry, while an old subject, has only recently recognized both its method and its task; as a science it is young and full of promise.

The history of chemistry always found a place among Cullen's introductory lectures and in later years, particularly after moving to Edinburgh, he expanded this section of his course considerably.[26] He wanted his students to realize that the subject he was revitalizing had a long and noble past, but unlike many other chemists who lectured on the history of their subject, he was careful not to smother the student's curiosity under a blanket of erudition. 'After the example of Dr. Boerhaave,' Cullen told his class, 'some other teachers constantly premise a history of chemistry, but I think a detail of the discoveries that have been made from time to time in chemistry, or an account of the several great men who have appeared at different times as artists in this way, is extremely improper to persons not acquainted with the art. I shall therefore not trouble you by offering such at present, but shall give only a very short account of the first beginnings & different appearances of the art since.'[27]

Of course Cullen wanted his students to discover more than the antiquity and diversity of the subject; he believed the history of chemistry should reveal the progress of the science. When Cullen began teaching in Edinburgh he emphasized that while his lectures on the history of chemistry were not the first to be delivered there, his use of these lectures to reveal the direction of chemistry's advancement was a significant innovation. 'According to the custom long established in this University, we shall begin with a history of Chemistry; and though this was introduced only probably to fill up the time allotted for public Lectures, yet it is of great service to the young students, if properly attended to: It shews by what means the art has been brought to its present state of perfection.'[28] In Cullen's mind there was no doubt that a knowledge of history made one aware of 'the rise, progress, and improvement of Chemistry'.[29]

Practical knowledge is the older of the two branches of chemistry, but since this knowledge was discovered by accident and transmitted by mere imitation, it failed to reach the level of consciousness which is the distinguishing feature of philosophy. 'As the operations on metals are really chemical, the antiquity of such operations allows the chemists to carry the antiquity of their art very high, but it must be observed that it is not a practice accidently discovered & even frequently imitated by others that establishes an art, but when such practices are reduced to rules which direct the variations that may be necessary in different circumstances.'[30] We do not know when the accumulated knowledge of practical chemistry was first given the form of an art, but certainly the process was a gradual one and drew upon the ideas of many contributors.

> The Chemists [i.e. the Paracelsians] refer the origin of this art to Egypt, and sometimes they call it the Hermetic art, from Hermes Trismigistus whom they suppose was the first inventor. But it is ridiculous to refer the invention of any general art to one particular person; it is always by the successive labors of Several that an art is brought to perfection, and instead of being able to refer the origin of any particular art to one particular person, we can scarce tell for certain who invented any particular art: who

knows the discovery of the Mariner's Needle, and what disputes there have been about the Air pump, Barometers, Thermometers etc. The Case is, a second improves upon it, a third puts it in execution; and all claim the honor of the invention.[31]

Theoretical knowledge, the other branch of chemistry, was first developed by the Greeks, but the speculations of Parmenides and Heraclitus 'were only bold assertions and rude conjectures,' and 'the Grecians in general were only intent upon Theory, without recourse to experiment'.[32]

The alchemists, who were the first true chemists, appeared in Europe during the last centuries of the Roman Empire. Although interested in attaining both theoretical and practical knowledge, these chemists failed to identify what could properly be expected of their subject and hence were led on by fruitless and delusive hopes. 'Till the restoration of learning in the fifteenth & sixteenth centuries, except Roger Bacon & a few others who applied chemistry to philosophy & arts, the whole of the chemists were employed in search of the philosophers stone—a set of fanatical visionary men without learning or candour.'[33] Although they did not lay firm foundations for the study of chemistry, the alchemists, like the practitioners and theoreticians who preceded them, made important, if unintended, contributions to the development of modern chemistry. 'The greatest use they have been of to Chemistry is that by their means the art was preserved and has since been applied to nobler purposes.'[34]

Pharmaceutical chemistry appeared shortly after alchemy. The Arabian physicians were particularly adept in this field, and their knowledge and that gained by the alchemists became known in Europe through the works of Paracelsus.

> This famous man, by applying himself entirely to search after empirical remedies, became acquainted with several prepared by chemistry, with which he performed many cures. His success, attended by a great deal of boasting, gained a reputation to chemical remedies, procured him many followers, & even introduced the chemical philosophy into the schools of medicine, so that, in the sixteenth century a great number of chemists appeared, some of them good operators, but all of them bad philosophers, mostly employed in the preparation of medicines but still professedly attached to the search after the philosopher's stone. The application then to chemistry was dangerous, as infecting men with very foolish visionary hopes which often engaged them in measures that ruined their fortunes.[35]

Although some of Paracelsus' heirs began to advance more reasonable explanations of chemical action, they claimed too much for their subject and hence, inevitably, called forth opposing theories. 'When the Chemists had once begun a chemical theory, they would allow nothing else to have a share in it. They endeavored to account for every disease from this doctrine of acids & alkalies, and to this they said was owing all the process of Digestion & Chilifaction as is shewn in anatomical treatises. Afterwards the royal Society, Boyle & Newton, introduced the Mechanical Theories, which seemed diametrically opposite to the Chemists.'[36] Cullen, made wiser by their error, hoped to avoid this confrontation by carefully distinguishing between the subject matters of chemistry and mechanics.

True philosophical chemistry first appeared in the early seventeenth century. 'It was then that Galileo, and Lord Bacon appeared and . . . laid the foundation of Natural Philosophy. Some have attempted a contrast between these two, but very unfairly, as they were employed in quite different pursuits. When Galileo was promoting mechanical philosophy, Bacon was laying the foundation of Chemical.'[37] Although Bacon's contribution was prescriptive rather than exemplary, his enemies are in error when 'they say that he made no improvements in Chemistry, . . . they ought to consider that he planned out the road by which we have made so great discoveries.'[38]

Robert Boyle emerges in Cullen's account as the first truly modern chemist. 'This great philosopher, by his ingenious labours & writings, first shewed that chemistry might be delivered in clear terms, that it makes a great part of Natural Philosophy, & is especially fitted for the improvement of arts & manufactures.'[39] Boyle is a good model for chemical investigators, yet a close examination of his work reveals that his critical writings were his most enduring. 'Though he rather objects to other systems than proposes any of his own, he shewed us the use and importance of experiments; he taught us that more Chemistry is to be learned in the shops of artizans than the laboured theories of Chemists: He likewise stript Chemistry of all its uncouth and mysterious language.'[40] After being refashioned by Boyle, chemistry 'came to be sufficiently established as a creditable, a solid, & usefull art when the great Sir Isaac Newton, both by practicing it & employing it in his reasonings, shewed his opinion of it'.[41] However, Cullen lamented, we may never know the full extent of Newton's interest in chemistry, for he 'has left several folio Manuscripts upon Chemistry, which I fear will never be published'.[42]

Since its foundation in the great age of English natural philosophy, chemistry has flourished in many lands. 'The physicians of greatest note, such as Bohn, Hoffman, Stahl, Boerhaave, Freind, have also been the most assiduous cultivators & improvers of chemistry.'[43] And in Germany the advance of chemistry has been favoured by unusual circumstances. 'Germany has always been famous for Chemistry, which is, as it were, its natural science. The country being full of mines, the people have been naturally led to Metallurgy & other arts. Hence we have had *Stahl*, who is full of improvements, and was the first that gave us anything of Chemical system.'[44] The establishment of national scientific academies has also contributed to the notable progress of chemistry. 'The members of the Royal Society, of the Academy of Sciences, & of every other society instituted for promoting natural knowledge, by their frequent labours in chemistry, have shewn us what may be expected from it.'[45] As a result of these developments, the importance of the science which Cullen's students were studying had finally been recognized. 'By the labours of many ingenious men for these hundred years past, chemistry has been cleared of its jargon & mysterious language, has been directed to its proper end, & is now considered by every man acquainted with literature & the foundation of useful arts as a considerable part of Natural Philosophy, capable of being applied to very important purposes of society.'[46]

Cullen's views on the uses of chemistry, his next topic, will hardly surprise anyone who has followed him through the lectures on the definition and history of chemistry. Just as the phenomena of natural philosophy that manifest qualitative change were assigned to chemistry, while those involving quantitative change were assigned to mechanics, so those arts that deal in qualitative change should look to chemistry for improvement while the others, which effect quantitative change, should look to mechanics.

> Thus the giving metals a particular form, as of wheels, springs, &c, & adapting several pieces so formed to particular purposes, is the business of the mechanician. But the mechanician without the assistance of chemistry would have no such metals to work on, but in their place stones & earth. It is the chemist who from these stones & earths processes malleable metals. It is the chemist who gives to the metals the degree of hardness, ductility, elasticity, or other property that fits the several purposes of [the] mechanician, from which you may see the connection of chemistry with the arts of life.[47]

If, with Cullen, we accept this division of responsibility, we can hardly deny 'that the chemical philosophy, tho less attended to, is however of no less extensive use than the mechanical'.[48] Chemistry, according to this scheme, has as important a place among the applied sciences as it has within natural philosophy. Cullen not only argued for the independence of chemistry, he sought to make it the most comprehensive of the physical sciences.

Such bold claims are exhilarating, but how, we want to know, is the chemist going to improve the arts? Cullen had no desire to deceive his students and he warned them that 'you will find many difficulties in practice of which theory is not aware'.[49] What he wished to assert was that if any science were to suggest innovations to the arts he called chemical, it would be the science of chemistry. 'In short,' he told his students, 'I need say but this one thing: Wherever any art requires a matter endued with any peculiar properties, it is the chemical philosophy which either informs us of the natural bodies possessed of these properties or induces such in bodies which had them not before, or lastly, produces new bodies endued with the necessary qualities.'[50] In seeking to make such improvements, the philosophical chemist acts as the assay master to the chemical arts. 'You know the use of the assay master to the workers in mines. The assay master, from a quarter of an ounce of any ore, in a short time & at a trifling expence can not only determine what quantity of metal it yields, but also the manner of working it to greatest advantage. The philosophical chemist is the assay master to arts in general. He brings all new projects to the test, & in the common arts he makes trials in small for their improvement, which could not be done in the large way of business but at the hazard of a great expence.'[51] Yet the analogy is loose and the original question remains unanswered.

The philosophical chemist brings two kinds of special knowledge to the study of the arts. Having mastered the foundations and methods of theoretical reasoning, he is prepared to make use of this abstract but potentially valuable mode of inquiry. Cautiously Cullen told his students that 'tho no body would recommend a wantoness of theory less than myself, yet I must be advocate for its utility under proper restrictions. It is a most powerfull means of exciting us to Experiments, and consequently the knowledge of facts. . . . The knowledge of facts in practice will be considerably enlarged by the means employed for theoretical enquires.'[52] But the more useful advantage available to the philosophical chemist, given the rudimentary state of chemical theory, is his familiarity with a great variety of chemical substances. He knows the 'physical properties & relations of bodies & their causes,' and 'the experiments that shew these,' all of which constitute a knowledge of 'the very arts of life, or facts that may improve them & furnish new arts'.[53]

To make this, the most useful knowledge presented in his lectures, accessible to his students, Cullen discussed each of the chemical arts under the heading of the most appropriate chemical principle. 'The course we are [to] give,' he said, 'is to be divided into five parts according to the five heads of salts, oils, waters, earths, & metals, to which almost all substances may be referred; & I shall now mention the chemical arts I speak of as they may be referred to these several heads.'[54] A natural history of the chemical industries thus became the key to their improvement through the application of philosophy.[55]

Cullen's lectures contain brief discussions of the various chemical processes used in British industry. 'To the head of salts belongs the manufacture of salt petre, which has lately been so much in demand, & the manufacture of salt ammoniac, so necessary to several tradesmen & particularly to the tinning of metals & the making of white iron. Both salt petre & salt ammoniac have hitherto been imported to us . . . but the materials of which they are composed are to be found every where.'[56] The raw materials used in bleaching are also mentioned in this first category. 'To the head of salts too belongs the making of potashes, for which we have so much use at present. The materials our country affords in plenty & the manufacture in small is perfectly well understood by the chemist. That it has not been practiced at large I can only impute to gentlemen's views not being sufficiently turned to these arts.'[57] Sugar manufacture also is included under the salt principle. 'Another manufacture belonging to this head is the refining of sugars, which I imagine stands in need of some improvement. In this I shall not pretend to make that improvement, but I must say that it is to be expected chiefly from chemistry, which first taught the art both of making & refining sugars.'[58] And the salting of fish and the use of lime in farming, both topics of particular interest to Cullen, were also listed among the first set of chemical arts. 'By attending to the chemical history of [putrefaction] we are directed how to avoid it & thereby to preserve most kinds of natural productions in their native perfection, the various advantages of which you can easily imagine. By the same chemical history of putrefaction we are also directed how to promote it, which is frequently required, as in furnishing the proper manures for the farmer.'[59]

The list of arts mentioned under each of the other chemical principles is less extensive than the first. Under the heading of oils and inflammable substances we find the arts of extracting and purifying oils, 'the art of making soap, entirely a chemical work & of the most important use,' and fermentation, the basic process in the brewing industry.[60] Under the waters are listed the arts 'of discovering & correcting the faults of common water, . . . the virtue of mineral waters, & the art of making salt water fresh.'[61] Cullen thought it 'unlucky that the Chemists should have bestowed so much praise on Mineral Waters,' which were credited with therapeutic powers, while they have 'done so little with respect to common' waters, which are used by so many industries.[62] The fourth category, the earths, includes 'the art of glass, of artificial gems & the curious art of enamels, both which owe their present perfection to chemists. To the head of earths too belongs the whole art of pottery, from the making of bricks to the manufacture of porcelane.'[63] Also under the earths he listed 'the arts of preparing quicklime, plaster, & cements'.[64] Finally, under the fifth heading of metals are listed the oldest of the chemical arts: 'You all know that the extraction of metals from their ores, the separation of the more valuable metals from the less so, the introducing [of] new qualities into metals, & the making [of] many artificial metals are all of them entirely chemical operations.'[65]

The ancient hope of transmuting base metals into gold remained to be considered. Cullen's first comment on the problem seems to indicate uncompromising opposition: 'Truly, my opinion is that the practice is impossible & the attempt ridiculous.'[66] But Scottish philosophers in Cullen's era allowed themselves very few absolute statements about the material world, and he immediately retreated from this outright rejection into an argument based on experience and probability. 'If the adepts [of transmutation] will be offended at the expression impossible, I will say that many arguments render it improbable, that the writings on the subject are so involved in jargon & riddles as to be unintelligible, the practices of the pretenders to this art have been so full of knavery, that upon the whole the pursuit must be in the highest degree imprudent.'[67] His conclusion was further weakened by the fact that theoretically, transmutation remained quite plausible. 'I must here allow what has been advanced by the more sober chemists: That the baser metals may be meliorated, that is brought nearer to the qualities of gold & silver; & that by certain secret operations of chemistry more gold & silver may be extracted from the baser metals than is done in the common way of business.'[68] When confronting difficult problems such as transmutation, Cullen did not try to hide from his students his own inability to resolve them.

Cullen had little more to add to his account of chemistry's task. He had argued that as a branch of philosophy it had to follow the one method by which valid explanations can be generated: generalization from empirical observation followed by verification through experimentation. He had also shown that as a body of useful knowledge, chemistry had to address itself to the needs and problems of the arts. But what are the criteria of usefulness? How can one judge whether or not a proposed innovation is valuable? The question did not trouble Cullen any more than it troubled his fellow improvers. 'It is the merchant,' he said, 'that must inform us what arts there is a demand for and it is the merchant who must inform us of the marketable qualities of manufactures.'[69] The merchant, the master of the marketplace, will guide the application of science to industry, but to do this he must be able to mediate between the demands placed upon the arts and the innovations suggested by philosophically trained chemists. 'It is the merchant of more liberal education & extensive views that must excite & direct the slow toiling industry of the artificer. Now I take the merchant to be, in respect to the artificer, what the physician is to the apothecary, & I imagine the physician, unless he knows some what of the apothecary's art, will make but a blundering prescriber. Neither can the merchant direct the hands of the tradesman without a general taste & some knowledge of arts.'[70] To become a successful 'physician of the arts', the merchant must acquire the liberal ideas which flow from the study of philosophic chemistry. It is the activity of the liberally educated merchant that synthesizes the practical and philosophic tasks of chemistry.

3. THE CHEMICAL PRINCIPLES

By beginning his course with a discussion of the definition, history and usefulness of chemistry, Cullen was able to introduce his students to his conception of the subject before getting involved in difficult theoretical problems. But after completing these initial lectures, he had to begin the formidable assignment of making chemists out of the novices enrolled in his course. 'As I suppose no knowledge of this kind previously acquired,' he told them, 'I propose to begin with the very rudiments of the art, & to avoid entering upon any doctrines before the foundations for such are previously delivered, & to give these things first that are necessary for the understanding of what follows; that your knowledge acquired from this course may not be such as that acquired from many others which have been exhibited, that is, a number of practices without connexion among themselves & without the foundations of any [of] them.'[71]

Cullen divided his chemistry course into two parts. 'The first shall contain the general doctrines & the general rules of practice, & the second shall contain the application of these general doctrines & precepts to particular subjects.'[72] The first part, on the general doctrines of chemistry, was itself 'divided into three heads: The first treating of the bodies of which the chemical philosophy treats or on which the chemical art operates . . . ; the second treats of the powers in nature by which the physical changes of bodies are produced; & the third treats of the method of art in directing & variously modifying these powers of nature towards producing the required changes of bodies.'[73] Cullen thus began his lectures on the doctrines of chemistry with 'an account of the bodies which are the objects of chemistry'.[74]

What is it that chemists study? This question could not be answered until chemists had purified the corrupt and repelling language employed by their predecessors. Cullen was particularly anxious that his students should not be put off by the abstruseness of chemical terminology.

> I would have you to observe that the difficulty & mystery which often appear in matters of science & learning are only owing to the terms of art used in them, & if many gentlemen had not been rebuted by the uncouth dress in which science was offered to them, we must believe that many of these who now shew an acute & sound judgement in the affairs of life would also in science have excelled many of those who are devoted to it & who were engaged in it only by necessity & a phlegmatic temper. This is particularly the case with respect to chemistry, which is as easy to be comprehended as any of the common affairs of life, but gentlemen have been kept from applying to it by the jargon in which it has been industriously involved. We propose to amend that & avoid terms of art as much as possible, but many of them are unavoidable & shall be explained, & I must beg your attention to our first lectures, for without your understanding our terms, all afterwards must be darkness & confusion.[75]

Although chemistry addresses itself to the practical arts, the philosophic study of the science can only be carried on with a precise language purged of all the obscure terms employed by craftsmen.

Cullen believed that any meaningful discussion of the objects of chemistry must begin with a description of the physical elements. Referring to his definition of the subject, he argued that 'as in our account of the nature of chemistry we told you that bodies for the most part become the objects of chemistry as they are mixts or compounds, that is, bodies consisting of different elements, so it is necessary in the first place to consider these different elements by the various union of which bodies become the objects of chemistry.'[76] The evidence and interpretations offered by the mechanical philosophers make it quite clear, Cullen believed, that the material world is compounded out of atoms having certain enduring qualities.

To introduce his students to the corpuscular notion of matter, Cullen presented the classic arguments developed during the preceding century.

> It has long been observed that bodies are divisible to a very great degree of subtility, of which many instances might be produced: As in gold when drawn out on silver, certain colouring substances dispersed in a great [quantity] of liquid & tinging it when there must be supposed a particle of tinging substance in every drop of the water, or, what is still more subtile, the effluvia of odoriferous substances dispersed over a very great space, when we must suppose that in every part of that space there is a particle of the odoriferous body & the whole so numerous that we can hardly express [it] & at the same time the part so divided is not of any sensible weight, which is a division that is minute to a very great degree, but however great, we must suppose it [to] have an end & that there are in nature a number of particles not farther divisible or changeable in their forms by any power but that of the Deity. We believe this from the steadiness in the appearances of material bodies, for so long as we have any records of the observations of men, we find the vegetable, animal, & mineral substances, tho daily destroyed & removed, have still appeared the same, whereas, if the principles or elements of which they are composed were changeable, so must the several compounds be. Thus in all the different parts of the earth where mines are wrought & for so long as they have been wrought, there have yet appeared but six genuine metals. Many of these mines have been exhausted, filled up & after a term of years wrought again, when the same metal is found as before, which all shews a sameness of the elements of which they are composed.[77]

If the students found these arguments unconvincing, Cullen suggested that they listen to 'the words of Sir Isaac Newton on this subject':

> It seems probable to me that God in the beginning created matter in solid, massive, hard, impenetrable, moveable particles; incomparably harder than any of the porous bodies compounded of them, nay so hard as never to wear or break in pieces, no human power being able to divide what God made one at the creation. While these particles continue entire, they may compose bodies of one and the same nature and texture in all ages, but should they wear away and break in pieces, the nature of things depending on them would be changed: water & earth composed of old, worn particles & fragments of particles would

> not be of the same nature now as water & earth composed of entire particles at the beginning. And therefore, that nature may be lasting, the changes of corporeal things are to be placed only in the various separations & new associations and motions of these permanent particles. These primitive particles, perfectly hard, indivisible & unchangeable in their forms by any power in our system, are often called Atoms.[78]

When describing the most elementary constituents of the material world, Cullen told his students that 'I think every circumstance seems to [be] most favorable to the doctrine of the corpuscularian philosophers.'[79]

Cullen opposed the atomic philosophers on only one point, their assumption that we can know anything about the ultimate constituents of matter. 'The real elements [are] not known because of their minuteness. . . . Because of their cohesion with each other we can only examine mixts.'[80] We can in fact do no more than analyze substances that possess properties we find interesting. Mechanical analysis divides bodies quantitatively, but it never reaches the atomic level. Chemists analyze bodies into qualitatively different constituents, which are no more elemental than those isolated by mechanical analysis. 'Chemistry,' Cullen concluded, 'resolves bodies into different parts [and] these have been supposed [to be] elements, but that [is] doubtfull & we give them only as the chemical principles, usefully known & now to be explained.'[81] The atomic theory may be the most probable theory of matter, but in the context of the Scottish philosophy it has no direct bearing on the explanation of natural phenomena.

Since for epistemological reasons the chemical principles could not be called elements in the sense that atoms were thought of as physical elements, their contribution to the understanding of chemistry remained a bit uncertain. Clearly their first task was to help organize the classification of chemical substances. Having lectured on botany and materia medica, Cullen was familiar with the methods and problems involved in taxonomy. Unlike Charles Alston, who taught these subjects in the University of Edinburgh, Cullen considered himself a disciple of Linnaeus.[82] Since chemistry and natural history were so closely connected on the level of subject matter, Cullen saw no reason not to classify them in the same way. 'In giving [our] account of bodies,' he told his chemistry students, 'Natural History & chemistry [are] mixed. Natural history [is] necessary, but [it] does not go far without chemistry.'[83] Since he found the distinction between natural and artificial substances quite meaningless, Cullen did not draw a sharp line between the objects of chemistry and those of natural history. He emphasized that 'in our account of bodies as objects of chemistry, the production of nature & art [are] necessarily mixed & this constantly practiced as . . . suitable to our purpose of giving a general view of chemistry.'[84]

Once one has established the propriety of classifying chemical substances in the same way that one orders the minerals, plants and animals studied by natural historians, the procedure to be followed is quite easily explained. In his paper on the classification of salts Cullen argued that

> it may be often useful to refer the productions of chemistry to their proper classes, orders, genera, species, and varieties as is commonly done with respect to the objects of natural history. This will lead us to distinguish between the general and particular properties of bodies. As species and variety in the history of fossils and in chemistry cannot be distinguished on the same principles as they are in botany and zoology, it is necessary for me to explain in what manner I use these terms. When any matter appears to be possessed of properties or of a combination of properties peculiar to it and competent to no other I consider it as a species. But when this matter, while it yet retains most of the properties which determined the species, either by a new modification of itself or by an addition of other matter, receives some variation of its properties or some additional ones, it is to be considered as a variety.[85]

Of course in carrying out this scheme of differentiation in the real world the chemist is bound to encounter difficult and ambiguous cases. Nevertheless, some system of classification must be employed and Cullen, drawing upon the chemical principles defined by earlier chemists, chose to use a system based on five principles. He justified this choice by saying that although 'it is indeed difficult, perhaps impossible, to find a proper order liable to no inconvenience, . . . we must take that which is liable to the least. As the productions of nature and art, almost all of them appear under either a saline, inflammable, watery, earthy, or metallic form, I think it more suitable to the purposes of chemistry to consider bodies first, rather under these titles than as they happen to be of vegetable, animal, or fossil origin.'[86]

But are these five chemical principles more than headings in a taxonomic scheme? At times it seems that Cullen, like the older chemists, treats their properties as the causes of chemical qualities. For instance, when characterizing the different principles he said that 'salt is a body soluble in water, fusible in the fire & sapid. Sulphur or oil is a body not miscible with water, but volatile & inflammable—

to be distinguished from mineral sulphur. Water is a body fluid with heat & volatile, but not inflammable & insipid. Earth a dry solid body not fusible, volatile or inflammable. Mercury is fluid, insipid, volatile.'[87] But are these definitive properties enduring qualities of five distinct kinds of substances? Does a sulphurous substance, when combined with some other substance, give the compound sulphurous qualities? These questions, which troubled all eighteenth-century chemists who tried to set their subject on firm theoretical foundations, were not clearly resolved in the chemistry system Cullen offered his students. Despite his disclaimers, Cullen tried to explain certain chemical phenomena in terms of the enduring properties of the chemical principles. The ambiguity which this approach introduced into his system reveals how far eighteenth-century chemists were from developing an adequate and acceptable theory of chemical elements.[88]

The difficulties Cullen encountered when classifying the objects of chemistry are well illustrated by his handling of two specific yet related questions: How should one distinguish between salts and earths and What are the generic properties of the calcareous substances? Cullen could not avoid engaging these two questions, both of which had long and intertwined histories, and the answers he arrived at deserve our careful consideration because of their influence upon Joseph Black's study of magnesia alba and fixed air. We shall, therefore, begin untangling the complex background to these questions by examining the problems involved in classifying the earths and salts.

According to the acid-alkali hypothesis favoured by the iatrochemists, all salts, to quote a definition published in 1666, 'can be divided into two parts, into alkali and acid'.[89] But since most of the salts formed by the combination of the mineral acids and the known alkalis volatilize when heated, this definition stood in direct contradiction to the older Paracelsian definition of salt as 'the principle of fixity and incombustibility'.[90] Most of the acid-alkali salts, which were paradigmatic for the iatrochemists, are soluble in water, however, and solubility therefore soon replaced incombustibility as the essential physical property of salts. During the latter part of the seventeenth century the identification of salinity and solubility was widely accepted. Nehemiah Grew attempted to quantify the study of salts by determining their particular degrees of solubility in water and Boerhaave described Grew's experiments in his lectures on chemistry.[91] Like Grew, Boerhaave believed that solubility was the chief mark of distinction between a salt, which he defined as 'a fossil body which melts both in fire and water,' and an earth, which, he said, is 'not soluble either by water or fire'.[92]

While the acid-alkali theorists were developing their doctrine of salts, Becher and Stahl were advancing an entirely different system of chemical principles, one which characterizes salts in quite a different way. As Cullen told his chemistry students, 'they make two primary kinds [of chemical matter], Water & Earth. The earth of three kinds; Vitrescible, Inflammable, & Mercurial, which are much the same with the ancient Salt, Sulphur, & Mercury. They come to much the same as our five [principles]. The Mercurial, Inflammable, & Vitrescible Earths & Water the same, only Salt not an element but a mixt of water & the first earth.'[93] The solubility of salts and the supposed insolubility of earths were matters of indifference to the Stahlians. Thus Stahl himself, when accounting for the causticity of quicklime, speaks of a soluble and an insoluble earth. 'In *Lime* newly calcined,' he asserts, 'there are two substances contain'd; the first [is] a subtile corrosive earthy matter which enters into Water, tho' it communicates no remarkable taste thereto, whence such Water dissolves common Sulphur, precipitates acid Solution, [and has the other properties of lime-water]. The second [is] a grosser matter which in the form of earth remains at the bottom of the Water and seems to be the other, tho' not pure, constituent *alkaline* principle.'[94] Cullen, despite his harsh judgment of Becher and Stahl—'they are not judicious philosophers, their doctrines [are] not consistent, they solve no difficulties'[95]—realized that the properties of the alkaline earths such as quicklime create problems for chemists who wish to distinguish between the earths and the salts in terms of their solubilities.

As Cullen's choice of principles shows, he believed that the earths could be distinguished from the salts, but when describing the properties of these principles, he tried to avoid placing too much emphasis upon the criterion of solubility. He defined salt as 'a body soluble in water, fusible in the fire, & sapid,' whereas earth, he said, is 'a dry, solid body not fusible, volatile, or inflammable'.[96] He had not defined earth as insoluble, yet the implication was inescapable: 'Salt,' he later admitted, is 'distinct from . . . earth [in being] miscible with water.'[97]

By the time Cullen took up this problem the term salt had been applied to a wide variety of substances. Not only were acid-alkali compounds called salts, but the acids and alkalis themselves were considered salts. But if one calls the three perfect alkalis, potash, soda and the volatile alkali, salts, should one also call all alkaline substances salts? Many chemists believed that the true alkali salts could be distinguished from the alkaline earths which, while largely insoluble, effervesce when treated with acid. Cullen tried to resolve this problem by distinguishing between the simple and the compound salts and, within the class of compound salts, between the perfect and the 'not perfect' salts.

> The Simple are of two kinds; that is, Acid or Alkaline. . . . The Compound salts have always as one of their constituent parts an acid. In the various compound salts the acid is joined either to an alkaline salt, to an earthy, or to a metallic substance. The compound salts formed of an acid and earth, or of an acid and metallic substance, are by some considered as not perfect salts, as one of the constituent parts only is saline, but whatever is in this, they cannot be properly considered under the head of salts but must be deferred till the differences of earthy and metallic substances are ascertained under their respective titles.[98]

Although willing to acknowledge the alkalinity of certain earths, Cullen continued to insist that earths and salts can be distinguished from each other.

Cullen, and Black too, well might have muddled along without worrying this question further had not a German chemist, Johann Pott, challenged the analysis of sea salt published by Cullen's friend Duhamel du Monceau. The subject of sea salt was a classic problem in chemistry; Robert Boyle had described a process for obtaining a pure alkali from sea salt, Stahl had suggested that the salt contains an alkali found in no other substance, and Boerhaave had told his students that 'no known experiment hath hitherto produced the least fixed alkali from sea-salt'.[99] Duhamel, in his memoir 'Sur la Base du Sel Marin,'[100] described the experiments that enabled him to identify the alkali in sea salt as soda (sodium carbonate). The analysis itself is convincing only if one assumes that salts are soluble and earths are insoluble. To determine whether the base of sea salt is a salt or an earth, Duhamel added an alkali salt to a solution of sea salt. Had the base in sea salt been an earth, it would have been displaced by the alkali salt and settled to the bottom as a precipitate. Since no precipitate appeared, the base of sea salt was, he concluded, a salt: 'Cette base ne se peut précipiter par les Sels alkalis, comme cela arriveroit vraisemblablement si elle étoit terreuse.'[101] Then, using two different series of chemical reactions, Duhamel identified the alkali salt by transforming sea salt into the well known fixed alkali soda.[102]

Johann Pott, who studied under both Stahl and Hoffmann, became an expert on earths while trying to discover the way in which porcelain was made at Meissen. In 1743, seven years after the publication of Duhamel's paper, Pott announced that he too had successfully analyzed sea salt and that the base was an alkaline calcareous earth, not a salt.[103] Accepting the criterion of solubility, Pott argued that the soda which forms the base of sea salt is insoluble and therefore an earth. To demonstrate its insolubility, he combined Glauber's salt (sodium sulphate), which contains the base of sea salt, with fixed alkali (potassium carbonate), thereby obtaining a precipitate which, he asserted, contains the insoluble base of sea salt. He was not sure whether this earth was calcareous or not for he was unable to synthesize sea salt from marine acid and lime. In fact, as Cullen was to demonstrate, the precipitate Pott obtained from Glauber's salt contained magnesium salts, which had contaminated his Glauber's salt. But until Pott's argument had been convincingly refuted, his challenge cast doubt upon one of the most notable advances in the analysis of salts.

Pott published his system for classifying the earths just before Cullen began teaching chemistry and Cullen used this system in his lectures on earths.[104] But in Cullen's opinion, Pott's errors in the study of salts could not be excused by appealing to his special knowledge of the earths, and on the question of sea salt Cullen supported Duhamel's analysis without reservation. 'It is but very lately,' Cullen wrote,

> that the ingenious and very knowing Mr. du Hamel du Monceau, unassisted as it would seem by the writings either of Boyle or Stahl, has set this matter in a full and proper light. He has supplied the defects of Mr. Boyle and corrected the mistakes of Dr. Stahl; and for the knowledge of the properties of the fossil alkali, the reasons for considering it as a species of fixed alkali, and the application of this in chemistry, I must refer to Mr. Du Hamel's memoir on The Basis of Sea Salt. . . . The utility and necessity of that memoir will appear especially from hence that even the disciples and followers of Dr. Stahl in Germany did not apprehend his doctrine on this subject. See in the *Miscellanea Berolinensia*, Tom. 7, what frivolous objections the otherwise knowing and experienced Pott throws out against the memoir of Mr. Du Hamel.[105]

Obviously, Cullen asserted, Pott has simply missed the point Duhamel set out to make. 'I call it frivolous,' he exclaimed, 'when Mr. Pott argues that the basis of sea salt is not an alkaline salt because it is not exactly the same with the vegetable alkali, which indeed it was the purpose of Mr. Du Hamel's memoir to shew.'[106] Clearly Pott failed to grasp the way in which Duhamel employed the criterion of solubility and he totally neglected Duhamel's use of the other properties by which one can distinguish between a salt and an earth.

> When Mr. Du Hamel says that the basis of sea salt is distinguished from an alkaline earth by this that it is soluble in water, Mr. Pott would avoid this by telling us that quicklime and chalk are in part dissolved in water by the means of long boiling. Every body I dare say will perceive that this does not properly apply here, or if it does it will equally destroy the distinction between the vegetable alkali and alkaline earths. The other parts of Mr. Pott's reasoning against Mr. Du Hamel are of a piece with these: he does not indeed touch the proper question whether or not the basis of sea salt as readily dissolving in water, as effervescing with acids, as precipitating earths and metals from these, and as shewing many other properties of the vegetable alkali does not deserve to be considered as a fixed alkaline salt, and in respect of its peculiar properties as a particular species of this, and lastly as a very different substance from any alkaline earth yet known.[107]

Cullen's defence of Duhamel's analysis was made convincing by his identification of the fundamental error in Pott's experiments. Pott was misled, Cullen concluded, because he used Glauber's salt prepared from the brine of sea water rather than the pure salt made according to the method described by Glauber.

> The salt now used under this name in medicine is rarely prepared by the original method of Glauber, but is commonly obtained by a proper crystallization from the bittern of the salt boilers. This may be supposed to answer the purposes of medicine as well as any other, but as it is composed of the vitriolic acid and the earth called *magnesia alba*, it is to be considered in chemistry as very different from the genuine neutral composed of the acid joined to the fossil alkali. Mr. Pott is therefore very much mistaken in adducing the former composition as a proof of his opinion concerning the basis of sea salt.[108]

Cullen pointed out that it is quite easy to distinguish between pure and contaminated Glauber's salt, for when potash is added to a solution of the pure salt no precipitate is deposited. 'To shew the difference it is enough now to observe that a solution of true Glauber's salt is not precipitated by the vegetable alkali, whereas the calcarious Glauber's salt has its earth precipitated copiously by the same alkali; as is now well known from the common method used in Scotland for obtaining the *magnesia alba*.'[109]

Cullen felt that Pott's confusion about Glauber's salt could be excused, but his ignorance concerning the true nature of alimentary salt, that is, sea salt, was quite indefensible. 'Now physicians may perhaps find little difference between the calcarious and the genuine

Glauber's salt, so that Mr. Pott may have had some foundation for alleging that Glauber's salt is composed of an acid joined to the *magnesia alba*, but when he asserts that alimentary salt is composed of an acid and earth his error must appear very gross. The appearances and effects of an alimentary salt that contains a perfect alkali and one that approaches to a calcarious salt are very different.'[110] And where can one find the difference between a salt containing a perfect alkali and one containing a calcareous earth delineated? In Brownrigg's book on salt boiling, Cullen assures us. 'The ingenious Dr. Brownrigg has fully explained the difference. There was hardly any thing wanting to render the useful work of this author very perfect but his not being as it would seem better acquainted with the nature of the fossil alkali [i.e., he failed adequately to emphasize Duhamel's distinction between soda, the fossil alkali, and potash, the vegetable alkali].'[111] Contrary to Cullen's flattering recollection, however, Brownrigg's characterization of calcareous substances in no way solves the difficulties raised by Cullen's need to distinguish between the salts and the earths.

Brownrigg did not try to hide his own confusion about the difference between earths and salts. When salt is made from sea water, a 'white subtile earth' collects on the walls of the boiling house, indicating, he thought, that 'sea water probably holds an earthy substance so very light and subtile that . . . it is elevated along with the watery vapours'.[112] According to all the accepted definitions of chemical principles, it is 'absurd to talk of a volatile earth'. Nevertheless, several chemical phenomena seemed to indicate that some earths are in fact volatile: 'Such are those volatile fumes that arise from quick lime when water is poured upon it; and such is the subtile alcaline earth of lime water, which arises with it [in the boilers of steam engines] in distillation.'[113] Solubility also seemed to be a property of both earths and salts. Consider 'this earth, to which the salt boilers have given the name scratch,' which 'separates from the brine before the salt begins to form in it; and is either taken out in the scratch pans in the form of a white powder, or else adheres to the bottom, and sides of the salt pan in a hard stoney incrustation.'[114] Brownrigg was puzzled because scratch, which is in fact a mixture of calcium and magnesium salts, 'is capable of being dissolved by water in very considerable quantities. . . . The great solubility of this earth in water, shews that it nearly approaches to the nature of salts. It is even found to enter the composition of perfect salts.'[115] In this reference to perfect salts Brownrigg is alluding to the fact that Epsom salts (magnesium sulphate) can be obtained by adding sulphuric acid to scratch. 'With [vitriolic acid] it [i.e., scratch] is converted into a neutral salt, which Dr. Lister ranks amongst the species of his calcareous nitre. This salt, in taste, very nearly resembles the bitter purging salt of Epsom waters.'[116] But as he looked more closely at this substance called scratch, Brownrigg noticed a new set of earthy properties. Sea shells, a common source of lime, seemed to be composed primarily of scratch, and the earth itself can be calcined. 'I found,' he reports, 'that stonescratch, after calcination by a violent heat in an air furnace, fell into powder in the open air, and had other properties of quick lime.'[117] Therefore, Brownrigg concluded, 'the marine scratch . . . seems to deserve the title of a saline calcarious earth.'[118] Cullen obviously had forgotten about Brownrigg's discussion of scratch when he told his students where to look for a clear distinction between the perfect salts and the calcareous earths.

Brownrigg's treatment of calcareous substances indicates that he had been deeply influenced by the German conception of earths. Like the German chemists, he considered magnesia alba one of the saline calcareous earths. Since Cullen recommended Brownrigg's book to his students and since Joseph Black was to make an intensive study of magnesia alba, we must now turn from the earth-salt problem to an examination of the various meanings attached to the term calcareous.

Historically, the terms calcareous, calcination and calx were first used to describe the change in physical appearance caused by roasting limestone to produce quicklime.[119] This meaning was still in use in the eighteenth century and Cullen referred to these distinctive physical changes when defining these terms for his students. 'There are,' he said, 'certain bodies which, tho ignited with the strongest fire we can apply, yet there parts are not dissipated . . . but have their texture quite broke, so that they fall down into a fine powder, which is called a calx, & such bodies are said to be calcined.'[120] But by the middle of the eighteenth century the term calcination had been applied to processes other than those used to make quicklime. In Cullen's opinion, any roasting process which reduces a substance to a fine powder can be considered a calcination. 'This happens to many bodies by a slight fire. If such are heterogeneous substances, . . . their Water, Oils & Volatile Salts are dissipated & there remains only their Fixed Salts & Earths. Several metals may be calcined by a slow & gradual fire.'[121] Thus for Cullen all fixed salts, fixed earths and calcined metals are to be thought of as calxes.

This generalization of the terms calcination and calx made it necessary to distinguish between calcinable substances having different properties. Of course, as Cullen knew, it is possible to speak of calcinations that cause an increase in weight and those that cause a decrease. 'Upon this subject,' he said, it is 'necessary to notice the weight acquired by calcined metals. Metals calcined by the fire acquire weight; lead in minium acquires 1/10.'[122] Differences in chemical qualities, however, provide a more adequate basis for distinguishing among the calcinable substances, and it was the increasing attention given to chemical properties which led to a redefinition of the term calcareous. Limestone, the original calcareous substance, becomes caustic when calcined, and while this ability to yield a caustic calx was not at first considered an essential property of calcareous substances, gradually the notions of calcareous and caustic-when-calcined came to be identified with each other.

As soon as chemical properties are emphasized in the definition of calcareous substances, however, one is once again embroiled in the theoretical debates that so confused the distinction between the earths and the salts. Is causticity caused by the presence of an alkali salt or is it the generic property of a certain class of alkaline earths? It is the alkaline properties of the calcareous substances which suggest this question. As Cullen pointed out to his chemistry students, 'the Calcareous Earths have somewhat of an Alcaline Quality; they effervesce with Acids, they acquire this more after calcination. They approach to the Saline as being more diffusable, if not soluble, in water.'[123] According to Cullen's description, calcareous substances have a distinctive set of properties; they are alkaline, they are saline and they effervesce with acids. In his opinion these properties all tend to confirm the truth of the salt theory of causticity. It was this theory, however, that Black had to abandon before he could make sense out of the properties of the calcareous earth called magnesia alba.

The salt theory of causticity has a long and complex history of its own. Van Helmont's formulation of it represents the thinking of the acid-alkali theorists. 'In Lime,' he wrote, there are 'two salts, the one a *lixivial Alcali* salt, and the other, an acide or sharp one.'[124] John Mayow modified this hypothesis somewhat, arguing that when quicklime and potash are boiled together to produce caustic lye, the alkali in the potash attracts the neutralizing acid salt in quicklime, thereby exposing the intense alkalinity, that is, the causticity, of its alkali salt. It is this alkali salt in calcareous substances which gives caustic solutions their power: 'Since . . . the alcaline salt of quicklime is extremely acrid and fiery,' Mayow concluded, 'it necessarily follows that the lixivial water impregnated with it becomes in the highest degree biting, caustic, and fiery.'[125] Stahl also explained caustification in terms of intense alkalinity, but unlike Mayow he believed that the alkali in potash, rather than that in quicklime, is the source of causticity in lye. During the formation of caustic lye, he argued, 'the most subtile, fix'd earthy substance of the Quick-lime unites to the sulphureous part of the alkaline Salt, and absorbs the volatile, saline, acid one thereof; by which means it renders the Alkali [in the potash] more fix'd, and more penetrating [i.e., caustic] in its action.'[126] And in 1724 Charles DuFay claimed that the nearly inert powder he obtained by evaporating lime-water contained the salt which gives quicklime its characteristic properties: 'Si j'ose avancer mon sentiment,' he wrote, 'je crois qu'on peut regarder cette croûte [i.e., the residue left after lime-water has been dryed] comme un Sel extremement impur, absorbé dans beaucoup de terre insipide.'[127]

While all chemists agreed that the calcareous substances are alkaline, the agreement did not help clarify the distinctive features of the calcareous earths. In the late seventeenth century alkalis were frequently defined as the class of substances that effervesce when combined with acids, while the class of calcareous earths was even more loosely defined: 'What in the seventeenth century was described as a "calcarious earth" may have consisted largely of chalk, but on the other hand it is possible that it comprised any of the earths later called *magnesia, alumina* and *baryta*.'[128] In the middle of the eighteenth century Cullen still treated effervescence as a defining property of alkalis. When discussing borax, for example, he said 'it effervesces with acids & changes vegetable juices green, & we shall see other reasons for considering it as an alkali.'[129] He also seems to have accepted Pott's use of *terra alcaline* and *calcaria* as equivalent designations for the first class of earths, for we find Cullen writing that 'the first [class], or Alkaline Earths, may be burnt to a Quicklime & may be dissolved in acid Menstruums with Effervescence.'[130] But so long as Cullen wished to distinguish between salts and earths, he was reluctant to think of causticity as an essential property of the alkaline earths. He believed, rather, that alkalinity is a property of salts, and since he thought of causticity as an intense form of alkalinity, he was convinced that all calcareous substances contain a perfect alkali salt.

The salt theory of causticity was very much alive during the middle of the eighteenth century and several of Cullen's most respected contemporaries were busy looking for the salt in quicklime, although they were not all convinced the salt was a perfect alkali.

In 1745 Paul Malouin published a thorough study of the memoir in which DuFay claimed to have isolated the salt in lime-water.[131] He concluded that the substance DuFay had isolated was indeed the active component in quicklime and he identified it as the vitriolated salt that, a few years previously, Boulduc had called the *sel sélénitique*. 'Il est resté au fond du vaisseau,' Malouin reported, 'des crysteaux . . . qui me paroissoient . . . être formées de ces aiguilles que j'ai reconnu être un sel calcaire ou sélénitique qui est le vrai sel de la chaux.'[132] There were at least two possible responses to Malouin's identification of the salt in quicklime. One could say that if the salt in quicklime is a compound of a calcareous earth and the acid of sulphur, then it is not, according to Cullen's terminology, a perfect salt and hence not really a salt at all. Or one could ask why gypsum and limestone have such different properties if they contain the same essential substance, namely, *sel sélénitique*. It was this second question that attracted the attention of two of Malouin's chemical colleagues in the *Académie*.

Pierre Macquer set himself the task of clarifying the chemical relationship between quicklime and gypsum.[133] He accepted Malouin's identification of the salt in quicklime and tried to determine whether the differences between quicklime and gypsum were caused by different concentrations of the *sel sélénitique* in the two substances. Several experiments convinced him that their differences could not be attributed to different concentrations of the salt and he next asked if perhaps the salts in each substance contain different proportions of the vitriolic acid and calcareous earth. Although his experimental tests of this hypothesis were a bit more encouraging, Macquer did not really settle the question, even in his own mind. The slight differences he detected in the concentrations of acid and sand in quicklime and gypsum were simply not enough to account for the differences in their properties.

Duhamel du Monceau was less impressed than Macquer by Malouin's analysis of quicklime. Having spent years thinking about the origins of causticity, he had accumulated a vast store of experiments and ideas, which he arranged and published as a response to Malouin's memoir.[134] From Cullen's point of view, the theoretical asides sprinkled throughout Duhamel's memoir must have seemed entirely regressive. When he analyzed quicklime made from a variety of substances, Duhamel found slight chemical differences, which led him to conclude that there are in fact many kinds of quicklime, an assumption implying that causticity is a generic rather than a specific property.[135] He evaporated lime-water to obtain DuFay's salt but, he argued, 'il n'est pas probable qu'une aussi petite quantité de sel donne à la chaux les propriétés qui la rendent si approchante des sels alkalis. C'est neanmoins ce dont il est question, quand on parle de sel de la chaux.'[136] In fact, he asserted, the acid in quicklime is its distinctive ingredient. 'On a obligation,' he wrote, 'à ceux qui ont eu la sagacité de nous démontrer la présence des acides minéraux dans les chaux qu'ils ont examinées.'[137] And it is the acid in quicklime, he argued, that makes the alkaline earth soluble and therefore active:

'Il y a dans l'eau de chaux un acide qui tient une terre en dissolution, & qui, s'étant uni au sel alkali, avoit formé avec lui les cristeaux dont nous venons de parler: je soupçonnois que ces cristeaux étoient vitrioliques.'[138] Clearly among the chemists of the *Académie* the search for the salt in quicklime had moved far away from Cullen's conception of a salt.

As we have seen above, Brownrigg believed that at least some calcareous substances are both earthy and saline. Like Duhamel he believed that there are many earths that are calcareous, each yielding a distinct kind of quicklime. 'I am far from thinking,' he said, 'that all earths which come under that general denomination [i.e., calcareous], are exactly of the same nature, and agree in all their properties. On the contrary it is more reasonable to believe that there are several kinds of calcareous earth, which are more or less subtile, more or less corrosive, and have other specific differences.'[139] Magnesia alba, 'a kind of calcareous earth, very nearly related to the scratch of sea water,'[140] is, Brownrigg assumed, one of the saline alkaline earths which becomes caustic when calcined.

When describing the properties of magnesia alba, Brownrigg referred his readers to the pioneering studies of Frederick Hoffmann.[141] Hoffmann isolated the magnesia alba contained in the water of a mineral spring and demonstrated that it can be distinguished from the alkaline earth, which is the base of lime. He also demonstrated that magnesia alba and the calcareous nitre previously described by Martin Lister are identical and that the earth in magnesia alba is the base in Epsom salt.[142] Despite the thoroughness of Hoffmann's analysis, however, at least one problem remained. Hoffmann's magnesia alba came from mineral spring water whereas the magnesia alba available in Scotland was made from bittern. If calcareous substances differ according to their natural source, how can one be sure that Scottish magnesia alba will have the same properties as that studied by Hoffmann? When, a few years after the publication of Brownrigg's book, Joseph Black began his experimental study of magnesia alba, he took as his first problem establishing the chemical identity of his magnesia alba and that studied by Hoffmann.[143]

In opposition to all those who ignored the distinction between alkali salts and alkaline earths and in the face of the prevailing belief that there are specific differences between the various calcareous substances, Cullen continued to search for the perfect salt of quicklime. As part of his classification of the simple salts he argued that 'in concluding these observations on alkaline salts, I must observe that the saline substance obtained from quicklime deserves perhaps to be considered as a distinct species of alkali, and I shall willingly add it to these above. I omitted it, however, as I could have done no more than mention it.'[144] And rather than surrender the distinction between earths and salts, he chose to postulate a kind of transmutation, which links the two principles.

> The varieties to which alimentary salt is liable depend more upon the different methods of preparing it than upon the different sources from which it is taken. It is well known that the vegetable fixed alkali by repeated solutions and calcinations may be converted almost entirely into an earth. The fossil alkali is still more easily changed in the same way, and therefore it is that alimentary salt by long boiling is changed almost into an earth or into a calcarious salt. It is on this account too found of very great variety, as it may either consist of the acid joined to a perfect alkali or of the same acid joined to a substance that recedes more or less from the nature of the alkali and approaches to the nature of an earth.[145]

Cullen's system of chemical principles was beginning to dissolve into a flux. Even potash, he admitted, 'appears to me to be always in some degree varied by the species of acid with which it has been combined'.[146] Although eager to classify the objects of chemistry, Cullen had chosen a set of principles and emphasized a set of properties which prevented him from distinguishing clearly between the calcareous earths, the alkali salts and the alkaline earths. Only after abandoning the search for a perfect alkali salt in calcareous substances and only after determining that very few chemically distinct substances can be made caustic could Black say to Cullen 'I see no reason for your saying that the alkalies may acquire different properties from having been previously combined with different acids.'[147] The major achievement of Black's investigation of magnesia alba was the order it brought to the classification of the calcareous substances and the new foundation it provided for the theory of causticity, and the magnitude of that achievement can only be appreciated when Black's treatment of the subject is set against the taxonomic and theoretical confusions from which it emerged.

In attempting to reformulate the received theories of chemistry in terms acceptable to the Scottish theory of knowledge, Cullen took on a task, which could not be completed in a single generation. His conservatism, one aspect of which was his unwillingness to surrender the older materialist explanations of chemical action, can be clearly seen in his decision to include fire and air among the chemical principles. After recapitulating the distinctive features of the five principles used in classifying the objects of chemistry, Cullen went on to say that there are 'two other principles of bodies: fire & air'.[148] They are, he argues, two substances whose properties cause the changes that chemists study. Fire, he said, is present

> in almost all bodies & increases their weight. With respect to it, different opinions of Boyle & Newton, of Boerhaave & others. Boerhaave does not [assert] directly [that] it is the sulphur of the chemists, but the French & German chemists do & this must be left to the particular consideration of Fire. With respect to Air; from Mr Hales experiments [it is apparent that] it does enter into the composition of bodies & [is] in general to be noticed, but its effects on the qualities of bodies [are] not known & it may be only a different arrangement [of particles].[149]

Cullen was clearly unable to carry through a complete Humean subordination of explaining to ordering. In fact, his determined defence of the salt-earth distinction reflects, in the realm of the chemical principles, his continued attachment to the acid-alkali theory of chemical action. Even in his last years as a chemistry teacher he suggested that there may be a primal acid and a primal alkali which lie behind the various chemical phenomena we study. 'From [a] comparative & general view of the acid,' he reasoned, 'we are led to conclude that the opinion of some is not improbable who suppose that there is only one primogenial acid in nature, and that the different species that we can examine are nothing more than various modifications of this acid with foreign matters. . . . Some chemists have [also] imagined, not improbably, that as there is only one primogenial acid, so there is but one primogenial alkali, different modifications of which appear to us as distinct species of alkali.'[150] If there is a primal alkali, then the alkalinity of salts and earths must have a common cause, and causticity, which is an intense form of alkalinity, is just another manifestation of the properties of this fundamental alkali. Surely Cullen should not be faulted for suggesting concepts that provide general explanations for the phenomena studied by chemists. But in treating such concepts both as substantial principles and as abstract powers, he failed to distinguish between the world of nature and the constructs of the mind, a distinction which the natural philosophy of his time considered fundamental.

Cullen knew that something was amiss, for when restating the nature of the chemical principles, he claimed no more for them than the most sceptical of his contemporaries would allow. 'All these substances we call principles,' he emphasized,

> are really mixts & farther resolvible . . . & we may venture to say that neither in art or nature have we found a perfectly simple, homogeneous body. Water, however transparent & colourless,

> however insipid & inodorous, however durable & unchangeable in these & all other respects it may appear to us, there is reason to affirm, as you will find afterwards, that the purest [water] from nature or art that ever occurred is still associated with other bodies of a different nature. Mercury appears the most simple homogeneous body of any we can examine, but Mr Boyle could change a great part of mercury into water. It is the same with respect to every substance occurring to us, & tho there were substances occurring that appeared to us homogeneous, indivisible & unchangeable, we could well judge whither really they were so or [whether they] only appeared so from the imperfection of our art, as gold appeared a perfectly simple, indestructible body till experiments were made on it by the burning glass.[151]

The principles, being neither elemental nor pure, cannot possess enduring chemical qualities. They are, however, the most simple substances we can isolate by chemical analysis and hence they provide the best ground for classifying the objects of chemistry. 'Upon the whole then,' he concluded, 'the chemical principles really [are] mixts . . . [Although] they appear to be simple & perhaps the first degree of mixture, & [although] they are really the elements of compounds in Stahl's sense, & tho by chemical resolutions we do not arrive at the real & most simple elements of all bodies, yet it is usefull to know the principles of compounds.'[152]

The more Cullen addressed himself to the epistemological status of the principles, as opposed to the way he used them in his system of chemistry, the more sceptical his critique became. 'Tho we said that the knowledge of the composition of bodies is often usefull,' he continued, 'it must still be understood with two limitations: First, from the nature of the principles we cannot always conclude with respect to the mixt, [e.g.] in soap & glass, secondly, [neither] from the nature of the mixts can we conclude with respect to the nature of the ingredients. This observation was necessary as the chemists have reasoned very falsely, of which many instances might be given.'[153] Boerhaave was just one of 'the chemical philosophers [who was] faulty in this respect'. He wrongly attributed 'natural fluidity . . . to water, dryness & solidity to Earth, Inflammability to sulphur, [and] almost to every quality a distinct principle'.[154] While we know nothing of elements, Cullen continued, we are aware that frequently 'qualities are produced & destroyed by a change of texture without a change of principles—instance in Water; solid, fluid & elastic'.[155] Therefore, rather than seizing upon the obvious supposition that the elements differ, we should ask whether or not there is 'but one element & qualities may depend on texture'.[156] Had all chemists taken seriously the arguments contained in 'Mr. Boyle's two treatises: Sceptical chemist & Producibleness of the chemical Qualities,'[157] they would be trying to explain chemical phenomena in terms of textures and powers rather than elements and principles. Whether by doing so they would have advanced the science of chemistry is a somewhat different question.

Cullen could neither combine the seventeenth-century theories of matter with the Scottish philosophy of knowledge nor could he ignore the materialist aspects of the older concepts. Newton had not only propounded certain laws, he had described the nature of a level of reality that was, according to Cullen's philosophical colleagues, inaccessible to the mind of man. The new epistemology insisted that the elemental world described by Newton was of no use to chemists and that no new theory of primal matter could be postulated to replace it. Cullen considered chemistry a new science but he was reluctant to jettison all previous concepts just because they were philosophically tainted. Yet so long as he clung to Newtonian atomism he could not advance convincing arguments against those who used mechanical theories to explain chemical phenomena. He could not even lay the ghost of transmutation. 'Indeed,' he noted, 'upon [the] supposition of permanent [chemical] Elements we might contradict this Notion of Immutation a priori, but we have here endeavoured to prove it a posteriori or Experimentally'.[158] Cullen's belief in the existence of atoms rendered his chemical principles theoretically useless, first by requiring that to explain phenomena they, like the atoms, must possess enduring properties and then by undermining the possibility of attributing enduring properties to anything as complex as a chemical principle. When it actually came to explaining the phenomena studied by chemists, Cullen did little to bridge the gap between the received theories of chemistry and the epistemological position championed by his philosophical colleagues.

4. ELECTIVE ATTRACTION

Although Cullen often treated the properties of the principles as causative when explaining chemical phenomena, he never consciously surrendered his allegiance to Newtonian positivism. 'Theory consists of the doctrine of causes,' he asserted, and causes are the creation of a mind reflecting upon the reality revealed to it by the senses. Thus causes are real, in the sense that they can be correlated with the relations between objects in the world outside the mind, but they are neither things in themselves nor enduring properties of bodies. A knowledge of causes is worth attaining because it enables us to comprehend and hence master the material world, but the usefulness of this knowledge, like its validity, is in no way improved by associating causes with material substances. Therefore, when organizing his lectures, Cullen carefully separated his account of the causes of chemical change from his descriptions of the elements and principles.

There are, according to Cullen, four 'Primary Causes of the changes of bodies occurring in chemical operations: 1 – Common and Elective Attraction, 2 – Fire, 3 – Air, 4 – Ferments.'[159] All the changes characteristic of chemistry, changes such as 'combination or separation, condensation or rarefaction,' are, he believed, 'modifications of motion'.[160] The causes of these changes of motion can be grouped into two pairs of opposing forces. 'The first causes then seem to me [to be] attraction & Fire, & next to them, tho subordinate, air & ferments.'[161]

Attraction, the first of these causes, is also a model of what a cause should be. Cullen assumed that the notion of attraction has been accepted by all modern philosophers familiar with the Newtonian achievement. Furthermore, the search for the material cause behind the force of attraction has been properly given up as a false problem in philosophy. The reality of the force of attraction cannot be seriously questioned, for one can list numerous phenomena that reveal its effects even though its laws have not been adequately formulated.[162] Cullen, of course, recognized the truth and importance of the Newtonian theory of gravitation, but unlike Buffon and several other French chemists of his time, he did not believe that the theory of gravitational attraction provided an adequate conceptual basis for understanding chemistry.[163] As a model of theoretical explanation, the theory of universal attraction was unrivalled, but in the realm of chemistry one must account for the particular features of the relative attractions between bodies, and this could only be done by developing the theory of elective attractions.

The changes that chemists study are primarily, but not exclusively, changes of combination and separation. It was well known that 'some bodies will join, others will not'.[164] Among those that do form compounds, 'bodies that join with several others will join more readily with one than another'.[165] The relative stability of the compounds formed by these substances is termed their elective attraction and it is knowledge of this relation between the constituents in the compound that is 'of use in chemistry'.[166]

Naturally one does not ask what causes the elective attractions between chemical substances; 'what they depend on [is] not known'.[167]

The force itself is the cause and it is the general law governing the operation of this force that chemists attempt to determine. In this task, Cullen told his students, it is helpful 'to enumerate them [i.e., the elective attractions] & reduce them to general heads. This [was] attempted by Mr. Geoffroy. His table presents us at one view with the relations of a great number of bodies.'[168]

Cullen had copies of Geoffroy's famous table printed so that his chemistry students would not have to copy it from the board during his lectures.[169] Certainly one would have to consult a copy of that table to follow Cullen's comments about elective attraction. Column by column he marched his students through Geoffroy's table, explaining the terms used to denote substances, describing the reactions implied by the arrangement, and adding his own comments and corrections: 'The second column, . . . why silver stands here, why gold in the lowermost. Col. 4, why fixed alkali repeated here. Stahl's problem solved by Mr. Geoffroy. Applicable to nitre also.'[170] Geoffroy had successfully inaugurated the study of elective attractions and Cullen was merely building upon the foundations he laid without essentially altering his plan. The use of diagrams in the study of elective attraction, Cullen's one notable contribution to the study, was a didactic rather than a theoretical innovation.[171] So long as Cullen taught chemistry he continued to emphasize and improve upon the table of elective attractions, for he firmly believed that chemists were on the brink of determining the laws that govern this first cause of change in chemistry.[172]

5. FIRE

Fire, the second primary cause of chemical change, is, according to Cullen, both a principle and a power. As a principle, fire is to be considered as a substance possessing certain specific chemical properties. As a power, fire is defined as a force arising from the motion of particles. Cullen knew that by proposing this dual characterization of fire he could be charged with confusing causes and substances but, Newtonian that he was, he believed that the arguments for the materiality of fire, particularly those advanced by Dutch chemists and natural philosophers, could not be neglected.[173] There are, he stated, 'two opinions' concerning the nature of fire. That favoured by the English Newtonians sees 'fire consisting in motion which may be in common to all bodies,' whereas the other, the Boerhaavian opinion, defines fire as 'motion peculiar to one kind of bodies only'.[174] Each of these hypotheses can explain only a few of the phenomena associated with fire. 'The English must allow that the phenomena [of fire] belong to a peculiar modification [i.e., only a few substances are inflammable]. But if a peculiar substance, or a peculiar modification, how do we come to find this every where & diffused over all matter [i.e., all substances will absorb and transmit heat].'[175] If chemistry and mechanics are to develop as independent branches of a coherent system of natural philosophy, the concepts they employ, such as those used in the theories of gravitational and elective attractions, must be compatible. This necessary compatibility clearly was not a feature of the conceptions of fire Cullen was attempting to combine. 'Fire is a substance' and 'fire is a power' are statements which reflect entirely different assumptions about the nature of the physical reality described by natural philosophy. The need to clarify the idea of fire thus forced Cullen to confront conceptual conflicts which could not be resolved by eclectic accommodations.

Since there was no consensus concerning the real nature of fire, Cullen devoted most of his attention to the 'history of phenomena'. A truly comprehensive theory could emerge only after all the manifestations of fire had been noted and classified. Once the relevant phenomena had been ordered, philosophers could begin abstracting from the data of the senses and speculating about the relations between fire and the changes it causes. Cullen never doubted that 'fire [is] a primary cause of motion & change of bodies,' an assumption he considered 'obvious upon the slightest view, especially in chemistry'.[176] But he was not at all sure what fire was and how its effects were to be explained.

How could chemists determine which changes are caused by fire? One can begin by describing the most distinctive manifestations of fire, identifying their essential properties, and then treating the presence of these properties as the sign of fire's action. Consider, for instance, the phenomena of heat, light and burning. Each of these was thought to be caused by fire and yet they all have heat as a common property. 'Light, tho often without heat, yet condensed gives heat; heat increased produces light. These three then all referred to the same.'[177] Thus heat seems to provide the best means for detecting and measuring the action of fire. But 'our sensation of heat [is] relative, we want another mark that may be more universal. Perhaps there is none such, but the most so expansion.'[178] The argument Cullen presented was the common eighteenth-century justification for using thermometers, which measure expansion caused by heat, as detectors of the presence and action of fire. 'Expansion can be perceived where heat & light are not. Expansion can be measured more exactly than any other effect of Fire.'[179]

Before discussing the various thermometric substances and scales used by chemists, Cullen cautioned his students against believing that expansion can be taken as a measure of the absolute quantity of fire in a body. 'We cannot,' he told them, 'say with Boerhaave [that the relationship between expansion and fire is] peculiar, universal & inseparable, but [it is] the most proper mark & measure, & therefore [is] applied.'[180] Even if the same temperature scale is used, different thermometric substances will show different degrees of expansion when warmed by equal quantities of fire. 'Fire does not expand all bodies equally . . . but a body of steady nature will be equally affected at all times & such may serve to compare heats.'[181] Thus the measurements afforded by the thermometer are entirely relative. When we use thermometers, 'we do not thus acquire a knowledge of absolute heat nor the proportion of one heat to another, but only the arithmetical difference.'[182] To develop an adequate theory of fire one must do more than simply record the temperature differences indicated by the thermometer.

Cullen begins his 'history of the phenomena' with an account of 'the laws according to which Fire passes from one body to another'.[183] The first of these laws states quite simply that 'Fire [is] distributed equally around'.[184] If this law is understood to mean that heat flows constantly from warm bodies to cold bodies until they reach the same temperature, then Cullen is willing to accept it without correction: 'All bodies in the same room [are] of the same heat, all bodies [are] of the heat of the ambient fluid.'[185] But if this law asserts that the quantity of fire is the same in all space, then Cullen has grounds for questioning it. It is true that if a pound of water at 80 degrees is mixed with a pound of water at 40 degrees the temperature of the mixture will be 60 degrees, but if different substances at different temperatures are mixed, the law, in its simple form, does not seem to apply. 'A heated body imparts its heat to any other till both are equally hot, [e.g.] cold & hot water mixed. The same does not hold of water & Mercury by Boerhaave's account, but [his conclusion can be] doubted.'[186] Cullen is alluding to an experiment performed by Fahrenheit and described by Boerhaave. By mixing together 'equal bulks' of water and mercury having different initial temperatures, Fahrenheit found that 'if the water were hotter than the mercury . . . the degree of heat arising from the mixture, was always more than the half which was expected. On the contrary, if the quicksilver were hotter than the water, and equal bulks of each were intermixed, the temperature produced by them was always less hot than the half of the difference.'[187] Boerhaave, in an effort to maintain some connection between his theory of fire and the physical properties of bodies, concluded that this experiment indicates that heat is distributed in bodies 'in proportion to their bulk, or extension, and not of their density'.[188] Cullen, however, seems to have realized that Boerhaave's interpretation contradicts the very experimental facts he was attempting to explain. Thus the way in which the quantity of heat in a given space distributes itself among several different kinds of bodies remained, for Cullen, an unsolved problem.

The problem of the relationship between physical properties and heat properties appears again in Cullen's discussion of the rates at which bodies heat up and cool off. It was well established that when exposed to the same source of heat, 'all bodies [are] not equally heated in the same time'.[189] There are, of course, many physical properties which affect the rate at which a body is warmed, and Cullen listed several of them: 'A. bodies heated sooner or slower according to their surface, b. bodies heated according to bulk, in similar bodies in the ratio of the diameters, c. bodies heated by light according to their color . . ., d. bodies heated according to their density. . . .'[190] Of all these factors, the relationship between density and rate of heating troubled Cullen the most. If heat is the motion of microscopic particles, then the densest substances should be those which heat up most slowly. As a theoretical statement, 'Boerhaave affirms it universally'.[191] There are, however, several experiments which indicate that the relationship between density and absorption of fire is much more complex than Boerhaave supposed. 'Dr. Martin's experiments shew it otherwise: quicksilver [heats up] sooner than water, spirit of wine & oil heat sooner than water; but not in the inverse ratio of the densities.'[192] Martine's experiments obliged Cullen to admit, although reluctantly, that 'bodies heat not according to density'.[193]

Cullen's lectures on fire probably owe a great deal to George Martine's *Essays Medical and Philosophical*.[194] In essays on 'The Construction and Gradation of Thermometers,' on 'The Comparison of Different Thermometers,' on 'The Heating and Cooling of Bodies,' and on 'The Various Degrees of Heat in Bodies,' Martine clearly summarized the works of the leading heat theorists of the early eighteenth century and presented original experiments designed to test the truth of the theories he described. In the experiments which Cullen referred to, Martine compared the rate at which the temperature increases in equal volumes of mercury and water when they are placed before a common source of heat. He used two thin-walled glass vessels, one filled with water and the other with mercury, each having a sensitive thermometer suspended in it. When the two containers were placed an equal distance before a strong fire, the temperature of the mercury rose twice as rapidly as that of water. This and other similar experiments led Martine to a conclusion, which he stated forcefully and which his successors could scarcely ignore. 'I know,' he said, 'no stronger instance than this of the weakness, or if I may venture to say so, of the presumptuousness of the human understanding, in pronouncing too hastily concerning the nature of things from some general preconceived theories. That rule of bodies acquiring given degrees of heat or cold *caeteris paribus*, in times just reciprocally proportional to their densities, tho' plausible enough in theory, I cannot find to be true in fact.'[195]

The heating and cooling of bodies is accompanied by other phenomena which were so far beyond the grasp of eighteenth-century heat theory that Cullen could do little more than list them. Boiling, melting, inflammation, vitrification; all these physical changes affect the rate at which bodies gain and lose heat. Newton's law of cooling provided a reasonable first approximation to the rate of cooling when high initial temperatures were involved, even though 'experiments don't shew it [i.e., the temperature change, to be] according to this theory'.[196] Beyond that, there were no laws worth describing. In his search for the laws governing the rise and fall of temperature, itself an effect of the flow of fire, Cullen, like the best natural philosophers of his age, could only catalogue the various and seemingly independent phenomena involved.

In the second part of his history of the phenomena of fire Cullen describes 'the effects of its presence,' the first of which is expansion. With the addition of fire, 'all bodies [are] expanded,' but the degree of expansion 'is different in different bodies'.[197] Drawing upon the work of Musschenbroek, Cullen reported that the relative expansion of various substances, as 'measured by the pyrometer,' is 'iron 80 degrees, steel 85, copper 89, brass 110, tin 153, lead 155'. And like Musschenbroek, he concluded that the expansion caused by fire is 'not inversely as the densities—example in iron & lead—nor [is it] inversely as the cohaerence, for the cohaerence of iron is to lead as 450 to 29 1/4, [whereas] the expansion [is] 80 to 155.'[198] Thus expansion, like the rate of heating, was an aspect of the theory of fire which could not be reduced to the kinds of mechanical explanations available in the middle of the eighteenth century.[199]

Fusion, the melting of a solid into a liquid, is 'the second effect of the presence of Fire in bodies'. 'Many bodies, as they are expanded, become soft, ductile & flexible, & at length quite fluid, & the same bodies, upon removing the Fire, become consistant, solid & rigid.'[200] Once again, Cullen reports, the facts, though well known, have not been knitted together by an adequate theory. Mercury has never been solidified, even when exposed to the most intense cold known. Oils, acids and alkalis change from solids to liquids at various temperatures, while 'water impregnated with salts, neutral or compound, resists cold more than water alone'.[201] Substances that are normally solid also challenged Cullen's understanding of fusion. 'Salts by themselves melt in this order: Vitriol, Alum, Nitre, Glauber's Salt, Sal Ammoniac, Borax, Fixed Alkali, Vitriolated Tartar. With respect to these, one fusion depending on their water & another after their water is dissipated. . . . The Metals: First Tin, then lead, Gold, Silver, Copper, Iron.'[202] Even the earths are 'vitrescent, that is, may be rendered fluid, only they require a greater heat than culinary fires can give. Few of them resist the burning glass & those that do, for ought we know, might be vitrified by a still stronger heat, for to this we can't yet set bounds.'[203]

Although the temperature at which substances melt cannot be correlated with their physical or chemical properties, every substance melts at a single fixed temperature. And since the temperature at which a substance changes from solid to liquid, or from liquid to solid, does not vary, the action of a single cause will suffice to explain this change. Thus, Cullen argues, the existence of a fixed point of fusion renders the supposed existence of frigorific particles superfluous, for the effects of fire alone can adequately account for the changes of state involved.

> As all bodies, tho at different points of heat, may be rendered fluid & in a less degree become solid & approach to fluidity or return to it as the heat increases or diminishes, may not we then conclude that the fluidity of bodies depends on the action of Fire in them & nothing in the figure & determine that question, whether the freezing of water depends upon the mere absence of cold [sic] or upon the admixture of what are called frigorific particles. As all bodies are always firm or fluid at a precise point of heat, as water for instance at 32 in our thermometers, & only then, we must either say that these frigorific particles are present at all times & places & farther their action depends upon the heat or cold, which is resolving the action of freezing into the same.[204]

While it is true that 'upon the whole, as people employed in thermometers constantly find, the freezing point [is] the same,' Cullen clearly goes to far in saying that 'no one observation has ever been adduced to shew ice above 32 or water under it'.[205] Certainly he knew of Fahrenheit's discovery of supercooling, in which water remains liquid at a temperature well below its freezing point.[206] Supercooling, however, was one of the facts about the behaviour of heat that remained entirely anomalous until Black formulated his theory of latent heat, and Cullen obviously suppressed it, whether consciously or not we cannot tell, while arguing against the hypothetical frigorific particles.

The change from liquid to vapour is the next topic in Cullen's history of the phenomena. 'The third effect of Fire is ebullition & dissipation. This we said happens to all fluids at the boiling point.'[207] Again he attempts to link the effect of fire to the physical properties of the substances being altered.

> To [explain] this dissipation it is necessary to suppose that the parts of the fluid are rarefied so far as to be specifically lighter than air, so mercury must be rarefied more than 11 200 times & considering that the air too in retorts is rarer than the external air, the rarefaction must be still vastly greater. What it may be in different bodies has not been examined, but from an experiment of Dr. Newentytts with the Aeolipile it appears that water, in boiling & [being] converted into steam, is rarefied above 13 000 times & which Dr. Desagulier, considering the allowances made, thinks it may be supposed 14000.[208]

Cullen evidently believed that in being vaporized substances are transformed into elastic fluids having mutually repelling parts. 'All bodies in being thus rarefied are changed into elastic fluids, sometimes of incredible force.'[209] In other words, when explaining boiling under normal circumstances, Cullen accepted the Newtonian hypothesis which asserts that elasticity is a property of a state of matter rather than a property of a particular substance such as air.[210]

When heated beyond the boiling point, most bodies either become red hot or burst into flame. 'Another effect of Fire in all bodies that are not soon dissipated is that by the continued action of the Fire they become luminous or shining, . . . that is, either they [become] red through their whole substance or they have a luminous vapour which we call flame playing upon their surface.'[211] There was little Cullen could say about these effects of fire. 'What is the degree of heat at which bodies take the red heat or at which they break out into flame is not exactly known.'[212] A few facts concerning inflammation were, however, well established:

> It is only a certain number of bodies, to wit alcohol & oily bodies, that can be brought to flame, that is, bodies that by heat send out an inflammable vapour, & to the flaming of this the access of air in motion is absolutely necessary, but not to produce the red heat above mentioned. . . . Under this ignition more of the parts of bodies are dissipated & in a particular manner, for in the former method of dissipation the rarefied substances, again condensed by the cold, resume their former

> appearance, & partly the dissipations [are] in this way still, but here the parts thoroughly ignited are dissipated so as to have their texture absolutely broke, so as they can by no means [be] collected in their former shape.[213]

Thus burning, unlike boiling, has an irreversible effect upon the properties of the substances being changed.

Some bodies, when intensely heated, 'are rendered fluid & in cooling again are formed into glass. . . . Others are not rendered fluid or vitrified but have their texture quite broke so that they fall down into a fine powder which is called a calx, & such bodies are said to be calcined.'[214] The calcination of metals involves a variety of phenomena which Cullen could only describe:

> Several Metals may be calcined by a slow & gradual fire, but by a greater are turned into glass. Upon this subject [it is] necessary to notice the weight acquired by calcined metals. Metals calcined by the fire acquire weight—lead in minium acquires 1/10. Calcined in close vessels they do so, calcined by the burning glass they do so. Therefore Fire adhaeres to bodies & increases their weight. But Boerhaave found the weight of a bar of Iron heated & cold not to differ in weight. This may be owing to the increased bulk, giving less weight in the same air, as lead & tin, which it should have done if the Fire did not compensate the increased bulk. All this [is] uncertain & it may be that Fire in an elastic state does not gravitate but like air may be fixed & adhere to bodies, increasing their weight.[215]

The earths that can be calcined also have distinctive properties: 'The Calcareous Earths have somewhat of an Alcaline Quality; they effervesce with Acids, they acquire this more after calcination. They approach to the Saline as being more diffusible, if not more soluble in water. Joined to the others they promote vitrification.'[216] With this description of vitrification and calcination Cullen closes his history of the effects of fire's presence in bodies.

Having given a history of the communication and effects of fire, Cullen turned to a consideration of its sources. The first of these, 'the great source of light, heat & fire in this Earth,' is the sun.[217] Had he set out to write a comprehensive history of fire, Cullen would have had to survey the entire field of optics, but since he was lecturing on chemistry alone, he limited his description of light to the heat which can be obtained from it. 'The effects of the Sun are communicated to us in Light, which is a particular modification of Fire whereby it moves with incredible velocity in strait lines. The various affections of Light, tho they may seem to make a part of the history of Fire, yet as they relate so little to our subject, for this & many other reasons

I shall not meddle with it & only notice under what circumstances Light comes to produce Heat & its effects above mentioned.'[218] When concentrated by lenses and mirrors, sunlight provides the chemist with one of his most powerful tools. 'For the particular accounts of experiments with the burning glass,' Cullen told his students, 'I must refer you to the French Memoirs & other authors. It is enough to our purpose to have mentioned the general effects as a proof that the light of the Sun, collected & condensed, will produce all the effects of heat & Fire.'[219]

'A second source of heat & Fire is the attrition of hard bodies.'[220] The widespread acceptance of the mechanical philosophy had made the production of heat by friction the most commonsensical of all the sources of heat. 'With regard to all this it seems necessary that the quickest vibrations must be excited. Thus a nail driven into wood, while it is driving & the wood yields to it, no heat appears, but when driven to the head, two or three strokes heat it up strongly; & that this vibration must be produced by the hardness of body or the swiftness of the stroke with regard to the heat produced by attrition, as it does not occurr in soft bodies, so by the interposition of the softest bodies, fluids, it is prevented. Hence the use of greasing the axles of wheels.'[221] Perhaps, Cullen suggests, the electrical glow produced by shaking the barometrical phosphorus, 'a third method of producing fire,' is simply another manifestation of the release of fire by friction. 'The electrical experiments [are] now so common they need not be described & everybody knows that in many cases it appears luminous, which is one proof of Fire, & you know too that it can be applied to kindle spirit of wine, &c. It is always & only excited by friction, which makes me place it here next to the last described manner of producing Fire.'[222]

Cullen's fourth source of heat, 'the intestine motion of the small parts of bodies,' seems to be as mechanical as friction, but in fact he is referring to the heat given off by chemical reactions. As 'one instance or two' of this source of fire he points out that 'Oil of Vitriol & water raises a considerable heat, oil of turpentine added to the acids of vitriol & nitre raises a considerable flame, fermenting substances raise heat, putrefying substances raise actual flame, animal heat of this kind.'[223]

The fifth source of fire and heat, 'much a kind to the last, is the production of heat merely by the application of air to bodies'. Here Cullen has in mind the phosphors, 'such bodies as are luminous of themselves or become such merely by the access of the air. . . . These are of different kinds: one kind are luminous with out heat, as Bologna stone, &c, one kind are luminous with heat & burning, viz.

Kunckels, a third are luminous, hot, burning, & exploding violently, as Geoffroys Reguline Phosphorus.'[224]

While discussing the burning of inflammable bodies, the sixth source of fire, Cullen touches on some of the most difficult aspects of fire theory. After describing the inflammable substances and the process of burning, he asks what in fact is the fuel being consumed. Are all particles of matter capable of becoming fire if adequately agitated or is there a special substance in nature which alone has the essential properties of fire?

> Inflammable bodies are often called the pabula or aliment of fire because by their means fire can be long continued in one place & by collecting a great quantity of inflammable matter & inflaming it at once, a great force of fire may be produced. But here Dr. Boerhaave is at pains to caution us least in terming these bodies the pabulum of fire we may think that these bodies supply fire by being themselves converted into fire. This the conjecture of Sir Isaac Newton. Boerhaave's objection from the increase of the fire in the universe not to the purpose, for if fire consists in the motion of bodies of a certain size or with a certain velocity, any bodies that may be reduced to the size or be moved with the velocity & so become fire, but so soon as being joined with others they become larger or their velocity diminishes, they are no longer fire.[225]

Although Cullen could not resolve the differences between the Boerhaavian and the Newtonian theories of fire, he hoped that by including more facts in his history of the phenomena he could begin to move towards a resolution. Clearly the chemical composition and behaviour of inflammable substances deserved close examination. 'Inflammable bodies,' Cullen told his students, '[are] not simple but compound, consisting of water, salts & earths. Acids [are] present in all inflammable substances & [are] necessary to most heating mixtures. Flame only ignites oily vapour & that vapour [is] necessary to flame, which therefore appears only on the surface—flame blown away; live coals extinguished in spirit of wine. Air necessary to flame & burning, fire extinguished in vacuo, fire extinguished in small quantity of unrenewed air. Hence the opinion of air containing a pabulum of fire—air that has passed thro fire extinguishes other fire.'[226] Naturally Cullen's emphasis upon the chemical aspects of burning was not simply a response to the facts of nature. He had completed his history of the phenomena of fire and was preparing to reveal, although 'with doubt & diffidence,' the theory that, in his mind, connected all these phenomena to the fundamental nature of fire. It was, of course, a chemical theory.

After cautioning his students that 'our conjectures [are] to be received as such,' Cullen presented a series of hypotheses which he believed would be of use to anyone attempting to understand the phenomena of fire. His first four assumptions link the theory of fire to the theory of the aether: '1. Fire [is] a fluid, 2. Fire [is] an elastic fluid, 3. Fire pervades all bodies & keeps their parts asunder, 4. [We know there is] an aether in bodies from the reflexion &c of light, from electricity. The same with fire & present everywhere.'[227] The fifth statement identifies repulsion as the essential property of the aether: '5. Two great principles, the attractive & repulsive, the source of motion & change.'[228] In the nine succeeding hypotheses Cullen explains several phenomena associated with fire, including specific differences in rates of heating and changes of state, in terms of the density of the aether contained in the bodies undergoing the change. '6. Fire [is] a distinct substance but present in all bodies, 7. Fire must be distributed equally around, 8. Like vibrations will be excited in all contiguous bodies, 9. The aether of different densities in different bodies—from the fluidity of bodies, from the heating of bodies, from heating & cooling mixtures, 10. These bodies whose attractions are greater than the expansive force of fire in their pores will be perfectly hard & will not yield to any pressure, but the sudden motion communicated to the aether occasioning a condensation of it may break them, 11. These bodies whose attractions are less than the expansive force will be soft & ductile & so thro different degrees till, 12. The attractive forces are equal to expansive & then they will be fluid, 13. Thro all these degrees the aether [is] increasing in density, the body will be more & more expanded, 14. If the expansive force exceed the attractive, the body will become elastic, & such is the vapour arising from heated bodies when the rarefaction is such as to render them specifically lighter than the ambient air.'[229]

Cullen then turns to the relationship between light and the aether: '15. Is not light the vibration of this aether & [is] this aether not fit for such vibration till it is of a certain density, & therefore [do] only bodies heated to a certain degree, that is accumulating aether till it is of a certain density, become luminous & shining, 16. Does not the consuming of bodies by fire depend on the density of the aether or the increased expansive force & does not this appear from the condensation of the rays of light & of all other fire, 17. Is not light in material bodies produced by collision of two elastic fluids—of light & rarefied vapour—do not electrical experiments shew this—the phosphorus of Kunckel & Homberg, 18. May not light be produced in consequence of increased density or in consequence of collision.'[230]

In his next hypothetical statement Cullen asserts that the attraction and repulsion between bodies varies with the ratio of the densities of their material particles to that of the aether contained in their pores. '19. As the density of aether is different in different bodies & must be rarer in the empty spaces between bodies than in their pores, we may easily conceive how larger masses do not unite. If the density of the one is not so much greater than that [of] the other as to overcome the attraction of its parts, the aether in each upon their approach will repell each other. If the density is greater, they will be attracted & join each other. As the aether will always be rarer in the empty spaces between bodies than in their pores, even homogeneous bodies will repell each other at some distance, but pressed closer till that denser aether is driven out, they will cohaere strongly. Bodies dissolved in fluids by the above theory must be equally dispersed in them, as [may also be seen] from the transparency.'[231]

Cullen then applies his aether density argument to the optical properties of bodies. '20. In the parts of fluids the aether must be equally dispersed. In solids it may be unequally & they may have pores in which it will be of a different density—such solids must be opake. Bodies of very different densities will probably reflect the most part of the light falling on them. Most fluids [are] transparent from the uniformity of the medium, but mercury not so from reflexion.'[232] The relationship between optical properties and aether density also enables him to explain one of the chief differences between calcined and vitrified bodies. '21. The coherence of the most part of large masses [is] owing to fluids present in them. If these [are] dissipable while fire not retained, the bodies [are] calcined, 22. If the uniform mass [is] rendered fluid by fire, it must form a glass.'[233]

Turning next to the sources of fire, Cullen offers two statements concerning the importance of motion in any theory of fire. '23. Does not the generation of fire by attrition shew it [to be] an elastic fluid under strong vibration? Must not the hardest bodies be fittest & soft or fluid bodies prevent it? Does not the swiftness of motion necessary shew the same? Does not the density of the air necessary shew that strong oscillations are necessary to fire? 24. Are not inflammable bodies such as contain the densest aether while they resist its oscillations the least?'[234] But this emphasis upon the mechanical theory of fire does not lead him to neglect the chemical aspects of this problem. '25. Are not acids the chief ingredients in such [i.e., inflammable bodies] & this [is] confirmed by their dissolving most substances.'[235]

The final two assertions of this aphoristically argued theory connect the properties of the aether with those of the air and the acids. '26. Is not the aether present in a fixed state as air? 27. Do not acids set this aether loose as it does air in heated mixtures?'[236]

Cullen never became overly attached to his air-acid-aether theory of fire. He did not claim to be its author, he only proposed it because the dominant mechanical theories of natural philosophy were of so little use in chemistry. 'This theory is not new,' he told his students, 'you may collect it from Sir Isaac Newton's own works but more particularly from Bryan Robinson's treatise on the Aether of Sir Isaac Newton. It is the most plausible scheme of chemical philosophy and will at least check the false theories of the corpuscularians.'[237] While optimistic about the future of chemical theory, Cullen reminded his students that the laws describing the action of the aether had not yet been determined. 'Tho we see so far in general, yet do we not know the laws according to which the action of the attractive & repulsive powers are directed, & therefore [we] should be often puzled.'[238] Since chemistry seeks to explain the specific properties of bodies, the aether theory must meet a demanding set of criteria. For example, one 'difficulty is to find the cause of *Elective* attraction, or why Aether does not admit of an equal union with all bodies'.[239] But if the chemical notion of fire could be successfully subsumed into the broader concept of the aether, then chemists could hope to construct an attraction-repulsion theory which would address the particular phenomena they study while drawing upon the conceptual resources of an overarching natural philosophy. Cullen insisted that in proposing his hypotheses he had only gone so far as to 'point out a path in which the theory of fire is to be pursued & to support what I let out on this subject, viz. that attraction & fire were to be considered as the primary causes of motion & consequently of the changes'.[240] To be a philosophical chemist in the eighteenth century he had to go as far as that.

6. AIR

After having presented his thoughts on attraction and repulsion, the primary causes of chemical change, Cullen turned his attention to the chemical properties of air. He did not doubt air's involvement in chemical processes but he was not at all sure how to account for its action. 'From Mr. Hales experiments' he learned that air 'does enter into the composition of bodies & [is] in general to be noticed, but its effects on the qualities of bodies [are] not known & it may be only a different arrangement.'[241] Unlike Hales, he did not believe that air is the only substance that can be made truly elastic. 'Air has been considered generally as such,' Cullen acknowledged, but he believed that 'its chief properties as an elastic fluid manifestly depend on the presence of fire.'[242] Air occupies a middling position in Cullen's chemical theory. 'As air is a very universal agent,' he argued, 'it deserves to be particularly considered & the more so as there is otherwise no place assigned to it in our course, as there is for the two other essential instruments, water & earth.'[243]

In calling air a 'universal agent' and 'an essential instrument', Cullen was adopting the Boerhaavian scheme in which certain substances are identified as the causes of certain chemical changes. And yet he wished to restrict his discussion of air, as he had restricted his discussion of light, to those properties which cause specific effects. 'With respect to air,' he said, 'these general properties considered in every course of National Philosophy [are] not to be mentioned here & only those properties that relate to chemistry' will be mentioned.[244] Cullen's discussion of air is therefore marked by a tension created by his desire to treat air as both a general cause of change and as a specific substance. In Cullen's lectures this tension is creative, encouraging a general rethinking of the fundamental conceptual problems involved. But for Black, who was less committed to the discovery of general causative substances than his master, the concept of chemical instruments was philosophically meaningless. In his own lectures Black said that

> under the title of the Apparatus, or instruments, I do not propose to take under consideration the nature and properties [of fire], of water, of air, and of earth, as Dr. Boerhaave has done, who has given elaborate treatises upon these four subjects, under the title of instruments of chemistry. It is true that we apply heat to produce changes and effects upon different bodies; and that when we employ it in this way, it may be called, in metaphorical language, an instrument in our hands. It is also true that we make use of water on many occasions; as an assistant by which we can reduce some bodies into a state of fluidity, and mix them more effectually together. But if, upon this account, we must treat of fire and water as instruments, for the same reasons, and with the same propriety, I might bring most of the objects of chemistry under the same denomination. For, after we have learned by experiments the powers or qualities which belong to those bodies, we take advantage of this knowledge, and employ them accordingly to serve our purpose on different occasions.[245]

By following Cullen's advice and concentrating upon the particular changes caused by the fixation of air, Black was able to inaugurate the experimental study of the chemical properties of specific airs.

Cullen grouped the chemical effects of air into four categories: '1. These depending on its constant oscillations, 2. These depending on its dissolution [of] almost all bodies heated to [a] certain degree, 3. These depending on its being impregnated with various matters & insinuating these into other bodies exposed to it, 4. These depending on its being lodged in a fixed state in bodies & being thence restored to an elastic state.'[246] The first of these effects is the cause of many important chemical processes. 'The effects depending on its [i.e., air's] constant oscillations are almost all these changes of bodies that are produced by the intestine motions of the parts of bodies, such as solutions & its frequent attendant effervescence, fermentation, putrefaction & even the burning of bodies.'[247] Phenomena such as the motion of dust in a shaft of sunlight reveal that atmospheric air 'is scarcely for two moments in the same state, but is constantly expanding & contracting & [it appears that] it must be constantly vibrating & oscillating & in some measure must affect all other bodies placed in it in the same manner & thereby increase their intestine motions, especially these of fluid bodies & even solids, by occasioning a constant oscillation of fire in their pores.'[248] There are many experiments, Cullen assures us, that demonstrate the influence of the air's motion upon bodies.

> No effervescence happens when the air is compressed & prevented from oscillation & producing the like in bodies, thus no liquors effervesce in glasses hermetically sealed. In the same manner, fermentation & putrefaction are prevented unless exposed to the open air, which shews the oscillations of this to be necessary. . . . Compare this with the production of flame by the collision of electrical bodies & it will confirm our theory that the oscillations of the air contribute & are necessary to the burning & flaming of bodies.[249]

If one accepts Cullen's assertion that these chemical changes can be explained in terms of the motion impressed upon the constituent particles of bodies by the ambient air, then there can be no doubting the importance of air in chemical processes.

The next set of properties is more properly chemical than the first. 'The second head to which the effects of air [can be] referred,' Cullen told his pupils, 'is the solvent power of the air, upon which we suppose the evaporation of water & other fluids depends.'[250] Cullen presented his theory of evaporation under this heading. He begins by describing the theory that Desaguliers published in 1729.

> The evaporation of water & the ascent of vapours, as connected with the theory of meteors, has been often considered & deserves our notice as concerned in many chemical operations. Several hypotheses, the last [of which] is that of Desagulier in the *Transactions*, 407. He shews the fallacy of the former opinion, & which we shall suppose done, & consider only the Doctor's, which [states] that as fire rarefies bodies & water & at [times] changes [them] into an elastic fluid, so soon as it is expanded so

much as to be specifically higher than air, it will rise in it & according to the degree of expansion it will always ascend till it meets air of the same specific gravity.[251]

Cullen was willing to believe that during evaporation a substance, through rarefaction, becomes lighter than the surrounding air and floats upwards, but he found the arguments offered to support this theory questionable and the theory itself inadequate. He believed that Desaguliers' theory was founded upon a single experiment and an erroneous set of assumptions. 'From an experiment,' Cullen reported, 'he finds that water, by the boiling heat converted into steam, is expanded to a bulk 14000 [times] greater than that of cold water.' But since 'the specific gravity of air is to that of water as 800:1,' water will evaporate long before turning into steam.[252] 'Now Desagulier,' Cullen continued, 'using the scale of Sir Isaac Newton's thermometer where the heat of boiling water is marked 34 & upon the same scale the mean heat of summer is 5 & the mean heat of spring or autumn 3 & the mean heat of winter 2, he supposes the rarefaction will be in the same proportion, so if the rarefaction by 34 be 14000, the mean heat of summer will give 2058, the heat of spring & autumn 1235 & the mean heat of winter 823, which you see is sufficient when the density of the air is 800.'[253] But Cullen knew that

in all this there is a considerable mistake. Desagulier considers these numbers on the scale as expressions of absolute heat & that the heats expressed by these will be in the same proportion to one another. But we have observed before that the beginning of our scales is arbitrary & the numbers don't express the proportions of one heat to another. Sir Isaac Newton begins his scale at the freezing point, Fahrenheit his [at] 32 degrees below that & with as good reason he might have gone 40 or perhaps more farther. But according to these different scales Desagulier's calculations would come out very differently.[254]

Desaguliers has not only misunderstood the nature of thermometric scales, he has totally ignored the discontinuities which accompany changes of state. As Musschenbroek's experiments indicate, 'we don't find that double the heat applied to bodies gives double the expansion . . . & with respect to water, this expansion is not remarkable till it comes to the boiling point. The body of the water heated from 53 to 212 is only expanded 1/85 of its whole bulk so that here in particular it would seem as if the rarefaction was not in proportion of the numbers marked on the thermometer, since all at once & only then it becomes remarkable.'[255] If all these criticisms of Desaguliers' theory 'were not thought sufficient,' Cullen continued, 'I could produce several other objections.' But he limits himself to one more instance of Desaguliers' errors, one which calls into question his competence as an observer of nature. 'We don't find,' Cullen states, 'the evaporation of bodies to be in proportion to the heat & density [of the air as much] as to the dryness & moisture, so that it is often greater with cold & dryness in winter than with heat & moisture in summer.'[256]

Cullen would not have lingered so long over Desaguliers' errors had he not had a better theory of evaporation to offer. 'Upon this subject of the rise of vapours,' he told his students, 'I had a hint from Mr. Henry Home that the action of the air as a solvent of water was to be considered & the more I consider it the more I am of that gentleman's opinion.'[257] Home suggested that evaporation is caused by an elective attraction between air and water. Cullen thought this a likely hypothesis and argued that 'an attraction betwixt water & air is sufficiently manifest since we never find them absolutely separate from each other; there is no air without water nor no water without air, as is well known from experiments with the air pump & [those of] Mariotte.'[258] Furthermore, he said, 'while the heat continues the same, water [is] dropt according to the rarity of the air.'[259] This rather cryptic statement alludes to an argument developed at greater length in Cullen's commentary on Home's essay. 'Air,' Cullen told Home,

> according to its density dissolves more or less of water. As the air is exhausted from the Receiver of the Air pump a cloud is formed in it & upon farther exhausting, the inside of the Receiver is bedewed & dimmed with Moisture which upon readmitting the Air disappears again. Now this is not to be accounted for upon the hypothesis of Rarefaction; the falling indeed may but the taking it up again while no change of heat occurrs cannot. To this purpose you will find an Elegant Experiment in Nieuwentyt's Religious Philosopher that is worth your attention.[260]

Being a chemist rather than a mechanical philosopher, Cullen obviously enjoyed putting forth a theory of evaporation which depended upon elective attraction as well as on hydrostatic displacement. He respected Desaguliers as a natural philosopher and followed him in rejecting Hales' assertion that air is the only 'amphibious' substance, yet he was no doubt pleased to demonstrate again why chemists had to formulate their own theories rather than simply apply the laws discovered by the mechanical philosophers.

Cullen next turned his attention to 'the third head to which we referr the effects of air, [namely] these depending on its being impregnated with various matters & insinuating these into other bodies exposed to it.'[261] Whereas the second cause of air's effects was its ability to dissolve substances through the action of elective attraction, this third cause is, like Desaguliers' theory of evaporation, entirely mechanical. There are, according to Cullen, three types of fumes that float in air. First, there are

> the bodies with which the air may be impregnated, [i.e.] all these that may be taken up with water in the way of solution. Secondly, [there are] all these that by fire may be converted in[to] an elastic vapour specifically lighter than air. Tho we alledged that the ascent of water was commonly & ordinarily in the way [of] solution, yet we must allow that this other too comes in for a share, as we see the evaporation is more copious at the boiling point & as we see that heated bodies send up fumes in vacuo. Thirdly, [there are] all these bodies that by fermentation, putrefaction & effervescence may be rarefied in the same manner.[262]

Having already dealt with evaporation, Cullen began a discussion of the fumes given off by fermentation, putrefaction and effervescence. 'We find,' he said, 'that fermenting &c substances constantly produce a quantity of air, sometimes permanent & sometimes quickly destructible. They are fumes of the same kind in so far as our smell discerns & there effects are in other instances the same.'[263] These fumes are not only chemically identical, they are the same as 'the fumes of the phosphorous, of the electrical matter of fire & in short of all bodies very thoroughly ignited & nearly consumed, [such as] charcoal & other fewel'.[264] Normally these fumes are masked by the vapours released during burning, 'but when from burning bodies every substance is dissipated but the most fixed, then the fumes are of this kind, & if we examine them we find them a very subtile acid, at least when joined to water it appears so.'[265] Cullen asserts that this acidic fume released by inflammable substances is the substance that deprives both aether and air of their elasticity and fixes them in various solid and liquid bodies. 'May we say then,' he asks, 'that this is the substance that chiefly absorbs & fixes elastic air? May we say too that this same substance fixes the aether of Sir Isaac, which we alledged to be the matter of fire? Perhaps it is but one & the same operation in both cases. If it is so, while air then surrounding bodies is filled with these fumes, its elasticity must be destroyed, but more accumulated, they may even act on the burning body & fix aether too upon elastic oscillations, on which the burning of the body depends, which must therefore cease.'[266]

Here we have the acid-air-aether theory used to explain both the presence of heat and flame during burning, in terms of the release of aether, and the need for a steady supply of fresh air, in terms of the absorption of the fixing acid given off during the release of aether.

Furthermore, if air is fixed when fire is freed, then air must be freed during the fixation of fire. 'With respect to the causes of [air's] restoration,' Cullen argued, 'they are chiefly Fire, fermentation & the mixture of certain substances, viz. all these that effervesce. Fire not only detaches the air from all bodies but even converts most bodies into such a fluid. Sir Isaac's opinion with respect to air makes this easy.'[267] But can we talk of the fixation of fire in reactions, such as those between acids and alkalis, which release elastic air in the form of effervescence? Yes, Cullen replies, if we understand 'that as different substances fix & detain in their pores more or less of the aether, so upon any new mixture, if the mixture is capable of containing more of the aether than the bodies separate, air will be absorbed, if less, air will be generated.'[268]

After describing the fixation and release of air, Cullen was ready to 'go on to consider the fourth head, or these effects depending on the airs being fixed in bodies & thence restored to an elastic state'. Although the experiments of Boyle and Hales have established beyond doubt 'that many bodies absorb elastic air,' Cullen could only report, 'that in such [bodies] it shews none of the ordinary effects of its elasticity'.[269] Because they had focused exclusively upon the elastic property of air, previous natural philosophers had totally ignored the chemical properties of fixed air. Cullen could do no more than acknowledge this problem and leave the task of first describing some of the qualitative 'effects depending on the airs being fixed in bodies' to his pupil Joseph Black.

Cullen was, however, somewhat more informed concerning the effects of air being 'restored to an elastic state'. This release 'causes effervescence & explosion, for effervescence is nothing but the air restored to its elastic state, gathering in larger bubbles & rising to the surface of the liquor & there bursting with noise. The explosion of bodies is the same restoring of air to an elastic state, which when done in a large mass containing a great quantity & let go all at once, it must violently compress & move the air & every other body in its way with great force.'[270] Black also believed that chemical explosions were caused by the rapid release of fixed air. In the course of describing his 'Experiments upon Magnesia Alba, Quick-lime, and other Alcaline Substances,' he argued that the explosive quality of fulminate of gold should be attributed to the fixed air that the metal absorbed from the alkali used in its preparation. 'From our experiments,' Black wrote,

> it seems probable, that the increase of weight, which some metals acquire, by being first dissolved in acids, and then separated from them again by alkalis, proceeds from air furnished by the

> alkalis: and that, in the *aurum fulminans*, which is prepared by the same means, this air adheres to the gold in such a peculiar manner, that, in a moderate degree of heat, the whole of it recovers its elasticity in the same instant of time; and thus, by the violent shock which it gives to the air around, produces the loud crack of fulmination of this powder.[271]

Cullen's later presentations of the air-acid-aether theory also deserve attention, for as he explored its implications and reflected upon the advances in chemical knowledge, he modified the theory in revealing ways. Consider the discussion of calcination he presented when lecturing in 1757–8. He first distinguished between the calcination of calcareous substances and the calcination of metals. 'There are,' he said, 'two kinds of Calcination: 1st When the Body is calcined by the dissipation of the cementing particles &c, Secondly when it is reduced to a powdery form by the Addition of something that it receives from the fire.'[272] He then attributed the weight gain that accompanies the calcination of metals to the addition of an acid. 'All metals,' he said,

> during their calcination, acquire a great increase of weight. Thus 100 pounds of lead affords 110 pounds of minium. In other instances the increase is near 1/2, and this is not owing to any heterogeneous particles either from the fewel or air, for the experiment succeeds in closs vessels and with the rays of the sun—hence Boyle wrote a treatise on the Pondersity of Fire & Flame. But I would not conclude that it was fire that gave this increase of weight, but probably it is a substance that enters the composition of the Aether; but what it is we are yet uncertain, as we have not been able to procure it separate. But I should conjecture that it is of the nature of an acid.[273]

To support this hypothesis Cullen emphasized the following facts: '1. We can reduce metals very readily to the state of calces by the addition of Acid & Sulphur, 2. metals when calcined, either per se or with an acid, may again be reduced back to their metaline form by such substances as attract [acids] more powerfully than metals do, such as Alkali, calcarious Earths, [and] Phlogiston.'[274] If one thinks of phlogiston as something that is consumed during burning, one will agree with 'the chymists [who] have been led to imagine there is a Phlogiston in metals, of which they are deprived when calcined, where as by our hypothesis,' Cullen points out, 'calcination is only the addition of a subtile acid, and to reduce the calx again to a metalline form we need only add any substance that will attract the acid from the metal. In this manner we separate the acid . . . by means of a Phlogiston.'[275]

Although Cullen adopted the term phlogiston and used it in his later lectures, he clearly never fundamentally altered his air-acid-aether theory. Phlogiston for him was simply a new name for fire and aether; it was not a new chemical principle. He chided other chemists, such as Macquer, for considering phlogiston a distinct chemical substance and for using the term to designate 'the inflammable parts of bodies without being obliged to tell whether these be oil or not'.[276] But Cullen also realized that the basic weakness in the more abstract theories of fire advanced by chemists such as Boerhaave and Homberg was their failure to distinguish 'between fixed & movable Fires, or Fire at rest and Fire in motion'.[277] Cullen held 'the intermediate opinion,' believing 'that there are only two elements, one of them the gravitating matter, the other a subtile aether'.[278] According to this hypothesis, 'its probable that light and all other phaenomena of Fire depend upon this Subtile Aether in all Bodies, but this will be hid till the parts are agitated in a particular manner'.[279] Furthermore, he said, 'I think its probable that inflammable Bodies are of such a particular texture as to recover these particles of Aether in a great measure.'[280] Thus for Cullen, phlogiston is the aether that has been captured by inflammable bodies. Clearly the concept of phlogiston never seriously disturbed Cullen's understanding of the forces which cause chemical change.

By the 1760s Cullen had again modified the way in which he presented his theory of calcination and inflammation. During the latter half of the eighteenth century natural philosophers in Scotland developed a distinct distrust of aether theories,[281] and Cullen too found the concept hard to pin down. Acids and airs at least have specific properties, which reveal their presence and assure one of their existence, but the aether seemed to be no more than a name for the power of repulsion. Of course when Cullen first formulated his air-acid-aether theory, none of the chemical properties of the airs had been described, but during the 1750s Black's research on fixed air provided the chemical concept of air with an experimental foundation that made the notion of aether seem, by comparison, even more hypothetical than it had been in the 1740s. Thus in lectures delivered in the 1760s we find Cullen no longer invoking the aether and speaking not of air in general, but of the mephitic air described by Joseph Black.

According to this revised theory, phlogiston is a combination of an acid and fixed air. 'The inflammability of all bodies,' therefore, 'seems principaly to be the Effect of an Acid & Mephitic Air, whence it is not improbable to suppose, that the three forms of inflammables depend upon a simple Phlogiston formed of those two matters Combined in different proportions: Viz. that in alcohol there is the greatest proportion of Mephitic Air, that in Oils the acid predominates, and that in Sulphur it predominates still more.'[282] During burning this phlogiston is chemically resolved into its constituent parts. 'From what has been said,' Cullen told his students,

> it appears that the subject of inflammation is a vapour arising from certain bodies, that this vapour must be raised before inflammation can take place, that this vapour is the phlogiston of chemists composed of an acid & mephitic air, that inflammation consists in the resolution of this phlogiston and the absorption of its mephitic air by the surrounding atmosphere, and upon the whole one may conclude that inflammation is an instance of the generation of heat by combination or mixture. That the mephitic air here mentioned is a part of the phlogiston appears from its effects. The acid is not so very obvious, yet it may be demonstrated in some cases, particularly of the burning phosphori which have all an acid for their basis.[283]

This same phlogiston can also be used to extract the fixed air from calcined metals. 'Reduction of Metals,' Cullen argued,

> is affected by the addition of fatty and unctuous inflammables, among the most powerful of which is charcoal. Chemists suppose that the Calcination . . . of metals [is] occasioned by the separation of their phlogiston, and that a sufficient Quantity being imparted to them by animal or Vegetable fuel, they again assume a Metaline form, but to obviate this theory we need only Observe, that calcareous Earths mild will reduce Metals as well as Charcoal or Oily inflammables. Is not the matter seperated fixt Air?[284]

Seen in retrospect, this final version of Cullen's air-acid-aether theory seems to be a brilliant anticipation of Lavoisier's oxygen theory of calcination, combustion and acidification. Although Cullen's suggestion can in no way detract from the uniqueness or importance of Lavoisier's achievement, the similarity between the relationships postulated by Cullen and those which lie at the centre of Lavoisier's new theory of pneumatic chemistry may help explain why Cullen, Black, and their students were not nearly as impressed by the 'French Chemistry' as their contemporaries and subsequent historians of chemistry expected them to be.[285] By the 1760s Scottish chemists knew that there are a variety of airs having different chemical properties, they knew that fixed air has acidic properties, they knew calcined metals gain weight because air is fixed in them, they knew that elastic air is a compound of fire and particles of matter which were previously fixed, and, knowing all this, they concluded that anyone studying airs should concentrate upon the change involved in going from the fixed to the elastic state. And this focusing upon change of state further emphasized the role of heat and helped to make it the central topic for theoretical investigations among Scottish chemists during the latter half of the eighteenth century.

7. HEAT AND MIXTURE

In the course of organizing the facts of chemistry into a systematic presentation Cullen could not avoid using traditional terms such as salt, earth, oil, fire, acid and alkali. These words, of course, signified more than arbitrary categories, for they were also the names of concepts incorporating a variety of connotations acquired through previous usage. In the seventeenth century it had been widely assumed that the observable properties associated with these terms, such as the inflammability of oils, were the essential properties of distinct substances. In the eighteenth century the arguments which the mechanists had used to cast doubt upon the reality of chemical qualities were generalized in such a way that many philosophers came to believe that the perception of a phenomenon tells one nothing about the material reality which gives rise to that perception. One could, of course, accept this redefinition of the relationship between perception and material reality and try to redefine the vocabulary of chemistry accordingly. Cullen frequently moved in this direction, declaring that chemical principles are useful for taxonomic purposes alone and that fire, for instance, is merely another name for the sensation of heat. Frequently, however, his thoughts followed roads opened by earlier natural philosophers, as when he attributed substantial reality to the atoms and postulated the existence of a primal acid. Instead of treating these instances of conceptual conservatism as irrelevant to Cullen's achievement as a teacher, we should recognize them as indications of his commitment to chemistry, a commitment that obliged him to resolve the differences between the received ideas of chemistry and the criteria of the philosophy of his time. But even after recognizing the task Cullen set himself, one must still ask whether he seriously tried to reconceptualize his key ideas according to the dictates of eighteenth-century philosophy. In other words, in moving from taxonomy to theory, did Cullen rework those concepts which violated the philosophical canons of his age? The answer is that whatever his success, Cullen recognized and accepted his responsibility in this regard and he struggled to speak of textures rather than principles and powers rather than substances. While his understanding of chemical phenomena drew heavily upon out-dated notions of essential properties and mechanistic causation, he carefully avoided advancing new chemical theories that employed terms his philosophical colleagues might dismiss as indefensible. While not always fruitful, the concepts he did use justified his claim that chemistry as he taught it should be considered one of the modern sciences.

Chemistry is concerned with specific properties and chemists investigate these properties by analyzing the substances which possses them. But what does chemical analysis reveal? Is the causal property in the compound a constituent substance or is it a form, a peculiar structure possessed by the original compound? In Cullen's opinion, texture, which he defined as 'a particular arrangement of particles,' is the cause of all specific properties. In the course of a discussion concerning the medicinal properties of plants he said,

> when a body is distilled by itself the process is called *Analysis Chemica* from a notion the chemists had that it divided the body into constituent parts. But experience has taught us that this opinion is false, for whatever plant we distill, we obtain the matters—viz. Phlegm, acid Volatile, Alkali, Oil, and a Caput Mortum of Earth—And these are the same in proportion from whatever plant it is produced, only from some we obtain a greater quantity of Volatile Alkali, from others a greater proportion of Acids. The French Academicians tryed many fruitless experiments of this kind and their Secretary justly observes that to endeavour to find out the medical virtues of plants by this chemical analysis is like thinking to find out the elegance and use of the severall parts of a fine house by taking it to pieces and considering the several parts by themselves, whereas the parts of the house & the plants derive their particular properties from a particular arrangement of their particles.[286]

In chemistry, as in pharmacy, Cullen believed that it is the texture of the compound, and not the essential qualities of its constituents, which gives it its particular properties.

Cullen advanced a complementary notion of power to explain qualitative change. As he told his medical students when lecturing on neurology, rather than discarding the older terms, we should redefine them more abstractly. 'To avoid determining anything with regard to these opinions' on the nature of the nervous fluid, he said, 'I have used the term "nervous power". . . . Whatever I say in joining with [those who] regard the soul as having a power in beginning motion, yet in matters of physic we must entirely abstract from it. [While] we must keep to the *language* of Whytt, Boerhaave and Haller [and while] it is true that the language of Boerhaave and such will seem to be the same as the language of the materialists, a very little explanation will show the difference.'[287] The two powers that cause chemical change are elective attraction and repulsion and the task of the theoretical chemist is to link these two powers to the changes of texture which he is able to observe.

To forge this link a series of middle-range hypotheses were needed, hypotheses which postulated specific correlations and suggested experiments by which one could test them. In his air-acid-aether theory Cullen argued that changes in texture are caused by the attractive and repulsive powers which manifest themselves in the fixation and release of elastic vapours; that the cause of heat is the cause of elasticity, the flow of heat being proportional to changes in temperature; and that a decrease in heat indicates that the cause of heat has been fixed by an acid. For the experimentalist the key assertion here is the middle one, the belief that heat flow and temperature change are proportional. Identifying the material basis of repulsion was not one of Cullen's problems; he was quite willing to say that repulsion is caused by aether or fire or phlogiston or the motion of particles. And the acidic quality of the fixative substance was not, from the experimental point of view, an essential part of his theory. Cullen simply thought in terms of the acid-alkali theory. When lecturing on medicine he said, 'from the chemists Boerhaave took the doctrine of acid and alkali, but he very properly limited and corrected it,'[288] thereby reconstructing and reviving the traditional theory in a way Cullen hoped to emulate in chemistry. In terms of observable phenomena, the most important property of the fixing substance was not its acidity but its ability to counteract the power that causes expansion and heat, thus bringing about a loss of elasticity and a fall in temperature. Consider, for instance, Cullen's account of the way in which the acidic property of saltpeter manifests itself: 'One of the properties of this [substance] is its producing cold in any liquors it is mixed with & this we know depends on its acid spirit, by which a greater degree of cold can be produced than at any time nature or art otherwise can shew.'[289] Cullen believed that by recording temperature changes, which reveal the way heat flows as its cause is fixed and freed in various substances, chemists could begin to formulate an experimentally based theory of the repulsive power, the power which, like elective attraction, is one of the fundamental causes of change in chemistry.

As a first step towards the formation of a well-founded theory of repulsion, Cullen examined the temperature changes that accompany the changes of state that occur in chemical solutions.[290] He believed that solution, like evaporation, could be attributed to the action of two causes, the mechanical and the attractive (i.e., the chemical). He said that solution is 'mechanical in consequence of the tenacity of the fluid, so by increasing the surface in proportion to the quantity of matter, any body, however heavy, may be suspended in any fluid. But solution [is] not always performed in this way, but by attraction [also]. For the body [to be] taken up at first by attraction, the body [must be] divided equally.'[291] The distinction seems clear enough, although neither Cullen nor Black was ever able to convince Lord Kames that a solution of clay in water is a mechanical suspension and not a case of elective attraction.[292] Cullen, however, pressed ahead, arguing that the term solution should be applied only to those cases in which the solid is dissolved by elective attraction and that chemical reactions only occur in solutions of this kind. 'Solution properly' applies, he said, to 'the attractive [situation in which] the solvend [is] equally divided; this only gives mixture or combination.'[293]

By defining solution as the result of elective attraction, Cullen made it a subject which falls within the realm of chemistry. He did not, however, believe that all chemical solutions are identical. Certain substances, if added to certain kinds of solutions, cause reactions which lower the temperature of the liquid whereas other combinations cause a rise in temperature. Cullen studied these reactions and arranged them in what he may have thought of as tables of elective repulsion. 'In particular,' his biographer tells us, 'he endeavoured to arrange, under a few general heads, in tables of heating and cooling mixtures, the variations of temperature that result from different chemical combinations.'[294] Cullen finally classified these reactions under two headings: those which cause a rise in the temperature of the solution and those which cause a fall. This arrangement led him again to narrow the meaning of solution. As he told the Literary Society of Glasgow: 'In the combination of bodies, I think we can observe two different kinds of it, one of which is properly mixture, the other solution only.'[295] These two kinds of reactions can be distinguished by the heat flow which accompanies them. 'In every instance of mixture,' Cullen reports, 'heat was generated, and in every instance of solution cold.'[296] Finally, to explain these changes in temperature, Cullen suggests that 'in every mixture there is a condensation, and in every solution a rarefaction'.[297]

Cullen believed that in this correlation of heat and mixture he had established a law firmly grounded upon experiments and of considerable generality in chemistry, and one which met the criteria of contemporary natural philosophy. With it he could, for instance, account for a peculiar difference between the crystalline and the deliquescent (anhydrous) states of salts. He begins his explanation with a description of the most readily noticeable differences between crystalline and deliquescent Glauber's salt (sodium sulphate).

> Glauber-salt is a neutral, consisting therefore properly of an acid and an alkali only. But to get this salt in its crystalline state, a certain quantity of water must be united with the acid and alkali. In this state the concretions of the salt are of a determinate form, and it has a certain degree of firmness and transparency. But if, on the contrary, the quantity of water necessary is wanting, the salt, though it still contains the usual proportion of acid and alkali, does not concrete at all, or if it does, concretes in no determinate form, absolutely without any transparency, and with very little firmness of cohesion. As the salt in this state readily absorbs moisture from the air, and becomes in some measure fluid with it, I named it the deliquescent state.[298]

The temperature change caused by dissolving Glauber's salt in water depends upon the state of the solid salt. 'In the deliquescent state,' Cullen said, 'it generates heat, in the crystalline state it generates cold.'[299] Cullen then asserts, as a first conclusion, that the rise or fall of the temperature does not depend upon the properties of a matter of fire or a matter of cold. 'As in both experiments it is plain that the same materials, water, acid, and alkali, and these only, are employed; so, I think, it is plain that the effects of heat or cold cannot be imputed to the quality of the matter, that is, either to calorific particles in the one case, or to frigorific in the other.'[300] As a second conclusion he asserts that 'the effects we speak of [are] to be imputed to the state of aggregation, which is manifestly different in the two cases'.[301] He explains this difference in aggregation by saying that 'when a deliquescent salt is added to water, a part of this [water] is united with it in the way of mixture, but . . . when a crystalline salt is added to the same, it is united with it by solution only'.[302] Thus the behaviour of the two states of the salts in solution confirms both his theory of heat and mixture and his notion that chemical properties arise from the state of aggregation of the substances being examined.

Cullen was also able to account for another apparent exception to his law, 'the combination of acids and volatile alkalis, which, though undoubtedly an instance of mixture, is, however, in all the accounts of experiments of this kind, constantly mentioned as a mixture generating cold.'[303] Here, he argues, 'a double effect seems to have occurred, that is, the generation of heat by the mixture of the acid and alkali, and the generation of cold by the solution of the alkali in a watery fluid. The last, however, being the most considerable, appears as the result of the whole experiment.'[304] Thus he concludes 'that the real tendency of acids with volatile alkalis is to generate heat, and that thereby I have removed seemingly the most considerable objection against the general position, that *every mixture generates heat, and that every solution generates cold.*'[305]

The phenomena described in Cullen's later essay on 'The Cold Produced by Evaporating Fluids' were brought to his attention by a student who was performing a series of experiments designed to extend this theory of heat and mixture. 'A young Gentleman,' Cullen tells us,

> one of my pupils, whom I had employed to examine the heat or cold that might be produced by the solution of certain substances in spirit of wine, observed to me: That, when a thermometer had been immersed in spirit of wine, tho the spirit was exactly of the temperature of the surrounding air, or somewhat colder; yet, upon taking the thermometer out of the spirit, and suspending it in the air, the mercury in the thermometer, which was of *Farenheit's* construction, always sunk two or three degrees. This recalled to my mind some experiments and observations of *M. de Mairan* to the same purpose; which I had read some time before. *See Dissertation sur la glace, edit.* 1749. *pag.* 248, *& seq.* When I first read the experiments of *M. de Mairan* in the place referred to, I suspected, that water, and perhaps other fluids, in evaporating produced, or as the phrase is, generated some degree of cold. The above experiment of my Pupil confirmed my suspicion, and engaged me to verify it by a variety of new trials.[306]

Having been reminded of the cooling that accompanies evaporation, Cullen developed a set of techiques that enabled him to compound the effect. He found that 'by repeated dippings [of the bulb of the thermometer into alcohol], the cold produced might be rendered very remarkable,' and that the fall in temperature could be further increased by moistening the bulb with a single drop of liquid and by blowing air past it to hasten the evaporation.[307] 'By taking these methods,' he reports, 'I have by spirit of wine made the mercury in the thermometer sink from 44 degrees to below the freezing point; and, by employing some other fluids to be mentioned by and by, I have produced a sinking of the thermometer much more considerable.'[308]

Cullen's essay contains a list of the several fluids he examined while attempting to determine their relative powers to remove heat during evaporation. Included in his list are: The *quick-lime spirit* of *sal. ammoniac*, The *aether* of *Frobenius*, The *nitrous aether*, The *volatile tincture of sulphur*, *Spirit of wine*, *Spirit of sal. ammoniac* made with the fixed *alkali*, *Brandy*, *Wine*, *Vinegar*, *Water*, *Oil of turpentine*, *Oil of mint*,

Oil of pimento. Just as one can refer to tables of melting temperatures and to tables listing the degree to which bodies expand when equally heated, so too, Cullen asserts, one can determine 'the comparative power of these fluids in producing cold, by the order in which I have set them down; having mentioned that fluid first which seemed to me to occasion the greatest sinking of the thermometer, and the rest follow in order as they seemed to occasion less and less.'[309]

Such tables merely order the phenomena however, without telling us why the cooling powers of the fluids follow this particular order. Earlier in the eighteenth century natural philosophers had argued that more heat was needed to cause an equal temperature change in a dense body than in one that is less dense, but Martine's experiments had shown this argument to be false. In the case of cooling power however, Cullen believed that the specific chemical property involved could be explained in terms of a physical property. 'From the above enumeration' of experiments, he said, 'I imagine it will appear, that the power of evaporating fluids in producing cold, is nearly according to the degree of volatility in each.'[310] On the basis of this correlation, which reinforced his original identification of the cause of cooling, Cullen concluded 'that *the cold produced is the effect of evaporation*'.[311]

There was, however, one case which seemed to indicate that the cold produced by evaporation could not be incorporated into Cullen's general theory of heat and mixture. 'When the ball of the thermometer is moistened with any of the fossil acids,' Cullen reports, 'a considerable degree of heat is produced.'[312] But in this case, just as with the similar challenge to his account of the heat flow in solutions, Cullen was able to explain the apparent anomaly and thereby turn a fact that threatened to falsify his theory into an additional confirmation of it. 'We know,' he wrote, 'that these acids attract water from the air; and also that these acids, mixed with water, always produce heat: it may therefore be supposed, that the heat produced, by moistening the ball of the thermometer with these acids, is to be imputed rather to their mixing with the water of the air, than to their evaporation singly.'[313] After experimentally demonstrating the plausibility of this explanation, Cullen restated his theory in its full generality: 'The experiment, as it stands, tends to prove, that the heat produced by acids, applied to the ball of the thermometer, is owing to the mixture of these with the water of the air; and therefore, it is still very probable, that all fluids, which do not immediately affect the mixture of the air, will, in evaporating, produce cold.'[314]

If all evaporation produces cold and this cold is caused by the liquid being dissolved in air, then the elective attraction between air and the liquid must be necessary for evaporation. In other words,

Cullen's experiments on cooling seemed to confirm Kames' attractionist theory of evaporation, an implication Cullen had good reason to reject. 'When I had proceeded thus far,' he said,

> I began to consider, whether the cold produced in the above experiments might not be the effect of the mixture of the several fluids with the air; and that therefore, to a list of cooling mixtures and solutions which I was then making up, I should now add the several so[lu]tions made by the air. By one who supposes the evaporation of fluids to depend upon the action of the air as a menstruum, this would be readily admitted; but, as I knew that fluids evaporate *in vacuo* as well as in the air, I resolved to suspend my opinion, till I should repeat my experiments in an exhausted receiver.[315]

His subsequent experiments on evaporation *in vacuo* focused Cullen's attention upon a series of phenomena, which he described but did not attempt to explain. Even before introducing liquids into the vacuum chamber he observed the effect of adiabatic expansion and compression. 'A thermometer,' he wrote, 'hung in the receiver of an air-pump, sinks always two or three degrees upon the air's being exhausted. After a little time, the thermometer *in vacuo* returns to the temperature of the air in the chamber, and upon letting air again into the receiver, the thermometer always rises two or three degrees above the temperature of the external air.'[316] The first effect Cullen noticed after creating a vacuum around his thermometer-alcohol apparatus was rapid boiling under reduced pressure. He reports that 'when a vessel containing spirit of wine, with a thermometer immersed in it, is set under the receiver of an air-pump; upon exhausting the air, the mercury in the thermometer sinks several degrees. It becomes more especially remarkable when the air in any plenty issues out of the spirit of wine. As the spirit continues long to give out air, so it is long before the thermometer immersed in it returns to the temperature of the external air.'[317] After the evacuated chamber had reached a stable temperature, Cullen found that he could obtain an extraordinary fall in temperature by evaporating the alcohol on the bulb of the thermometer. 'When,' he tells us, 'after some time it [i.e., the temperature] appears stationary, if it [i.e., the thermometer] is then drawn up out of the spirits and suspended in the vacuum, the mercury sinks very quickly eight or nine degrees; a good deal farther than it would have done in the same circumstances in the air. In the same manner as in the air, the thermometer *in vacuo* may be made to sink lower by repeated dippings into the spirit of wine.'[318] Thus, in contradiction to his solution theory of cold, Cullen found that the cooling caused by evaporation is more extreme when no air is involved. 'This experiment with spirit of wine was often enough repeated,' he assures us, 'to shew clearly, that the evaporation of the spirit *in vacuo* produces a greater degree of cold than the evaporation of the same in the air.'[319]

Other liquids, when subjected to the vacuum, exhibited similar effects. 'With regard to the different fluids,' he reports,

> the *phaenomena* were much the same. Upon exhausting the receiver, the fluid gave out a great quantity of elastic air; and while this happened, the mercury in the immersed thermometer sunk very fast and to a great length. In our trials, it generally sunk below the scale applied so that we could not measure exactly how far. In one experiment before exhausting the receiver, the thermometer had stood at 50 degrees, and, after exhausting, we could judge very certainly that it sunk to below 20. In another experiment made with nitrous aether, when the heat of the air was about 53 degrees, we set the vessel containing the aether in another a little larger, containing water. Upon exhausting the receiver, and the vessel's remaining for a few minutes *in vacuo*, we found the most part of the water frozen, and the vessel containing the aether surrounded with a thick and firm crust of ice.[320]

Cullen's theory of heat and mixture could not explain these phenomena, and he therefore, with admirable candour and restraint, closed his paper without elaborating upon his discoveries. 'Such a means of producing cold, and to so great a degree,' he said, 'has not, so far as I know, been observed before, and it seems to deserve being further examined by experiments. Till that is done, I do not chuse to give any account of some other remarkable *phaenomena* that have occurred in the above experiments, nor to enter into several speculations that the subject seems to suggest.'[321]

Evaporation in a vacuum involves a change of state and a temperature change that Cullen could not explain. Previously, when investigating the heat flow in solutions, he had come across a similar set of phenomena. While constructing his table of heating and cooling mixtures, Cullen had asked himself whether the melting of ice in water was a case of cooling caused by dissolution. He tried the experiment but obtained results which he found 'extremely curious; for, if the water be heated to 50°, and the ice be at 32°, the thermometer will sink to 32° during the solution.'[322] This experiment posed a fundamental challenge to one aspect of his theory. One could maintain that the melting of ice in water is a case of cold produced by solution, but one could no longer assume that the flow of heat always causes a change in temperature. In this case the ice had to be absorbing heat from the water, yet its temperature did not rise. Clearly the investigation of heat and chemical change involved more complex relationships than those suggested by Cullen's theory of heat and mixture.

There can be little doubt that Joseph Black was familiar with Cullen's theory of heat and mixture and that he knew about the anomalies that had cropped up during his master's research. He was also, no doubt, aware that the two anomalies described above appear during changes of state that do not cause other changes in the chemical qualities of the substances involved. As ice melts and becomes water there is a flow of heat, which does not cause an increase in temperature, and as water goes from liquid to vapour there is a cooling, which cannot be attributed to the chemical solution of water in air. Neither Cullen nor Black believed these problems called for the rejection of the theory of heat and mixture; the awkward facts simply indicated that more experiments and further conceptual refinement were needed.

Early in 1755, while still in Edinburgh, Black wrote to his former teacher saying:

> I have looked over your paper [on the cold produced by evaporating fluids] with the highest pleasure, and wish with all my heart you would prosecute the subject: it seems to be a copious and important article in the history of Heat. It immediately brought to my mind the sensation of cold which spirit of wine, but especially aether, gives when dropt upon the skin. As for improvements, I assure you I see none to be made in it, but what must necessarily follow the trying of new experiments, and this is a labour in which I cannot engage myself at present, having already bestowed too much time on my own chemical inquiries.[323]

It was not long, however, before Black found time to take up the experimental study of heat. As is well known, the investigations he carried out after being appointed professor in Glasgow resulted in the conceptual clarification and quantification of specific heat and latent heat, theoretical breakthroughs, which helped explain the problems raised by Martine's experiments on the rates at which bodies heat up and those posed by Cullen's experiments on the flow of heat during change of state. Thus Black's contributions to the theory of heat gain their proper historical setting when seen as brilliant extensions of Cullen's general theory of heat and mixture.

PART TWO

Joseph Black and the Advancement of Philosophical Chemistry

Dr Cullen in his lectures on Chemistry, will always be heard with attention and applause. Those experiments, however, which he is almost daily pointing out to us [the medical students in Edinburgh], as necessary to be made, for attaining farther excellence in that science, require, for their performance, such a long and unwearied attention, as is perhaps only consistent with an almost total abstraction from other avocations. How much Dr Cullen does in this way, even amidst the constant distractions of other business, is not easy to say. But the prosecution of the science still more amply upon that plan, may, with confidence, be expected from Dr Black, a man whose name as a Chemist is justly celebrated throughout all Europe, and whose reputation, at present, supports the fame of an inferior, but rival University. In him, the vigour and enterprise of youth, is conjoined to a well known and peculiarly preferable attachment to Chemistry. He, we apprehend, is the only person, who, in the Chemical Chair, can acquit himself with the same applause, that the present Professor has done for a number of years. [From the 'Address of the Students of Medicine . . . of Edinburgh, 1766'. Thomson, *Cullen*, 1.157.]

[7]

Education and Apprenticeship

Joseph Black was born in Bordeaux on 16 April 1728.[1] He was the fifth son and ninth child born to John Black and his wife Margaret Gordon and the fourth of their boys to survive the rigours of infancy. This large and healthy family continued to expand long after Joseph's arrival. Margaret Black gave birth to fifteen children between 1717 and 1736 and thirteen of them were still living in 1751.[2] A decade later John Black could still point with pride to his 'numerous issue of eight sons and four daughters yet alive'.[3]

Although large, the Black family was always close. Their sense of community and their concern for one another, evident throughout their later correspondence, was doubtless in part a response to their alien status in France. Protestants in a Catholic country, they lived with the realization that complete assimilation without conversion was impossible.[4] In France at that time, a child's birth was not legally registered until he was baptized into the Catholic faith. It was therefore not unusual in the eighteenth century for French Protestants to arrange a Catholic baptism for a child. John Black apparently did this for five of his sons, but not for Joseph.[5] However, we know from family records that Joseph was christened by a Mr Lothian nine days after his birth.[6] We also know that John Black considered all his children 'natural born French subjects'.[7] But the legal threat remained. In 1741, after Joseph had gone to Belfast, his father feared that rising intolerance might finally make his family's position in France untenable. He wrote to his brother saying that 'the old severities are begun again by forcing Protestant children into convents wich its' probable may soon make me resolve in sending Jenny and others of mine back again to Ireland as liberty is one of the choicest of blessings'.[8]

It would have been cruel to force John Black to choose between his Scotch-Irish ancestry and his life in Bordeaux for he was deeply attached to both. The son of a merchant, he had been reared in France and later apprenticed 'to learn trade' with his father-in-law in Dublin.[9] In 1699 he moved to Bordeaux to assist in the counting house of a distant relative. A few years later he began to build up a business of his own, acting as a factor in the expanding trade between Bordeaux, Great Britain and Ireland. As his business grew, so did his wealth. In 1751 he possessed 340,000 *Livres* worth of fixed and liquid assets as well as two substantial homes near Bordeaux, a house in Belfast and another in London.[10]

Affluence, however, had not undermined this cautious merchant's original view of the world. Pious and unpretentious, John Black found his greatest happiness in prudently providing for his children. In 1751 he wrote a characteristic letter to William Cullen about his son Joseph. 'I have told him, [that] I will ever, by the grace of God, endeavour to be a careful stewart of the small competency that kind Providence has entrusted me with . . . giving [my children] all to understand that it has, and ever shall be, my constant endeavour to lay a good and solid foundation; and that it is for them to contribute, by their piety, virtue and industry, to raise the superstructure, which I pray God may enable them to doe for his glory and their own happiness.'[11] John Black's children returned his heartfelt affection with lifelong filial respect. Joseph always retained a vivid memory of his father's 'good character and virtues'. He had, Joseph recalled, 'no ambition to be very rich, but was chearful and contented, benevolent and liberal-minded. He was industrious and prudent in business, of the strictest probity and honour, very temperate and regular in his manner of life. He and my mother, who was equally domestic, educated thirteen of their children.'[12] John's other sons, especially those who entered the world of business, were also grateful for his constant devotion to their professional interests.[13] John Black evidently earned and enjoyed his notable success as both a merchant and a father.

Joseph Black's mother was also from a family of merchants. Although commercial interests seem to have caused her father to move to Bordeaux, Jacobite sympathies probably also played a part. One would expect a Gordon from Aberdeenshire to favour the Stuarts, and a note by Joseph Black, tactfully purged in Ferguson's summary of it,[14] indicates that his mother's family shared this sentiment. This note also provides evidence of the tolerant view of religion shared by John Black and his famous Bordeaux friend Montesquieu. Joseph wrote:

> The Gordon of Halhead who was my mothers Father was a younger Brother of the Family who settled at Bordeaux as a merchant & made a Fortune with which he purchased the Estate of Halhead from his elder Brother (a bankrupt) and thus became Halhead himself & married a lady of the family of Byres of Tonley. His eldest Son (then of Halhead) was engaged in the rebellion of 1745 & after living concealed some time escaped to France and lived at Boulogne where he received

assistance & kind offices from my Father and it is to him that the President Montesquieu's letters refer.[15]

In 1716 John Black and Margaret Gordon were married in Bordeaux.[16] 'Equally domestic' Margaret Black helped her husband forge the family's sense of solidarity and purpose. Joseph's only surviving statement about his mother concerns his earliest schooling. 'I received,' he wrote, 'the first beginnings of Education from my Mother who taught all her children to read english.'[17] Evidently the size of her family did not keep Margaret Black from encouraging the individual interests of her children. Joseph, when nine years old, was described as 'a tall slender well shaped boy very attentive & good scholar & good memory to the Psalms-has a disposition to design'.[18] By the time he was twelve, his parents' instruction had prepared Joseph for more formal schooling in Belfast.

In August 1740 John Black informed his brother Robert that he had 'sent Joseph, Alexander [b. Aug. 1729] and Samuel [b. Nov. 1730], my three boys to the schools at Belfast where they are well arrived'.[19] James Blow, a family friend, met the boys in Dublin and accompanied them to Belfast[20] where they were to live with their aunt, John's 'sister Arbuckle'[21] and 'acquire the Rudiments of latin and greek'.[22] While it is not clear which school in Belfast he attended,[23] Joseph received the training required to continue on to college. 'After 4 years spent there at a private School,' he recalled, 'I was sent to the university of Glasgow.'[24]

Joseph's parents recognized and approved of his developing academic interests. While his son was still in Belfast, John proudly told his brother Robert that 'son Joseph proves an excellent scholar and of singular ingenuity,'[25] and in July 1745, after Joseph had completed his first year at Glasgow, John again mentioned his academic progress to Robert: 'It's well you approve son Joseph's inclinations; while virtuous they shall always be encouraged to him and all the family as much as lies in my power.'[26] Although towards the end of his life Joseph Black recalled that in Glasgow he 'attended all the Lectures on the languages & Philosophy in a regular succession,' the sequence of his studies cannot be definitely established.[27] The only university document containing information about Black's studies is the matriculation list for the Ethics class of 1746–7.[28] The normal course of study at Glasgow was 'first year, Latin; second year, Greek; third year, Logic; fourth year, ethics; fifth year, Physics.'[29] Since it is quite certain that 1744–5 was Black's first academic year at Glasgow,[30] he seems to have taken ethics during his third, rather than his fourth year. What this probably means is that he had a sufficient knowledge of Latin when he entered to be granted advanced placement, a standard procedure in the University of Glasgow during the eighteenth century.[31] During his last year as an arts student, 1747–8, Black studied physics under Robert Dick. Although there are no matriculation records for this class during these years,[32] we do know that Black 'was a favourite pupil of the professor of natural philosophy, Dr. Robert Dick, and the intimate companion of his son and successor.'[33] It thus appears likely that Black completed his arts course in the spring of 1748 and began studying medicine the following autumn.[34]

Upon completing his undergraduate studies, Joseph had to decide whether to continue as a student or leave the university, probably to enter the family business. 'Being then required by my father to chuse a Profession or employment,' he wrote, 'I chose that of Medicine.'[35] The practical aspects of the physician's life did not attract Black at all, but of all the professions, medicine was, as Robison remarked, 'the most suited to the general habits of his studies'.[36] By continuing his education as a medical student, Joseph could cultivate his interest in natural philosophy while satisfying his father's demand that he prepare to enter a profession. John Black evidently approved of his son's decision, for he informed his brother Robert that 'our physician Joseph at Glasgow [is] very well'.[37] Joseph therefore began his medical education 'by attending the Lectures of the Professor of Anatomy & of D. Cullen who was at that time professor of Medicine at Glasgow.'[38]

Almost as soon as Black had assured himself of a few more years in the university, his fundamental interest in natural philosophy reasserted itself.[39] Cullen's encyclopedic interests made it possible for Joseph to concentrate on chemistry while ostensibly furthering his medical education. Chemistry was one of several subjects that Cullen taught but it was the only one which really captured Black's interest. More than willing to accommodate his bright young pupil, Cullen eagerly encouraged Black's unofficial apprenticeship in chemistry. The paternal solicitude Cullen showed towards his student's self-realization was still vivid in Black's mind many years later when, while writing his autobiographical note, he recalled his early years as a medical student. 'D. Cullen about this time began also to give Lectures on Chemistry, which had never before been taught in the University of Glasgow & finding that I might be useful to him in that undertaking he employed me as his assistant in the laboratory & treated me with the same Confidence and friend-ship & direction in my Studies as if I had been one of his children.'[40]

Just when Black became Cullen's assistant cannot be established with certainty. The transformation from medical student to chemical apprentice was probably a gradual one, and in later life Black himself could not remember clearly when it occurred. When drafting his autobiographical note Black first wrote that he assisted Cullen for three or four years. He then crossed out the four and changed his statement to two or three years. Still unsatisfied, he crossed out the figure two and settled on three years.[41] Since we know that Black went to Edinburgh late in 1752, he apparently began studying chemistry with Cullen between autumn 1748 and autumn 1750. A letter from one of Cullen's young friends in Edinburgh indicates that the earlier figure is the more probable. In December 1748 Francis Hutcheson, son of the former professor of moral philosophy, wrote to Cullen and politely inquired about the progress of his chemistry course in Glasgow. 'I would have been extremely glad to have heard from you ere this how all was going on in Glasgow. No one has written me anything particularly of the classes. I judged, indeed, from a letter of Mr. Yates, that the chemistry was going on well, as it was so stuffed with chemical terms, mixts, compounds, &c. that I should have been obliged to call a consultation to unravel its textures, and extract the quintessence of it. J. Black might have answered me.'[42] Hutcheson would not have mentioned Black had he not been well known to Cullen both as a friend and as an outstanding chemistry student. Clearly by the end of 1748 Black had begun his long and rewarding apprenticeship under Cullen.

While Black and Cullen explored the philosophical questions raised by the phenomena of chemistry, John Black patiently bore the expense of his son's professional education. Joseph saw no reason to disturb his father's understanding of the situation; surely he would eventually become a physician. Black therefore seldom wrote to his father during the early years of his graduate work, and when he did, he only alluded to his academic progress. In the autumn of 1750 John let slip a good natured complaint about Joseph's failings as a correspondent. Writing to his son Alexander, he reported that 'I have after about ten or twelve months philosophic meditations been favoured with an epistle from our dumb doctor Joseph who as his brother Robert lately from Glasgow to Dumfries bound home writes me was still the same studious man.'[43] A month later John wrote Alexander another letter, which indicates that he believed Joseph's medical education, then in its third year, was proceeding satisfactorily: 'Son Joseph, if application to his studies will do it, will, I hope, become no small proficient but some more learning will be necessary before he is entitled to a piece of the old Montpellier rag.'[44] Yet six months later John reported 'that our Dumb Doctor keeps at hard study. I have not one scrape of pen from him these many many months.'[45] Considering the 'Dumb Doctor's . . . backwardness in epistolizing,'[46] his father's ignorance of his educational progress is hardly surprising.

This ignorance was dispelled by one of John Black's infrequent trips to Belfast. In October 1751 Joseph sent his father a letter welcoming him to Ireland,[47] and the following month he travelled from Glasgow to Belfast[48] where father and son were reunited for the first time since Joseph's departure from Bordeaux eleven years earlier. John Black may have intended to question his son about his extended medical studies, soon to enter their fourth year, but on discovering what a 'generous education' Joseph had acquired in Glasgow, he was overwhelmed with happiness. Joseph evidently attributed every characteristic that his father found admirable to the influence of his teacher, William Cullen. The day before his son left Belfast to return to Glasgow, John poured out his gratitude in a dignified letter to the man who had guided his son as he grew from unformed youth into promising manhood:

> as I had nott for many years seen, or but seldom heard from your pupil, my son Joseph, his brother George at Belfast acquainting him of my desire to have a little of his good company, I was, about eight or ten days agoe, most agreeably surprized to see him, who, butt a young stripling boy when he left France, was become so tall and promising a personage, and above all soe properly endowed with principles of a generous education, for which, as he often testifies, he is in the greatest measure beholden to you. Permit me, dear and very worthy Sir, to express with him our most gratefull sense of all your goodnesse to him, which, I assure you, I and all mine will ever retaine a most thankfull remembrance of, humbly entreating the continuance of your soe honourable and distinguished patronage, which, as in duty, so I am persuaded he will ever study more and more to deserve, by a humble and obedient deference and respect for yourself, your good family, and all your commands.[49]

Joseph's visit with his father earned him one more year of study, but at the end of that year he was still not ready to receive his degree. Early in May 1752 John Black asked his son to spend the summer with him at the family home in Belfast.[50] Instead of going to Ireland, however, Joseph joined his father at Halhead, the Scottish estate owned by Margaret Black's father. John and Joseph Black then spent over two months together on a 'Northern Jaunt',[51] travelling first to Aberdeen and then touring the Highlands, Edinburgh and the Lowlands on their way back to Glasgow. They seem to have reached an understanding about Joseph's future. When, in the middle of August, they parted in Glasgow, John recorded that he 'left there son Joseph to finish his studys in Physick & Chymistry with profess. Doct. Cullen'.[52] Joseph had secured yet another year to complete his professional training. Clearly John Black expected it would be his last.

During the autumn months of 1752 Joseph Black decided to complete his medical studies in Edinburgh. We do not know why he made this decision but it seems likely that Cullen encouraged his move. Edinburgh offered the finest medical education available in Great Britain, one far more comprehensive and prestigious than that available in Glasgow. Perhaps Cullen also recognized that as long as student and master remained together, chemistry would not yield to medicine. To get his degree in Edinburgh, Black would have to apply his chemical knowledge to a medical problem suitable for a dissertation. Knowing this, Black selected a topic which suited his abilities and his needs before leaving Glasgow. The chemical problem concerned the relationship between fixed air and causticity; the medical problem concerned the action of lime-water upon urinary stones.

Robison's examination of Black's earliest notes leaves little doubt that he concentrated on fixed air and causticity as part of his drive to complete his medical training. Before resolving to leave Glasgow, Black focused his attention upon those theoretical problems, such as the flow of heat during change of state, which were central to Cullen's understanding of chemistry. But in the final months of 1752 he performed a set of experiments which clarified the relationship between fixed air, effervescence, and the causticity of lime. 'In the oldest parcels of these notes,' Robison reports, 'I find queries respecting the nature of cooling mixtures, and the cold produced by liquefaction; but it is not till some time after, I think not before the year 1752, that I can date any observation relative to fixed air.'[53] He then describes Black's experiments,[54] adding that 'immediately after [the account of the fixed air experiments in Black's notes], a medical case is mentioned, which I know to have occurred in November 1752.'[55] It thus appears that Black went to Edinburgh after having laid the theoretical foundation for his dissertation.

By the beginning of December 1752 Black had settled into his new surroundings. Having committed himself to the study of medicine, he knew his father would be pleased by 'the news of my having quitted Furnaces & Fumes and Sulphur & Nitre'.[56] Like his second cousin Adam Ferguson, Black found student lodgings with the family of James Russell, professor of natural philosophy.[57] Joseph found the professor and his wife, who was Margaret Black's sister, 'the most easy frank People imaginable,' and he told his father 'I am extremely happy in my present situation.'[58] Naturally, moving to the capital city exacerbated the chronic condition of students: 'I find,' Joseph wrote, 'my Edinburgh Expenses will far exceed those of Glasgow for the very fees & books for 5 classes will encrease them considerable.'[59] John Black, who certainly had the means, responded sympathetically now that Joseph was at last making progress towards his degree.

Not only had Joseph moved to Edinburgh, he had also, by December, gained enough confidence to describe his thesis topic to his father.

> I am now thinking upon a thesis for taking out my Degree & Propose to myself a subject which has lately from some new Discoveries of two of the Professors here turned out a very curious one—I mean the Properties & virtues of Lime Water in Dissolving the Stone in the Bladder which if it turns out as it promises will be one of the most usefull Medicines in the Materia Medica & particularly worthy of attention as it shows a tendency to remove one of the most excruciating Disorders that render Men Miserable—The Discoveries of Dr. White and Dr. Alston are chiefly from Experiments of its dissolving power upon the Stone taken out of the body & these are very curious indeed but not numerous enough & the two Professors disagree too about its effects & draw very different conclusions from their tryals so that nothing can be determined without a greater number of Experiments. Now these are all of the Chemical kind & here my Chemistry will be of very great service to me.[60]

How fortunate Black felt! Alston and Whytt were debating an important medical question which he could easily resolve by utilizing his superior knowledge of chemistry.[61] He hardly seems to have felt challenged. The problems he faced were technical and experimental, matters of detail, not theory. From the experiments Black performed in Glasgow, Robison reports, 'it would appear, that he had before this time [November 1752] suspected the real nature of these substances [fixed air and quicklime].'[62] This must have been Black's opinion also. Everything seemed to be breaking right for him at the end of 1752.

Despite his self-confidence, however, Black could not find the time he needed to complete his dissertation. After a few months in Edinburgh he found himself completely immersed in medical studies. His first letter to Cullen, written in February 1753, begins with a surprising apology: 'My time is so taken up with five or six hours' attendance at the colleges, and the shop, and the patients, that really I can very seldom indulge myself in the pleasures of reflecting on my friends in Glasgow.'[63] For the first time in his long medical education, Black was sincerely interested in the practical realities of his profession. 'Medicine has put on quite a new face to me since I came here,' he told Cullen, 'and I can now see an operation with as great ease as before I did a dissection. I go among patients groaning or gasping without any uneasiness, so that objection is quite vanished.'[64] Although his newfound interest in medicine forced him to postpone his dissertation research, Black argued that a greater knowledge of practical medicine would be useful in the long run. 'I have found no time to read on the pathology, nor one word on lime-water, so that I must, if possible, spend another winter here, which I believe, is worth my while, on account of the shop, infirmary, and private patients. These, I find make me read with much more keenness and to better purpose.'[65] In a more candid statement, Black admitted that his dissertation, and perhaps theoretical chemistry also, were no longer his sole interests. 'As for myself, I am kept so busy that I have almost forgot all my Glasgow projects, and as medicine is grown quite agreeable to me now, I have no reason to be quite so anxious about them as I was.'[66] Nevertheless, he still had to deal with John Black's insistence that he complete his professional training. He told Cullen that 'I had lately a letter from my father, by which I see he expects I will take my degree here this spring, and is very keen for it; but I must persuade him soon of the impossibility of it; or, if I cannot, will be obliged to take it in Glasgow if I can get it without a thesis.'[67]

One month after describing his new interests to Cullen, Black again appealed to his father for more time. He began his letter by gently criticizing John Black for drawing a sharp distinction between chemistry and medicine. 'You are so kind as to congratulate me on my Escape from fumes &c. I assure you I now reflect with a good deal of pleasure on the various experiments I have tryed for I find every day that Chemistry is a branch of natural Philosophy of the most extensive & solid use in all the arts but particularly in Medicine.'[68] He then told his father that he could not master both the practical aspects of medicine and the experimental knowledge needed for his dissertation in one academic year. 'I have so much Business on my hands with 5 classes besides the shop & the infirmary that it is impossible for me to write the thesis & prepare for my degree this Season without neglecting the attendance or reading necessary to make my self master of the several Branches I am attending & the subject I propose to handle will cost me a good deal of time as I can make nothing of it without trying a number of Experiments & comparing many authors on it.' To head off the suggestion that he return to Glasgow and take his degree without a thesis, Black expatiated on the advantages of studying in Edinburgh. 'This [delay] will probably surprize you as I came here with a resolution to take my degree but that I find impossible, & indeed I find my self so extremely well situated for further improvement here that I believe I can spend next winter better no where particularly on account of the Shop & infirmary & my intimacy with Mr. Russel who gives me an account of all his Practice as also an acquaintance with Dr. Clerk Physician to the Infirmary to whom Dr. Cullen has introduced me.' Knowing full well his father could afford to support him for yet another year, Joseph also appealed to John's sense of responsibility. 'I am sensible the Expence I put you to here is a little extraordinary but that is unavoidable if I would be perfect in what I profess & undertake the care of other People's health with a good conscience.' Joseph then went on to emphasize the common interests that united father and son. In requesting continued support, he emphasized, he was not asking that his interests be preferred to those of his brothers and sisters. He then flattered his father with a lengthy response to his questions about the use of quicklime as a medicine. Yes, quicklime is too corrosive to use directly, but when mixed with water it forms a most efficacious medicine. 'Mrs. Stephens medicine is now generally laid aside as excessively acrid & nauseous. . . . The only usefull ingredients in her composition were Soap & a Quicklime made of Egg Shells but as Quicklime can never reach the bladder in Substance They give only Soap & lime water.' Then, having completed his appeal, Joseph waited to hear whether his father would allow him to complete his studies in Edinburgh.

Joseph's own dissatisfaction with his endless career as a student probably troubled him as much as his father's expectations. From the slight evidence available, it seems that John was willing to support his son, providing money was available, as long as he was happy and progressing towards his medical degree. John's genuine affection for Joseph is apparent in his desire to stay abreast of his studies. He became anxious and a trifle censorious only when his son failed to inform him about his current interests. By surrounding his appeal for a second year of study in Edinburgh with family and university news, Joseph easily won his father's assent. At the end of March, John Black wrote to his son George about 'Jos who continues still very busy att his chymistry experiments & attendance att Edinburgh Physick Colledge hospitals operation & butt does nott expect or apply as yet for his passing examination in order to obtain his degree.'[69] Furthermore, Joseph was fortunate in that John received his appeal shortly before hearing that a recently delivered grandchild had been named Joseph. If his philosophical son was so loved by his brothers and sisters, how could his father deny him. John Black soon reconciled himself to giving Joseph all the time he needed. In April 1753 John wrote to a friend saying that 'James Burnett writes me of 20 ult^{o} [i.e., 20 March] that his Wyfe my dear Bella was that morning safely delivered of a sturdy brave boy & called him by his unckle's name, my Son, Joseph. Son Joseph is still pursueing att Edinburgh his Physicall studys as well as in chymistry as in the operations at the hospitalls greatly assisted therein by his cuzen german Mr. Jas. Russel a worthy & noted Surgeon Apothecary with whom he is happily lodged & in proper & due time to pass his examen & receive his diploma of Dr. in Physick.'[70] Two days after writing this letter, John Black 'wrote to Son Joseph on the being Godfather to dear Jos. Burnett &c.'[71] Before finishing his first academic year in Edinburgh, Black was assured of a second.

Although Joseph Black realized that Edinburgh had a better medical school than Glasgow, he had nothing but scorn for the chemistry taught there. He therefore never doubted that his superior knowledge of chemistry would enable him to complete the dissertation he had planned in Glasgow. From late 1752, when he arrived in Edinburgh as a medical student, until 1756, when he left to begin his professorial career in Glasgow, Black was keenly aware of the profound differences between Cullen's conception of chemistry and the chemical ideas held by the members of the medical faculty in Edinburgh. As Cullen's brightest chemistry student, Black enjoyed the unusual pleasure of knowing that in this field his knowledge surpassed that of his teachers.

Even if we discount many of Black's comments as flattery designed to please Cullen, his harsh judgments of the chemistry taught in Edinburgh still reveal his own sense of superiority. 'Though I attend so many classes,' he told Cullen, 'I suppose you will venture to believe me when I tell you, I have little satisfaction in any of them. Alston's is the newest to me, and pleases me most; but he has neither had practice enough, nor is he a good enough chemist to be perfect in his profession; and, by-the-by, I think there is no branch should be more cultivated in a medical college.'[72] He also echoed his master in saying 'I see no science that stands so much in need of chemistry as Materia Medica: every article offers some experiments,' and he exhibited as well the bright student's penchant for damning the present state of knowledge: '[The preparation of Ethers] is a curious and a very necessary subject of inquiry; considering the stupid manner in which it is ordered in the Dispensatories.' [73] He concluded a long discussion of a chemical problem by acknowledging Cullen's mastery of the subject: 'This is probably all very tedious to you, who have thought on the subject already, but you know how keen a chemist I am, and you must forgive me.' [74] Black also reported that the practical chemists of Edinburgh are inexpert—'I am told they are not very nice about evaporating pans here'—and that the furnaces he designed are better than any to be found in the laboratories of Britain's leading medical school: 'I wish to know how you are pleased with the new furnaces. . . . I have met with nothing in that way here. Dr. Plummer's are slovenly and ill contrived, and the laboratory ones are built so as that the utmost fire almost you can raise in them will be just sufficient for the process they are appropriated to, so that they have no dampers, nor do not need them.' [75]

Although he never lost confidence in Cullen's approach to chemistry or in his own ability, Black finally admitted that he was unable to resolve the chemical problem which lay at the heart of his dissertation research. His second academic year in Edinburgh began to slip away and once again it appeared he would not complete his studies. Unwilling to postpone taking his degree for another year, Black gave up trying to explain the action of lime-water and turned instead to describing the medical effects of magnesia alba. Putting the best face possible on this retreat, he proudly informed his father that at last his degree was in sight.

> I have now thrown aside all thoughts of drawing or other inferiour Employments & am preparing for receiving the Doctor's Cap and Gown with great pomp and solemnity. I found it proper to lay aside lime water which I had chosen for the subject of my Thesis. It was difficult & would have appeared presumptuous in me to have attempted settling some points about which two of the Professors themselves are disputing. I have taken one which has been little touched upon by any body else. It is acidity in the stomach & a particular sort of powder employed in the cure of it. As there is some Chemistry in it I propose to address it to Dr Cullen if you approve of it. It is from him I learned my Chemistry & I think he ought to be the patron of my first attempt in this way. [76]

John Black could not help but be pleased by this long awaited announcement. Realizing this, Joseph added a final request for money and support while interning in London. 'My Degree and the printing of my Thesis must run away with a good deal of money & my lodging tho so very advantageous & agreeable on other accounts, yet like all other tolerable lodgings in Edinburgh is none of the

Cheapest. However this is the last heavy article except some time of [sic] an Hospital in London which I find is absolutely necessary to a young Physician. If you think that proper I am satisfied, & shall stop there with Pleasure.' John had no intention of denying his son support just when his efforts were about to be rewarded. On 6 April he mentioned in his diary his approval of whatever expenditure was needed to obtain the degree: 'By letter 6 ult^o^ from son Joseph att Edinburgh he acquaints his being soon to be admitted a graduate Doctor of Physick which ceremony will require some expense extraordinar which I order him to manidge with decent oeconomy & call for from Mess. Alexander.'[77]

On 1 June 1754 Joseph again wrote to his father.[78] After thanking John for his 'most affectionate letter of the 8th of April,' Joseph happily reported on his further progress towards a degree: 'I am not yet installed into the order of the great wig, but have gone through all the ceremony, & that has been put off by the Professors to wait for some others that are to be promoted along with me 6 days hence.' He then explained in detail why he had decided to postpone his projected trip to London. In August, John Black replied to this letter, knowing that by then his son had finally completed his professional education: 'Wrote . . . to son Joseph att Edinburgh answering his acquainting me that he had happily past all Examinations in order to his being received Doctor of Physick.'[79]

Although his dissertation was published by the middle of June, Joseph did not send his father a copy until late November. 'I take this opportunity,' he finally wrote, 'to send you a few copies of my Inaugural Dissertation which it is the custom here for the Young Doctors to distribute very liberally among their Friends and fellow Students & this custom oblidged me to print off 170 Copies—a piece of expence attended with no manner of advantage to me but which could not be avoided.'[80] Realizing that his father would find the subject abstruse, and anxious to apologize for the weaknesses in his argument, Joseph explained how his dissertation attained its final form.

> When I first began to write my Dissertation I determined to examine a medicine called magnesia alba which is used only in case of a sourness in the stomach because it has like Crabs eyes, Chalk &c, the property of sweetening any sour liquor with which it is mixed, in an instant. This substance I chose to make the subject of my Essay chiefly because it gave me an opportunity of trying some new chemical Experiments a few of which I hoped might possibly be curious & make the Dissertation vary a little from the common beaten road. In this I find I have

> succeeded. It has been taken notice of a little more than is usual, & indeed Dr Alston the Professor of Materia Medica has done me the Honour in a second Edition of his Dissertation on Quicklime to pay me an extreamly civil & high compliment upon it.[81] But however, when I had finished my Experiments & began to lay them together I found that the whole would appear to belong too much to pure Chemistry & too little to medicine. I therefore thought of adding a very short account of the nature causes & cure of the Heartburn or acidity in the stomach because it is in this Disease alone that magnesia is used as a remedy, & the having premised a few things on the nature of the disease would enable me to speak more distinctly in treating of the virtues of the remedy & of the manner of applying it. You may easily imagine that I must have been a little hurried in putting together that part which treats of the Disease, as it was late before I thought of it. I did not get it finished off with the exactness I wished for & would have the second to be looked on as the principle part of the Dissertation.[82]

Joseph Black believed that despite its flaws, his inaugural dissertation contained important experiments. He therefore suggested that since John Black had returned to Bordeaux, he show the thesis to his good friend Charles Secondat, Baron de Montesquieu. 'Perhaps,' he wrote, 'it may be some entertainment to Monsr. de Secondat to glance it over as he seems to be fond of natural enqiuries.' John Black did present a copy of his son's work to Montesquieu during one of their frequent visits,[83] and the Baron's comment, made a few days later, was all a proud father could hope for: 'Mr. Black, my very good friend, I rejoice with you; your son will be the honour of your name and of your family.'[84] By January 1755 John Black could also record in his diary that the Academy at Bordeaux, probably seconding Montesquieu's commendation, had expressed approval of Joseph's dissertation: 'Wrote att length to Son Joseph att Edinburgh now received a Doctor of Physick answering his with inauguration dissertation approved att this Academy &c.'[85] Nine months later Joseph replied to his father's letter, thanking him for bringing his thesis to the attention of the leading natural philosopher in Bordeaux and mourning the passing of Montesquieu, who died in February 1755. 'I observed with gratitude,' he wrote, 'how carefull you have been to put my thesis into the hands of all such as you imagined would have any relish for it but the great name of Montesquieu among these renewed my melancholy. I had already heard of that Loss all the world has reason to mourn & it affected me the more sensibly as he was your intimate friend.'[86]

To John Black, the publication, rather than the contents, of his son's dissertation was the important achievement, for it marked his admission into the medical profession. But William Cullen was naturally eager to see what his former student's dissertation contributed to chemical understanding or medical practice. In fact, while preparing his thesis Black evidently continued to consider himself Cullen's apprentice in chemistry. 'Previously to the publication of his thesis,' Cullen's biographer writes, 'Dr. Black submitted it to the revisal and correction of his friend and former preceptor; and, from a fragment of a scroll, containing remarks by Dr Cullen on that dissertation, it is evident with how much attention he had considered every part of it.'[87] Much to Black's dismay, the pressure of time and the need to relate his experiments to a medical topic prevented him from making full use of Cullen's comments. Although he dedicated his dissertation to Cullen, Black did not expect it to be well received by his chemistry master. Shortly after its publication he wrote to Cullen saying:

> I sent you on Thursday last a dozen copies of my Thesis, which you will find to be less altered from the sketch you saw before than you expected. The truth is, I was quite tired of it before I gave it out of my hands. There are many things in the first part which do not at all please me, but I was obliged to write it, and I shall be satisfied if it passes without much notice. The style through the whole is excessively dry and awkward, but my subject did not allow much elegance; and, though it had, I confess myself so imperfect in the Latin, that I really could not have attempted to beautify it.[88]

Professionally, the dissertation served its purpose but chemically it failed to meet Black's own expectations.

Black chose to stay in Edinburgh for another year after finishing his dissertation. He told his father that he needed the time to gain clinical experience,[89] but he clearly also wanted some free time to prepare a paper on his theory of causticity. In June 1754 Black asked Cullen if he thought he should publish his experiments. 'I perceive that the Professors here think my experiments new, and seem pleased with them,' Black wrote, 'particularly Alston and Rutherford, the latter particularly, because he looks upon himself as the introducer of magnesia into practice here. What do you think of printing the experimental part in the Physical Essays here? If the experiments are worth any thing they will be stolen by others; but yet I would rather have it so than make them public, unless you think it would be of some sort of service to me.'[90] The subject no doubt came up again when Black visited Cullen in Glasgow in the autumn of 1754.[91] Black finally completed his essay during the next spring and it was read before a meeting of the Edinburgh Philosophical Society on 5 June 1755.[92]

The following six months were particularly hectic for Black. Although his essay had been read, he continued to revise it before submitting it for publication. In August he wrote Cullen, asking him to have an experiment performed with the superior vacuum pump he had at his disposal in Glasgow, and the results of this experiment were subsequently incorporated into the published version of the essay.[93] In September Black wrote to his father and told him of the progress of his studies. 'I have been advised,' Joseph reported,

> to publish my experiments upon magnesia in a Collection of papers printed here under the title of Essays physical and literary by a society of Gentlemen (mostly professors of the University). As most of the papers of this Collection are English I have accordingly revised & translated that part of my Thesis in the same language & added besides some experiments upon Quicklime which seem to throw a new light upon the nature of that usefull Substance & teach us to account for most of its properties without haveing recourse to the many groundless suppositions invented by Philosophers to explain them. These Experiments have been read before the society above mentioned, approved of & ordered to be printed in their next volume which is to appear in the beginning of winter. You will perhaps begin by this time to think me tout a fait souffleur & so very fond of Chemistry as to neglect every thing else. I must own this last essay on Quicklime has detained me here somewhat longer than I intended but if I may trust to the opinion of several good Judges my time has not been thrown away & my paper may be of some service in introducing me into the world.[94]

Having finished his chemical investigations, he continued, he was at last ready to spend some time in clinical studies in London. 'I now propose soon to set out for London,' Joseph informed his father, 'to attend the hospitals there which have universally the greatest character for variety of practice & are thronged with young physicians from all quarters.'[95] John Black quickly approved his son's plans,[96] but Joseph's departure was delayed by his involvement in the struggle over the succession to Andrew Plummer's chair of medicine and chemistry in Edinburgh.

When Andrew Plummer became seriously ill during the summer of 1755, several candidates eager to obtain his chair began to press their cases.[97] Although a possible successor himself, Black was primarily interested in seeing that Cullen received the appointment. He therefore kept himself and his former teacher well informed during the tedious competition which ensued. When Cullen finally obtained the appointment in late November, he invited Black to assist him in giving the lectures during the first year. He also immediately asked his many influential acquaintances to see that Black succeeded him in Glasgow. It is therefore hardly surprising that while preparing his first scientific memoir, getting ready for an extended visit to London and jockeying for a university appointment, Black had no time to reflect upon the theoretical questions raised by the interesting heat phenomena described in the essay which Cullen sent him during the latter half of 1755.[98]

Apparently Black did manage to spend a few months in London, for we next hear of him in 1756 when he became professor of anatomy and botany in Glasgow. This chair had been previously held by Robert Hamilton,[99] but in 1756 he transferred to the chair of medicine which Cullen had vacated the previous year. When Robert Hamilton died in 1757, Black became the professor of medicine, and Thomas Hamilton, Robert's brother, succeeded to the chair of anatomy and botany.[100]

Once installed as a member of the faculty in Glasgow, Black began to pursue the heat investigations which led to his second great set of discoveries.[101] Although troubled by ill health,[102] he also busied himself with his new responsibilities as a professor and as a practising physician.[103] His efforts as a teacher must have been appreciated, for in 1763 the university provided him with a new laboratory and lecture room in the neighbourhood of the physic garden, the laboratory alone costing the considerable sum of £500.[104] He also began to interest himself in practical chemistry, particularly bleaching, since his research on quicklime seemed to bear directly upon this subject. In response to his father's urgings and the requests of several interested parties in northern Ireland, he drafted a memoir on the use of lime in bleaching and, as one of his correspondents later informed him, his essay was instrumental in bringing about the repeal of the law against bleaching with lime.[105] During the period in which he was teaching in Glasgow Black also visited the recently established Carron Iron Works at Falkirk. Although he was well acquainted with John Roebuck, the 'brother Doctor & chemist' who played a leading part in this important enterprise, Black insisted, when questioned, that he knew nothing about the secret process Roebuck was using in his other great effort in industrial chemistry, the production of vitriolic acid at Prestonpans.[106]

During his decade as a professor in Glasgow, Black had the pleasure of seeing his reputation as a chemist grow considerably. As he told his father in 1766, Cullen and William Robertson, the Principal of the University of Edinburgh, had for several years been trying to bring him to Edinburgh.[107] In Glasgow his lectures became so famous that four of his professorial colleagues attended his final course.[108] Black was also flattered by the petition that the Edinburgh medical students circulated in 1766, and he promised to send his father a copy.[109] Thus it was with considerable pleasure that Black in 1766 returned to Edinburgh to become professor of medicine and chemistry. He was, of course, no stranger there, and even before delivering his first lecture in the university he acted as 'best man' at the wedding of his cousin Adam Ferguson.[110] And yet, although Black's move to Edinburgh was the culmination of his professional career, it also marked, for reasons that remain unknown, the end of his career as a creative philosophical chemist.[111]

[8]

Black's Study of Causticity and Fixed Air

I. THE THEORY OF CAUSTICITY: THE FIRST PHASE

Joseph Black's essay on the relationship between causticity, alkalinity and fixed air rises like an unscaled peak in the history of chemistry. When viewed from the vantage point of the chemical revolution, it appears to be the source of that mainstream of eighteenth-century innovation, pneumatic chemistry. In his essay Black described several properties of carbon dioxide, the first of the atmospheric gases to be identified as a distinct chemical substance, and thereby inaugurated a branch of chemistry which was subsequently advanced in Great Britain first by Henry Cavendish, then by Black's own pupil Daniel Rutherford and finally and most famously by Joseph Priestley. But if viewed from the prospect of the early eighteenth century, a position which can only be reached with difficulty, Black's essay appears to loom up beyond a wide and unmapped valley. There are, of course, a few worn paths that wind from the more familiar terrain up the sides of the distant peak, but they all peter out before reaching the top. One trail emerges from Hales' investigation of fixed air, a second springs from the older theories on the nature of causticity, and yet another begins with the medical debate over the use of lime-water. But all these initiatives tend in different directions and none of them will carry the historian to his goal. Clearly we must find a route which follows the natural contours of the terrain while always tending towards the summit if we wish to complete our ascent and finally see the problem of fixed air and causticity as Joseph Black saw it when he wrote his essay.

Black's theory of causticity matured in three phases. The first ended in 1752 when he left Cullen and Glasgow for Edinburgh and a medical degree, the second ended in 1754 when he finally submitted the dissertation required to obtain that degree, and the third ended in 1756 with the publication of his essay on calcareous and alkaline substances. This rather simple progression was complicated, however, by the medical demands placed upon Black during the years in which he was preparing his dissertation. Before leaving Glasgow he believed he had acquired an adequate understanding of the essential nature of causticity. But when he tried to apply this knowledge to the problem of finding a more effective internal solvent for urinary stones, he discovered that his theory contained a critical flaw. Black however, was anxious to complete his medical education and he therefore hastily composed a dissertation, which satisfied the rather vague demands of his medical examiners, before he had resolved the fundamental chemical problem that interested him. After obtaining his degree he spent another year refining his original theory so that in its final form it could explain both the general relationship between causticity, alkalinity and fixed air and the troublesome specific properties of magnesia alba.

After studying with Cullen for over three years Black must have been well acquainted with the various properties of calcareous substances. He knew that caustic substances yield alkaline solutions when slaked with water and that they are completely neutralized when exposed to the air. He also knew that magnesia alba, which certain doctors were prescribing as an antacid, was, according to Hoffmann and Brownrigg, a saline calcareous earth. He was also aware that his teacher believed that the active constituent in quicklime is a salt and that air, when fixed, has certain chemical qualities which merit investigation. But all these pieces of information remained unrelated in the works and teachings of his predecessors. Whether by chance or design, Black brought them together into a coherent theory of causticity shortly before leaving Glasgow in 1752.

Our knowledge of these early views on causticity and fixed air is derived from the experimental notes which John Robison summarized in his introduction to Black's *Lectures on the Elements of Chemistry*.[1] Although the notes themselves are tantalizingly brief, it is clear they are the record of a series of experiments which establish all the main points of Black's later theory. These experiments on alkalis, calcareous substances and fixed air also indicate that, as Robison correctly concluded, Black 'had before this time suspected the real nature of these substances'.[2] The experiments themselves were not exploratory. They were designed to confirm a chemical theory that Black had formulated to make sense out of the unconnected facts that he had encountered during his training as a chemist.

In the first of these early experiments Black demonstrated that causticity is not caused by a substance impressed upon the calx by the fire. 'I see,' Robison tell us, 'that Mr. Black had entertained the opinion, that caustic alkalis acquired igneous matter from quicklime. In one memorandum, he hints at some way of catching this matter as it escapes from lime, while it becomes mild by exposure to the air;—but on the opposite blank page is written, "nothing escapes,—the cup rises considerably by absorbing aire."'[3] Professor Guerlac has suggested a description of this experiment, which seems more accurate as an account of the apparatus used than as a guide to Black's expectations. 'We may conjecture,' he says, 'that Black placed the caustic lime in a dish or saucer floating on water and inverted a glass vessel over it. He may have expected to trap some igneous matter if it were given off, or—and this is more likely—to note a fall in the water-level. Instead, the water rose, carrying the cup with it, and suggested to Black that air had been absorbed by the lime.'[4]

Guerlac asserts that Black expected to obtain evidence that a fiery substance is released during slaking and that he only thought of the absorption of air when he saw the water level rise instead of fall, but I find this supposition unlikely. Robison's term 'entertained' was, I believe, carefully chosen, for it appears that Black only introduced the fire theory to refute it. Being Cullen's student he no doubt considered causticity an intense form of alkalinity. This chemical property is exposed when the neutralizing acidic substance contained in calcareous earths and alkalis is driven off. Thus in going from a caustic to a mild state, the calx absorbs something from the air, or the air itself, rather than giving up its acrid constituent to the air. Of course, had Black believed that causticity is caused by the addition of fiery or acidic particles, he could have argued against the results of his own experiment. Perhaps the fire particles released during neutralization escape through the glass or perhaps the acid given off is dissolved in the water. But Black ignored such tendentious hypotheses for he had no desire to defend the archaic fire theories favoured by such older chemists as Stephen Hales and Robert Whytt.[5] Black's experiment confirmed his belief that caustic substances are neutralized by the addition of air, and that was all he needed as a first step towards the establishment of his new theory.

The second set of experiments were designed to demonstrate that calcareous earths and fixed alkalis contain fixed air. Robison reports that 'a few pages after' his first experiment, Black 'compares the loss of weight sustained by an ounce of chalk, when calcined, with its loss when dissolved in spirit of salt.'[6] Although Robison does not seem to have appreciated the significance of this experiment, Black thought it so important that he presented it at greater length in both his dissertation and in his essay. Actually, it was not enough to demonstrate that the common alkalis and calcareous earths contain fixed air; Hales had already done that and the received theories of causticity had remained unshaken. To make his new theory convincing, Black had to demonstrate that the fixed air in these substances is chemically significant. This he did by criticizing the method of equivalent weights, one of the most popular and exact analytic procedures used by eighteenth-century chemists.

William Homberg developed the technique of equivalent weights as a means for comparing the strengths of different acids. He took a fixed weight of a certain alkali salt and measured the weight of acid needed to neutralize it. He assumed that the resulting neutral salt is a compound of the original alkali and the added acid, and argued that the strength of the various acids could be measured by comparing the weights of each needed to neutralize the fixed weight of alkali.[7] Black, however, had good reasons for believing that there are three chemically active substances involved in these neutralizations. In experiment xxii of his dissertation he demonstrated that when a solution of fixed alkali is neutralized by an acid, the final weight of the solution is less than the total weight of the original alkali and acid. The lost weight, he concluded, belongs to the escaped air. 'I put,' Black said,

> two drachms of sal tartari [potassium carbonate] and an ounce of water together into a large flask furnished with a long and slender neck: the flask with the salt and the water weighed two ounces two drachms: I then gradually dropped in oil of vitriol diluted with water to complete saturation of the salt, which was attained when I had poured in two drachms two scruples and three grains of the acid; the flask now weighted two ounces four drachms and fifteen grains; the loss of weight was therefore one scruple and eight grains; of this a very small part might be water, or something similar; the rest air.[8]

As Black then pointed out, this quantitative discovery invalidates all measurements based upon the method of equivalent weights. 'These things misled the celebrated Homberg,' Black noted, 'when he endeavoured to find the proportion of solid and liquid in different acids; he attacked the question by saturating pure alkaline salt with each of the acids, and observing what the salt gained from a definite quantity of each acid; but his calculation was necessarily erroneous, as the alkali loses no small part of its weight in the saturation.'[9] Clearly if one has to take account of air when using the technique of equivalent weights then one can hardly deny the chemical significance of the air fixed in those substances that can be made caustic.

In his third set of experiments Black demonstrated that fixed air can be transferred from one substance to another. 'When I precipitate lime by a common alkali,' he wrote, 'there is no effervescence. The air quits the alkali for the lime, but it is not lime any longer, but c.c.c. [i.e., calcareous earth. $Ca(OH)_2 + K_2CO_3 \rightarrow 2KOH + CaCO_3 \downarrow$].'[10] Adding potash to a solution of calcareous salts and obtaining an earthy precipitate was hardly a novelty; Cullen had previously used this well known reaction to purify a solution of sea salt.[11] But Black was the first chemist to realize that in this reaction the fixed air in the potash is being transferred to the lime in the lime-water, thereby neutralizing it and making it insoluble. By comparing Black's account of this reaction with that published by Charles Alston in his *Dissertation on Quick-Lime and Lime-Water*, we can see that Black made his discovery by concentrating upon the chemical properties of fixed air rather than by quantifying the reaction. It was Alston, not Black, who carefully determined the weight of quicklime which will dissolve in a given quantity of water.[12] Then, armed with this information, he added three spoonfuls of lime-water to a solution of potash and weighed the precipitate. Since the weight of the precipitate was more than twice the weight of the lime contained in the three spoonfuls of lime-water, he concluded 'that the Precipitation was chiefly from the fixed Alcali'.[13] How did Alston explain this weight increase, which in Black's view was such a clear indication of the transfer of fixed air? Since Alston never seriously thought of air as having chemical properties, he simply assumed that the dissolved alkali consists of two earths, one of which is captured by the lime during precipitation. 'Do not then these Experiments,' he asked, 'make it . . . more than probable, that Quick-lime rectifies or acts on alcaline fixed Salts, by separating from them some Part of the coarsest of their Earth, and substituting in place thereof its own most soluble and subtile Part, whereby their Strength, as a Menstruum, is considerable increased?'[14] Because Alston did not think of fixed air as a substance having chemical properties, Professor Guerlac's assertion that these quantitative experiments 'brought Alston to the brink of those discoveries so brilliantly carried through by his young medical student, Joseph Black'[15] seems dubious at best.

From Robison's description of the fourth and last set of experiments, we know that Black had, like Hales, collected the elastic air released from solid bodies and that he was aware of the acidic properties of fixed air. 'He had,' Robison reports, 'discovered that a cubic inch of marble consisted of about half its weight of pure lime, and as much air as would fill a vessel holding six wine gallons, and that it was rendered tasteless and mild by this addition, in the same manner as oil of vitriol is rendered tasteless and mild, in the form of alabaster by its combination with calcareous earth.'[16]

The experimental notes which Robison salvaged for posterity thus reveal that Black had developed his new theory of causticity before applying himself to the task of writing his thesis. He had demonstrated that the causticity of quicklime and the alkalis is an intrinsic property of these substances, that both in nature and in the laboratory this caustic quality is neutralized by fixed air, that this fixed air can be transferred from one substance to another without being released as effervescence and without losing its distinctive chemical qualities, and that fixed air acts like an acid when combined with other substances. He had not yet, so far as we can tell, determined that fixed air is also a chemically distinct substance in its elastic state. It seems much more likely that at this early stage in the development of his theory he thought, as did Cullen, that air obtains its characteristic chemical properties from the acid which combines with it during the process of fixation.[17] Nevertheless, Black knew enough about causticity to believe that he could write a worthy medical dissertation without really exerting himself. He thought that the lime-water debate being carried on between Alston and Whytt could be easily and significantly advanced by someone who, like himself, possessed the necessary chemical skill and knowledge. Thus it comes as no surprise to find that following the experiments on causticity, 'the experiments with magnesia are soon mentioned' in Black's notebook.[18]

Black did not attempt to hide his original reason for taking magnesia alba as the subject for his medical dissertation. 'The experiments of the celebrated Doctors Whytt and Alston, Professors in this University,' he wrote in his dissertation, 'have brought to light so much that is new and useful about quick-lime that this substance has now certainly become most worthy of the attention of all. I hoped, indeed, that I might find out a new quick-lime and lime-water, different from the common sort, and perhaps more powerful in dissolving the calculus. In this I was disappointed, but do not regret the labour.'[19] Black did not regret the labour because he was finally able to acknowledge the presence of certain troublesome properties in magnesia without having to abandon his theory of causticity; but when first encountered, the anomaly created by this earth seemed devastating. To understand why Black expected to obtain a quicklime and lime-water from magnesia alba, we must remember that originally his theory of causticity said nothing about the specific differences between the various calcareous substances. Since Hoffmann, Brownrigg and Cullen believed that causticity is an intense form of alkalinity, he assumed that any alkaline earth which could be calcined would yield a caustic quicklime and an alkaline lime-water. In other words, Black believed that causticity is a generic property of those earths that are alkaline in their reactions with acids and that contain air. Since magnesia alba is evidently one of these earths, he assumed it would yield a quicklime.

Black devoted little time or attention to his dissertation research during his first year in Edinburgh. From the hints available in his few letters to Cullen it appears that he continued to study the general properties of lime-water without being troubled by the peculiarities of magnesia. In February 1753 he wrote to Cullen saying, 'I am vastly pleased with the hopes of your experiments on the solution of calculi; I want to know something certain on the subject and would wish they were tried in a heat pretty near the animal.'[20] Perhaps Black hoped that by being more careful than Alston and Whytt had been about the temperature of his lime-water solutions, he could obtain more revealing experimental data than they had.[21] Further on in the same letter he reports that he has been using an experimental technique previously employed by Robert Whytt in his research on lime-water. 'I had not particular use for the Irish limestone,' Black told his former teacher, 'but that I thought it the purest of any I had seen, and intended to compare it with lime-shell; however, I am in no hurry about it. Mr. Russell has put me on an excellent way of comparing their specific gravities by glass-bubbles.'[22] But these dilatory investigations neither enabled Black to resolve the anomaly raised by magnesia nor suggested any significant contribution to the lime-water debate. Since Black had to complete his medical education during his second year in Edinburgh, he unhappily decided to write a dissertation marked more by tentativeness and evasion than by theoretical brilliance.

Although Black's dissertation did not particularly please him, it does reveal that by 1754 he had identified the central problem posed by magnesia alba. He had by then succeeded in driving the air out of magnesia alba, thereby obtaining the calx which he called magnesia usta, but he had also discovered that when heated in water, the calx exhibits none of the alkaline properties he had expected. In the crucial experiment, 'a drachm of Magnesia usta was gently heated with an ounce of water, in a closed flask for several hours, the water was then passed through bibulous paper, it was quite devoid of taste, and neither made solution of mercurial sublimate in any way turbid [Black's colour test for the presence of magnesia] nor did it change the colour of syrup of violets [the test for alkalinity]: the Magnesia itself, when well dried, weighed one drachm ten grams: mixed with acids it did not effervesce, nor did it produce any effect on lime-water.'[23] This experiment demonstrates with inescapable force the difference between the properties of calcined magnesia and those of calcined calcareous earth, a difference that Black found entirely anomalous. 'We see from the immediately preceding experiments,' he said, 'that quite a notable change is produced in Magnesia by burning: yet very different from that which calcareous powders and stones undergo, when so treated. The explanation of this for a long time seemed to me very obscure.'[24] Although he performed many experiments designed to illuminate this problem, Black was not able to dispel completely the obscurity created by this difference before his dissertation was called for.

Anyone familiar with Black's theory of causticity as presented in his essay in 1756 can find adumbrations and intimations of that theory in the dissertation of 1754. But the theory itself was not the focus of the chemical sections of the inaugural dissertation. The experiments presented in the dissertation were chosen and arranged, Black tells us, so as to emphasize the differences between certain earths which had previously all been considered calcareous. In other words, the chemical argument of the dissertation is taxonomic, not theoretical.[25] As an introduction to his chemical experiments, Black wrote that 'in those [experiments] that follow we see that four species of powders or earths are altogether different from one another, and yet chemists are wont to confuse them together under the common name of absorbents.'[26] And in the very first experiment he bluntly emphasized the distinction between magnesia and the caustifiable substances, the very distinction he had been so reluctant to accept: 'Magnesia,' he stated, 'most manifestly differs from the calcareous sort of powders or stones. All those substances may properly enough be called calcareous which, when strongly heated in the fire, are converted into quick-lime.'[27] In his dissertation Black admitted that not all alkaline earths that can be calcined are calcareous, but why this is so and what this fact meant for his theory of causticity remained a mystery until the publication of his essay two years later. Medically, however, the significance of this fact was inescapable; it meant that his theory of causticity was of no use to researchers looking for an improved treatment for the stone. Perhaps, Black may have pondered, the advancement of philosophical chemistry and the improvement of our practical knowledge are more loosely linked than Cullen had indicated.

2. THE INAUGURAL DISSERTATION

Black's dissertation is divided into two parts. In the first and shorter part he presents a medical theory which explains the origins and characteristics of the digestive hyperacidity commonly treated with magnesia alba. Had Black been able to make a lime-water with magnesia, he could have simply neglected its antacid properties. But since his experiments had not pointed the way to a new medical use of this earth, he could give his dissertation the required medical tone only by expanding upon its established uses. Thus, he wrote in his dissertation, 'when I considered what I had written on [magnesia alba], it did not seem to have such a relation to Medicine as the motive of the work required, and I accordingly decided to preface it with some notes, as short as possible, on the acid humour arising from food, for which alone magnesia serves as a remedy.'[28]

Historians have generally passed over this first section of the dissertation with an embarrassed haste its author would no doubt have recommended.[29] Shortly after completing the thesis, Black told Cullen that 'there are many things in the first part that do not at all please me, but,' he added, 'I was obliged to write it, and I shall be satisfied if it passes without much notice.'[30] Yet this much abused section on the acid humour arising from food contains several arguments which reveal just how thoroughly Black had assimilated Cullen's guiding ideas about air and chemical change. When Black states that 'the nature of almost all vegetable matters is such that, when macerated in water, or dissolved in it, and kept at the temperature of the human body, they spontaneously fall into certain internal motion, commonly called fermentation, at the same time emit a great quantity of air, and at last are converted, wholly or in part, into an acid humour,'[31] we recognize Cullen's voice in the emphasis upon particle motion and the suggestion of an air-acid theory of fixation. Black also demonstrates his indebtedness and his loyalty to Cullen's ideas when he distinguishes sharply between the chemical effects of elastic air and those of fixed air. When trapped in the digestive tract in the elastic state, he argues, air causes both mechanical and chemical disorders. 'The fermentation of the food,' Black asserts,

> is a source of many ills; for the air, of which there is a great quantity given off, in its wandering through the intestines, produces borborygmi and flatus, and sometimes inflations and enormous pains as the intestine, constricted on both sides, including the air, denies it a vent. And indeed it seems likely that the air will be in all respects like every other air arising from fermenting liquids, and be full of a vapour so noxious to life that it suddenly suffocates animals that have breathed it, and therefore the spasms of the fibres of the stomach and intestines with which patients are very often troubled may to some extent be produced by the action of this air on the nerves.[32]

But when fixed in food, air has entirely beneficial effects on the body. 'There is,' Black argues,

> indeed much of this air in food, as it is at first taken into the body, deprived, it is true, of elasticity, nor does it ever recover elasticity in a strong healthy body, but is carried together with the food into the blood, also along with it a quantity of air, by no means insignificant, entangled among the food in the mouth, and thereafter deprived of its elasticity by the digestive forces.

> There is certainly air in blood and in all liquid secretions. A great quantity can be extracted from the solid parts by means of heat, so that it must also be in the nutrient liquid: hence it appears to be necessary for the proper composition of all these things.[33]

And yet, though he knew that fixed air contributes to a healthy constitution, Black was unable to say exactly how it affects the chemistry of the body. 'It is not to be doubted,' he adds, 'indeed that this air, extensively united with every part of our body, serves many great uses, nor is it to be supposed that its absence could be borne without inconveniences: but we do not seem to know what its use is, or what are the inconveniences that would result from its absence. Perhaps some of the colour of the blood and the solidity of its globules may be due to it.'[34] Evidently in medical theory as well as in chemistry, Cullen and Black attached great importance to the investigation of the specific effects of fixed air.

In the second section of his dissertation Black presents a systematic survey of the specific properties of magnesia alba. It bears repeating that this section does not culminate in a presentation of the general theory of causticity and fixed air. Although Black shows that fixed air is lost when magnesia alba is calcined, his purpose throughout is to describe the specific properties of magnesia so that it can be distinguished from calcareous earth and the other absorbent earths. This distinction was rendered important by his need to establish the limits of his theory of causticity, and it was this larger significance which both required and enabled Black to perform the marvellously accurate and revealing experiments upon which so large a part of his fame rests. But in the dissertation itself, the theory and the full meaning of the experiments remains latent.

Black's examination of magnesia alba begins with a pharmaceutical discussion of its preparation and use as a purgative. Hoffmann, who was, according to Black, 'almost the only writer who deals with the subject' of magnesia,[35] obtained identical samples of the earth from two different solutions. Black examined magnesia obtained from four different salt solutions; 'from the saline liquor remaining after the crystallization of common salt [i.e., bittern], from the bitter cathartic salt [i.e., Epsom salt (magnesium sulphate)], from the more generally known Glauber's salt [i.e., impure sodium sulphate], and lastly from common salt [i.e., impure sodium chloride]. These all,' Black reports, 'on the addition of alkaline lye, give Magnesia, which is exactly the same from whichever of them it may come.'[36] Although Black was only able to examine one of the two solutions Hoffmann used when preparing magnesia, he found that the earth he obtained from it was identical to his other samples of magnesia. Yet Hoffmann reported that solutions of his magnesia alba 'throw down a white powder when mixed with oil of vitriol,' whereas Black found that when he added sulphuric acid to a solution of his magnesia alba, 'it always remained quite clear'.[37] Although Black clearly believed that Hoffmann had failed to remove all the contaminants from his magnesia, he did not dare criticize the master in his inaugural dissertation. He therefore rather lamely suggested that Hoffmann's magnesia differed from his own because it had been obtained from the bittern of salt springs rather than from the bittern of sea water.[38]

Black preferred to use Epsom salt when preparing large quantities of magnesia. To obtain a pure sample, he advised, one should

> take of the bitter cathartic salt and of potashes equal parts: let them be dissolved separately in a sufficient quantity of water, and filtered through bibulous paper; let them be mixed with very thorough agitation; then let them boil over the fire for a little: now let there be added twice the volume, or even more, of pure boiling water; when well mixed let them repose until the Magnesia has settled at the bottom; let the superfluous water be poured off and new hot added; let it again be poured off, and repeat this, with cold water, ten or twelve times, or even oftener, if you wish the very purest Magnesia, and therefore that most fit for chemical experiments; lastly dry it [$MgSO_4 + K_2CO_3 \rightarrow K_2SO_4 + MgCO_3\downarrow$].[39]

He then explained that 'in this process the alkali draws the acid to itself with such force that it compels it to abandon the Magnesia. This then, as it is by itself insoluble in water, is necessarily precipitated.'[40] At the same time the alkali and the acid form 'a new middle salt' which is only slightly soluble and therefore must be removed from the precipitate by extensive washing. 'Indeed,' Black concluded, 'in the preparation of Magnesia as it is usually sold, this point is not sufficiently attended to; for it often contains some of the salt, and sometimes even has an alkaline taste.'[41]

Pure magnesia alba is valuable as a medicine, Black argued, because 'it quickly draws to itself any acid, changes it, and therefore also very promptly absorbs the acid humour in the stomach. But its great merit above the other absorbent powders is, that after the change of the acid, it mildly loosens the bowels; whence, although it has no taste, yet it plays the part of an otherwise unpleasant purgative, and so carries off the impurities arising from the fermentation of food, which quickly turn new food into acid.'[42] Hoffmann attributed this purgative power of magnesia to the middle salt that it forms with acid in the stomach and Black confirmed this attribution with an experiment:

> I saturated Magnesia with distilled vinegar—a product namely of fermentation, in all respects very similar to the acid derived from food: I dried up the salt thus made, and gave six drachms of it dissolved in water, to be taken gradually, to an adult man, who, after drinking the third part of it, had four evacuations without the least inconvenience. I gave the rest to an active woman of rather full habit of body, and it purged her ten times, so that this salt, though mild to the taste, seems yet to surpass other purgatives.[43]

But these were not the properties of magnesia that most interested Black, and following a last long paragraph on the pharmaceutical properties of magnesia, complete with a deferential reference to Charles Alston, 'our most sagacious Professor of Botany,' Black took leave of the medical aspects of magnesia with evident relief. 'Having discussed as shortly as I could the history and the actions of Magnesia,' he concluded, 'I may now in a few words glance at its chemical properties.'[44]

Black prefaced his account of the chemical properties of magnesia with a frank statement of his expectations. He had originally hoped to make a new quicklime and lime-water by calcining magnesia and dissolving the calx in water, but he had not been able to do so. He therefore decided to present his experiments in a way which would enable other chemists to distinguish among the four absorbent earths.[45]

In the first four experiments magnesia was treated with four different acids, the resulting salts were described, and the differences between these salts and those formed by calcareous earth and the same acid were pointed out. Thus in describing experiment i Black wrote that 'Magnesia is dissolved quickly and completely by vitriolic acid, and in their union a great abundance of air is expelled from them: when the mixture has attained saturation, a salt most similar to the bitter cathartic salt is produced [$MgCO_3 + H_2SO_4 \rightarrow MgSO_4 + H_2O + CO_2 \uparrow$].'[46] He then added, 'here note that Magnesia most manifestly differs from the calcareous sort of powders or stones,' for if acid of vitriol is dropped on any of them, it unites with the calcareous earth and 'falls with it . . . to the bottom: but this does not happen to solution of Magnesia, for it, in union with the acid of vitriol, dissolves very readily in water.'[47] In experiments ii, iii, and iv Black treated magnesia with nitric acid, muriatic acid and vinegar, in each case comparing the crystals obtained with those formed by calcareous earth and each of these acids. Black then completed this set of observations by saying, in experiment v, that when magnesia is 'dissolved in all these ways, it is easily recovered by the addition of fixed or volatile alkali'.[48]

In experiment vi Black attempted to use his theory of causticity, which remained unstated, to resolve an anomalous property of volatile alkali. 'If Magnesia is put into a hot solution of sal ammoniac,' he reported, 'it drives the volatile alkali out of it, and this easily betrays itself by its peculiar odour [$MgCO_3 + 2NH_4Cl \rightarrow MgCl_2 + H_2O + CO_2\uparrow + 2NH_3\uparrow$]. If the mixture is gently warmed for some time, the Magnesia is completely dissolved, unless more is put in than suffices for the expulsion of all the alkali, and the saturation of the acid. It can be recovered again by the addition of any fixed, or even volatile alkali [$MgCl_2 + (NH_4)_2CO_3 \rightarrow MgCO_3\downarrow + 2NH_4Cl$].'[49] Black then identified the problem which troubled him: 'This experiment directly contradicts the preceding. For here [experiment vi] the Magnesia seems to have a more powerful attraction than the volatile alkali for acids, while there [experiment v] the volatile alkali has a more powerful attraction than the Magnesia.'[50] Why this reaction goes both ways, Black admitted, 'beats me, and I did not understand why the very alkali which Magnesia drove out from the acid does not, on the spot, again drive out the Magnesia.'[51] But, he added, 'I surmised, indeed, that the alkali underwent some new changes in distillation, which might be the cause of this difference; for it can never be prepared pure without distillation: and this idea did not seem to be very incongruous, as from the experiments of the celebrated Hales, it absorbs air during distillation, and is therefore really changed.'[52]

What kind of change takes place during distillation? Since according to Black's theory of causticity the fixed air, which all mild alkalis contain, is driven off during calcination, he probably believed that the volatile alkali in sal ammoniac loses its air during distillation and therefore differs significantly from the mild volatile alkali. Thus, he may have reasoned, volatile alkali, which takes fixed air away from magnesia, is the same caustic volatile alkali that absorbs air, according to Hales, when distilled from sal ammoniac. To test this idea, Black tried the experiment with a sal ammoniac that had not been distilled. He 'prepared, from volatile alkali and the acid of salt, a sal ammoniac similar to the common kind: to see if Magnesia would be unable to decompose this compound: but everything occurred with it as in Experiment VI.'[53] Finally, in experiment vii, he found that when one part magnesia and two parts of sal ammoniac are mixed together and distilled, they 'produce volatile alkaline spirit and salt, quite similar to those which are separated from sal ammoniac by means of fixed alkali.'[54]

Black then turned from the mysteries of volatile alkali, which he resolved in his later essay on quicklime, to the elective attraction of magnesia. He demonstrated, in experiment viii, that when magnesia alba is heated in a solution containing calcareous earth and an acid, the magnesia attracts the acid while the calcareous earth forms a precipitate (e.g., $CaCl_2 + MgCO_3 \rightarrow CaCO_3\downarrow + MgCl_2$).[55] In experiment ix he pointed out that the reaction between magnesia and mercurial corrosive sublimate yields a strongly coloured precipitate and can be used to detect the presence of magnesia in a solution.[56] Then, in experiment x, he mixed one part quicklime with three parts magnesia, added water, and slowly heated the solution. After cooling it he found that the water had 'no taste of lime, will not in any way trouble solution of mercurial sublimate, nor leave any residue when evaporated at a gentle heat [$CaO + MgCO_3 \rightarrow CaCO_3\downarrow + MgO\downarrow$].'[57] In experiment xi he repeated the previous experiment except that this time he made a lime-water out of the quicklime and water before mixing them with the magnesia. 'The lime-water,' he found, 'soon completely lost all taste, and was converted into pure water.'[58] He then poured off the original lime-water, 'added new, left it at the same temperature, and repeated this operation, until I had poured in seven pounds fourteen ounces of lime-water; then indeed lime-water newly poured on no longer lost its taste.'[59] When he examined the precipitate left by all these neutralizations he found that 'the well dried Magnesia then weighed four drachms one scruple fifteen grains'.[60]

Black's interpretation of the reaction between magnesia and lime-water contains a revealing error. The magnesia he used originally weighed three drachms one scruple and the precipitate obtained, when dried, weighed four drachms one scruple fifteen grains. Weight considerations alone should therefore have prevented Black from calling the precipitate magnesia, yet that is what he did. The precipitate was, of course, a mixture of two insoluble earths, lime-stone, which had regained its air from the magnesia, and magnesia oxide, which had lost its air to the calcareous earth. But Black had already determined that magnesia has a stronger attraction for acids than the calcareous earths have and, knowing that fixed air acts like an acid, he may have found it difficult to think of calcareous earth taking fixed air away from magnesia. Nevertheless, the fixed air had to come from the magnesia since Black had constructed the experiment in a way that prevented it from entering from the atmosphere. Whatever the reasoning behind Black's statement that the precipitate is magnesia, it is clear that at this stage of his research he had not sorted out all the implications of the differences between magnesia and the calcareous earths.[61]

This confusion of the two earths was then carried over into the next experiment. In describing experiment xii Black said that 'Magnesia prepared in the way given in exper. xi well mixed with an equal part of sal ammoniac and distilled, gave a spirit and salt similar to those prepared in exper. vii and not caustic.'[62] What had he expected? Did he think that by adding a magnesia that had been deprived of its air to a salt made with caustic volatile alkali he would obtain a caustic volatile alkali or a caustic magnesia salt? He was disappointed, of course, because he was using a combination of earths that contained a great deal of fixed air and because at that stage he knew too little about the properties of volatile alkali. Fortunately such disappointments did not undermine his entire dissertation, for he was ostensibly only describing unrelated experiments and cataloguing properties. But Black did not leave these problems unresolved, and when these reactions reappear in his later essay, they are no longer anomalies but brilliantly incorporated extensions of his theory of causticity.

In the next seven experiments Black examined the properties of calcined magnesia. His description of experiment xiii tells us how the calcination itself was carried out. 'An Ounce of Magnesia was heated, in a crucible covered with a lid, for about an hour, at a temperature sufficient for the fusion of copper: when the crucible had been cooled the Magnesia weighed three drachms one scruple. We may call this Magnesia usta.'[63] In experiment xiv he noticed that magnesia usta, when treated with the various acids, forms the same salts as magnesia alba, 'but what is chiefly to be noted,' he wrote, is that 'the solution takes place most quietly, and without the very least ebullition.'[64] Magnesia usta also resembles magnesia alba in forming a coloured precipitate with a solution of mercurial corrosive sublimate (experiment xv) and in expelling volatile alkali from a solution of sal ammoniac (experiment xvi). But it differs markedly from the parent earth in that 'it does not seem to trouble calcareous powder dissolved in acids' (experiment xvii), 'nor does it produce any change in lime-water' (experiment xviii). Finally, Black attempted to dissolve calcined magnesia in water (experiment xix). This was the crucial experiment in his attempt to apply his theory to the clinical use of lime-water and, as we know, it failed.[65] This failure not only forced him to abandon all hope of producing a new and better lime-water, it revealed that the properties of magnesia are not entirely analogous to those of calcareous earth. If this analogy does not hold, Black probably thought, just how reliable is the whole theory of fixed air and causticity?

Black's final five experiments on magnesia were designed to prove that fixed air is the primary substance lost during calcination. In experiment xx he collected the condensable vapours driven off from magnesia during calcination: 'I put three ounces of Magnesia into a glass retort, with a receiver attached, and placed on sand: I then applied fire, which was gradually increased, until the Magnesia was just obscurely red hot. When everything was again cold, the Magnesia weighed one ounce three drachms and a half: it still effervesced strongly when put into acids, although not so much as before. In the receiver I found a little whitish water, weighing five drachms.'[66] Although he had considerably reduced the original weight of the magnesia, Black did not consider the experiment definitive. 'In this case,' he said, 'the heat applied was much less than that required for its complete burning, and so it still effervesced pretty strongly with acids, and I could not avoid this, for I had not the proper apparatus for the purpose: I was therefore forced to leave the experiment imperfect, until an opportunity should occur of carrying it out more properly.'[67] Even though the calcination was incomplete, the results of this experiment enabled him to suggest with rhetorical caution what he had known all along. 'The weight of the liquid caught in the receiver was not the half of that lost; what then is it that has disappeared?' Black asked. 'Perhaps some water,' he suggested, 'but more seems to be air, a great deal of air must have remained with the Magnesia, for, after the loss of so much, it still gives off air on the addition of acids.'[68]

Since he could not collect the air driven out of the magnesia alba, Black used a series of displacement reactions to prove that the chemical difference between magnesia alba and magnesia usta is caused by the loss of fixed air. In experiment xxi he roasted a sample of magnesia and then treated it with sulphuric acid: 'Two drachms of Magnesia burnt in the way described in exper. xiii were reduced to two scruples twelve grains; when put into spirit of vitriol they quietly dissolved [$MgO + H_2SO_4 \rightarrow MgSO_4 + H_2O$].'[69] He then added a fixed alkali to the solution to recover magnesia alba: 'Magnesia was again liberated from this by the addition of solution of fixed alkaline salt; of which, to do this, a large quantity is required; it was then well washed and dried [$MgSO_4 + K_2CO_3 \rightarrow MgCO_3\downarrow + K_2SO_4$].'[70] The magnesia, he found, had regained all its original properties: 'The weight had now increased again to one drachm fifty grains, it acted on acids with great ebullition and energy; put into solution of mercurial sublimate it formed a red powder; it precipitated calcareous powders from acids and rendered lime-water insipid: in short it had now recovered all the qualities which it had lost in burning.'[71] Again he asked, 'what is it then that the Magnesia lost in this experiment and again recovered, which when it was present gave so many qualities, which its removal took away?' And again he answered, 'perhaps air: for the consideration of the twentieth experiment makes it very probable that the greatest part of the weight lost in burning consisted of air going away: and this indeed is confirmed by the fact that, after burning, no air is given off by the action of acids.'[72] He then added that in this experiment, 'is it not also likely that the air is again restored to the Magnesia from the alkali, at the time when it [the alkali] removed it [the magnesia] from the acid: for it is proved by the experiments of the ingenious Hales that fixed alkali certainly abounds in air, of which it emits a great deal when it unites with a pure acid.'[73] Of course the two cases are a little different. 'In this case, Black admitted, the alkali 'certainly unites with an acid, though not with a pure acid, and indeed the middle salt formed by their union is in quantity, properties, in every way the same, as if the pure acid had been added.'[74] But since 'only a very little air comes off' when fixed alkali is added to a solution of vitriolated magnesia, is it not most likely 'that the air is driven out of the alkali by the acid and joins itself to the Magnesia?'[75]

In the next three experiments Black used acid-alkali neutralizations to prove that magnesia usta is simply magnesia alba deprived of its fixed air. He began by demonstrating, in experiment xxii, that Homberg's technique of equivalent weights gives erroneous results because it does not take account of the large quantity of air driven off as effervescence when acids react with mild alkalis.[76] Then, in experiment xxiii, he neutralized a sample of magnesia with spirit of vitriol and found that 'there had been lost by effervescence, one scruple and sixteen grains.'[77] Finally, in experiment xxiv, he 'did all the same things over again,' except this time he used 'two drachms of Magnesia reduced by effective burning to two scruples and twelve grains.' He found that 'four drachms one scruple and two grains of the same acid completed the solution,' and that the magnesia usta 'lost only two grains during the solution, which may indeed consist of water.'[78] Thus, he concluded, 'this experiment seems to show why Magnesia, usta, when united to acids, produces salts similar to those which it produced before the burning: for Magnesia usta has lost its air before the encounter with the acid; but Magnesia non usta loses it in the encounter itself.'[79]

Black then turned his attention to the other absorbent earths. In experiment xxv he

> dissolved an ounce of chalk in spirit of salt; and again separated it, from the acid, by means of fixed alkali [$CaCO_3 + 2HCl \rightarrow CaCl_2 + H_2O + CO_2$; $CaCl_2 + K_2CO_3 \rightarrow CaCO_3\downarrow + 2KCl$],

> and washed away all salt by repeatedly pouring on and pouring off water. The chalk then, on chemical examination, did not seem to have been changed in any way, inasmuch as it effervesced strongly with acid of vitriol, but yet was not dissolved by it; burnt in a sufficient fire, it became true quick-lime, which gave lime-water, and drove out a spirit of very sharp odour from sal ammoniac.[80]

He performed this particular experiment, he tells us, because he 'wished to know, whether chalk could acquire the properties of Magnesia: or if Magnesia owed these properties to a previous union with acid, and then separation from it,' but he found that 'chalk is nothing changed by this process.'[81] Then, after describing the effects of adding a calcareous substance to a hot solution of alum (experiment xxvi), he examined the properties of calcined hartshorn. He treated it with spirit of vitriol (experiment xxvii), acids of salt and nitre (experiment xxviii), and vinegar (experiment xxix), and in each case described the substance formed by the combination. In experiment xxx he found that 'calcined hartshorn suffers no change when burnt in the most vehement fire; nor does it acquire any of the properties of quick-lime,'[82] and he asserted that it therefore, like magnesia, should not be classed among the calcareous earths. Alum was the last of the absorbent earths Black studied. In his account of experiment xxxi he recorded the results of treating alum with the various acids and by performing experiment xxxii, the last described in the dissertation, he established that alum is yet another one of the absorbent earths which does not yield a quicklime when calcined.[83]

The chemical experiments described in Black's dissertation clearly fulfilled the modest task assigned to them. One could hardly ask for more convincing proofs that magnesia alba contains a great quantity of fixed air, that the release of this air is the chief result of calcining magnesia, and that magnesia deprived of its air yields neither a quicklime nor a lime-water. One can, therefore, properly term Black's dissertation an essay in the chemistry of fixed air. But can one with equal justice call it the first essay in pneumatic chemistry? Clearly not. Black focused his attention upon the specific, that is the chemical, effects of fixed air. He was not interested in the properties of this air in the elastic state nor did he suspect that there are a variety of chemically distinct airs in the atmosphere. Given his particular theoretical focus, he attached no significance to a chance observation he made before completing his dissertation. In January 1754 Black wrote to Cullen saying:

> I fully intended to have wrote last post, but really I happened to be intent upon something else at the proper time, and forgot it.

> It was, indeed, an experiment I was trying that amused me, in which I had mixed together some chalk and vitriolic acid at the bottom of a large cylindrical glass; the strong effervescence produced an air or vapour, which, flowing out at the top of the glass, extinguished a candle that stood close to it; and a piece of burning paper, immersed in it, was put out as effectually as if it had been dipped in water: yet the smell of it was not disagreeable.[84]

Had Black been interested in the chemical properties of elastic vapours, he would have been more than amused by this discovery. Yet he let it pass without even mentioning it again in the letter he wrote to Cullen twelve days later.[85] For Black the main problem remained the theory of causticity, and the anomalies he had encountered by 1754 had not fundamentally challenged the conception of fixed air that he had acquired during his years of apprenticeship with Cullen.

3. THE ESSAY ON QUICKLIME

Black's account of his 'Experiments upon Magnesia alba, Quicklime, and other Alcaline Substances' was read before the Philosophical Society of Edinburgh on 5 June 1755. During the year that elapsed between the printing of his Latin dissertation and the presentation of his English essay, Black managed to resolve most of the outstanding anomalies in his theory of causticity. Thus, while Part I of his essay is devoted to the experiments on magnesia previously presented in the dissertation, there is a significant improvement in the level of understanding. But it is Part II of the essay, the theoretical section, which makes that memoir more than a revision and translation of the dissertation. There, in a brilliant presentation of his theory of causticity, Black provided a model of scientific investigation, one which has been quite properly compared to Newton's optical treatises.[86] But in addition to the timeless logic of his argument, Black's essay contains statements that tell us a great deal about his own theoretical views and about his vision of the future of philosophical chemistry. And these statements suggest that Black saw his discovery of the chemical role of fixed air in a way that had little in common with the pneumatic chemistry of his successors.

Black was much less tentative about distinguishing between magnesia and calcareous earth in the essay than he had been in his dissertation. Magnesia's failure to yield a quicklime and lime-water was no longer treated as a reluctantly acknowledged cause for disappointment, it was accepted as a fact of nature. He reported that calcareous earth and vitriolic acid combine to form an insoluble salt which, according to a recent memoir by Marggras, 'possesses the properties of the gypseous or selenitic substances'.[87] It seems likely, Black added, that Hoffmann had not removed all the calcareous earth from the solutions he used when preparing magnesia alba and that 'the coagulum thus formed in the mother of nitre, may be owing to a quantity of quick-lime contained in it'.[88]

In the essay Black also distinguished between the two substances which precipitate when magnesia alba and lime-water combine ($MgCO_3 + Ca(OH)_2 \rightarrow CaCO_3\downarrow + MgO\downarrow + H_2O$). 'The powder,' that is, the precipitate, he reported, 'after being perfectly well dried, . . . did not dissolve entirely in spirit of vitriol; but, after a brisk effervescence, part of it subsided, in the same manner as the calcareous earths, when mixed with this acid.'[89] That part of the powder which does not effervesce or form an insoluble salt with vitriolic acid is, Black realized, magnesia which has been deprived of its air. Since he had learned that magnesia without fixed air is insoluble, Black was able to explain this reaction in a straightforward manner. 'We have already shown by experiment,' he said, 'that *magnesia alba* is a compound of a peculiar earth and fixed air. When this substance is mixed with lime-water, the lime shows a stronger attraction for fixed air, than that of the earth of *magnesia*; the air leaves this powder, to join itself to the lime. And as neither the lime when saturated with air, nor the *magnesia* when deprived of it, are soluble in water, the lime-water becomes perfectly pure and insipid, the lime which it contained being mixed with the *magnesia*.'[90]

Black appended a curious paragraph to his account of the reaction between magnesia and lime-water. 'When I first tried this experiment,' he wrote, 'I was at the trouble of digesting the mixture in the heat of boiling water, and did not then know that it would succeed in the heat of the air. But Dr Alston, who has obliged the world with many curious and useful discoveries on the subject of quick-lime, having had occasion to repeat it, I learned from him, that heat is not necessary: And he has moreover added an useful purpose, to which this property of *magnesia* may be applied; I mean the sweetening of water at sea, with which lime may have been mixed, to prevent its putrefaction.'[91] Perhaps Black was being perfectly honest about his indebtedness to Alston. Perhaps he was toying with Alston for having accepted his initial misidentification of the precipitate formed in this reaction.[92] But it seems most likely that Black, like Cullen,[93] had come to realize that Alston had held an essentially correct view of quicklime all along. Unlike Whytt, Cullen and Black, Alston had from the outset believed that quicklime is a distinct substance with a fixed, and quite small, solubility in water. His theory of caustification was erroneous and his general approach to chemistry was old-fashioned and narrow, but he had been right about the identity of the calcareous substances all along and Black probably wished to acknowledge the acuity of that original insight.

The reactions between magnesia and volatile alkali salts were also handled much more successfully in the essay than in the dissertation, primarily because in the essay Black could be more forthright in his use of arguments based upon elective attractions and heat.[94] He performed an elegant experiment to clarify the reactions which occur between these substances: 'Three drams of *magnesia* in fine powder, an ounce of salt-ammoniac, and six ounces of water, were mixed together, and digested six days in a retort joined to a receiver.'[95] The apparatus was arranged so that initially a dry volatile alkali salt was collected in the receiver, but after further heating this salt was dissolved in the water, which also condensed in the receiver ($MgCO_3 + 2NH_4Cl + H_2O \rightarrow MgCl_2 + 2NH_3\uparrow + CO_2\uparrow + 2H_2O$; $2NH_3 + CO_2 + H_2O \rightarrow (NH_4)_2CO_3$ (the dry salt in the receiver)). 'During the whole time,' Black pointed out, 'the neck of the retort was pointed a little upwards, and the most watery part of the vapour, which was condensed there, fell back into its body. In the beginning of the experiment, a volatile salt was therefore collected, in a dry form, in the receiver, and afterwards dissolved into spirit.'[96] After the retort had been allowed to cool, he found that it contained a saline solution, some unchanged magnesia alba, and some crystallized sal ammoniac. 'The saline liquor was separated from the other two, and then mixed with the alkaline spirit. A coagulum was immediately formed, and a *magnesia* precipitated from the mixture [$MgCl_2 + (NH_4)_2CO_3 \rightarrow MgCO_3\downarrow + 2NH_4Cl$].'[97]

The explanation offered for these reactions, which have little to do with causticity, is simple and direct. 'We learn,' Black said, 'by the latter part of this experiment [which was performed without heating], that the attraction of the volatile alkali for acids [i.e., the muriatic acid salt (chlorine)] is stronger than that of *magnesia*, since it [the volatile alkali salt] separated this powder [magnesia] from the acid to which it was joined. But it also appears, that a gentle heat is capable of overcoming this superiority of attraction, and of gradually elevating the alkali [i.e., ammonia], while it leaves the less volatile acid with the magnesia.'[98] This explanation is as elegant as the experiments it explains and it can be taken as a paradigm of Black's approach to the formation of chemical theories. The elements of the theory are heat, elective attraction and change of state and we will see these elements reappear in Black's chemistry much as we found them presented time and again in Cullen's lectures. These concepts, founded upon the notions of attraction, repulsion and the state of aggregation, rather than upon the qualitative properties of the earths and the airs, provided the foundation for his understanding of chemical processes.

Although he was no longer surprised that a caustic volatile alkali cannot be obtained from the reactions between magnesia alba and sal ammoniac, Black did not neglect this rather peculiar substance in his essay. He found that in addition to being caustic, 'the volatile alkali, when deprived of its air, . . . discovers . . . its natural degree of volatility, which was formerly somewhat repressed, by the air adhering to it, in the same manner as it is repressed by the addition of an acid.'[99] If one simply takes this specific property into account, then the caustic qualities of volatile alkali can be easily explained in terms of his general theory of causticity. One can prepare a solution of caustic volatile alkali by distilling an ammoniacal salt, as Marggras has done,[100] or one can use another alkaline substance to remove the fixed air from volatile alkali. 'A caustic volatile alkali may also be obtained,' according to Black, 'by mixing salt-ammoniac with half its weight of a caustic fixed alkali, or of *magnesia* which had been previously deprived of its air by fire; and then submitting these mixtures to distillation.'[101] Once again by focusing his attention upon the chemical transfer of fixed air, Black succeeded in incorporating a complex and anomalous set of chemical phenomena into his theory of causticity.

'In reflecting afterwards upon these experiments [on magnesia],' Black wrote at the beginning of the second part of his essay, 'an explication of the nature of lime offered itself, which seemed to account, in an easy manner, for most of the properties of that substance.'[102] The phrasing is graceful and the transition effective, but surely Black has purposefully rearranged the order of the ideas which led him to this first full statement of his theory of causticity. Of course we may, with Lord Brougham, call Black 'the historian of his own discoveries,'[103] if we choose to adopt the eighteenth-century notion of natural history. Black presented his theory of causticity as a natural consequence of his investigation of magnesia because that was the sequence which conformed to the prevailing notion of how science progresses. From unprejudiced empirical investigation one advances by cautious experimentation towards those general statements which codify the laws of nature. It matters little to the pedagogue how the scientist actually proceeded in making his discovery. Since Black was writing an essay in chemistry, not in intellectual history, we should expect to find that he presented his subject in whatever way made his conclusions appear most natural. This, I believe, is what he did in both his essay and his lectures.[104]

Black began his exposition of his theory by describing the four distinct experimental phenomena to be explained. These four phenomena, although presented in a different order, are essentially the same as the four facts established by the experiments on quicklime he performed before leaving Glasgow. The first phenomenon to be noticed is that all the earths he will be examining contain a great deal of fixed air in their mild states. 'It is sufficiently clear,' he argued, 'that the calcareous earths in their native state, and that the alkalis and *magnesia* in their ordinary condition, contain a large quantity of fixed air; and this air certainly adheres to them with considerable force, since a strong fire is necessary to separate it from *magnesia*; and the strongest is not sufficient to expel it entirely from fixed alkalis, or take away their power of effervescing with acid spirits.'[105] His second experimental fact was that fixed air acts like an acid in combination. The large amount of fixed air in mild absorbent earths and alkalis 'led me to conclude,' he said,

> that the relation between fixed air and alkaline substances, was somewhat similar to the relation between these and acids: that as the calcareous earths and alkalis attract acids strongly, and can be saturated with them, so they also attract fixed air, and are, in their ordinary state, saturated with it: and, when we mix an acid with an alkali, or with an absorbent earth, that the air is then set at liberty, and breaks out with violence; because the alkaline body attracts it more weakly than it does the acid, and because the acid and air cannot both be joined to the same body at the same time.[106]

The third fact to impress itself upon his mind was that causticity is an intrinsic property of caustifiable bodies and is not caused by an acrid substance obtained from the calcining fire. 'I also imagined,' he wrote,

> that when the calcareous earths are exposed to the action of a violent fire, and are thereby converted into quick-lime, they suffer no other change in their composition, than the loss of a small quantity of water, and of their fixed air. The remarkable acrimony which we perceive in them after this process, was not supposed to proceed from any additional matter received [from] the fire, but seemed to be an essential property of the pure earth, depending upon an attraction for those several substances which it then became capable of corroding or dissolving; which attraction had been insensible as long as the air adhered to the earth, but discovered itself upon the separation.[107]

And finally he asserted that fixed air will combine chemically with lime and saturate its causticity. His experiments, he said, led him to see 'crude lime . . . as a peculiar acrid earth, rendered mild by its union with fixed air; and quick-lime as the same earth, in which, by having separated the air, we discover that acrimony or attraction for water, for animal, vegetable, and for inflammable substances.'[108]

To explain these phenomena, Black advanced a theory which connects the chemical properties of the substances he was examining to their elective attractions for air and water. 'According to our theory,' he said,

> the relation of the calcareous earth to air and water, appeared to agree with the relation of the same earth to the vitriolic and vegetable acids. As chalk, for instance, has a stronger attraction for the vitriolic, than for the vegetable acid; and is dissolved with more difficulty when combined with the first, than when joined to the second; so it also attracts air more strongly than water, and is dissolved with more difficulty when saturated with air, than when compounded with water only [i.e., $CaCO_3$ is less soluble than $Ca(OH)_2$].'[109]

He then described the various phenomena that can be explained in terms of lime's greater attraction for fixed air than for water. If lime-water, for instance, 'be exposed to the open air, the particles of quick-lime which are nearest the surface, gradually attract the particles of fixed air which float in the atmosphere.'[110]

Knowing that Black understood the relationship between quick-lime, air and water in terms of elective attractions enables us to see more clearly why he was so reluctant to grant that magnesia will not form a quicklime or lime-water. The relationship between calcareous earth and magnesia was not simply one of analogy, it was a matter of essential identity. If causticity is simply a strong attraction for fixed air, then magnesia deprived of its air ought to be caustic. And if the attractions for water and fixed air differ in degree only, then magnesia deprived of its air ought to be soluble. At first Black thought he had discovered a law which applied to all the earths commonly called calcareous, but by the time he wrote his dissertation he had been forced to accept one of those difficult truths, which must be faced at some point by every ambitious young philosopher, namely that his new theory was not as comprehensive as he thought. Nevertheless, he accepted the restriction and, as we have already seen, in his essay he acknowledged the fact that magnesia deprived of its air is not soluble in water.

After describing how quicklime makes alkalis caustic by removing their fixed air, Black turned his attention to five propositions, which he formulated to test his theory. He said that his theory of lime and alkali 'recommended itself by its simplicity, and by affording an easy solution of many *phaenomena*; but appeared, upon a nearer view, to be attended with consequences that were so very new and extraordinary, as to render suspicious the principles from which they were drawn. I resolved however to examine, in a particular manner, such of these consequences as were the most unavoidable, and found the greatest number of them might be reduced to the following proportions.'[111]

The first proposition asserts that 'if we only separate a quantity of air from lime and alkalis, when we render them caustic, they will be found to lose part of their weight in the operation, but will saturate the same quantity of acid as before, and the saturation will be performed without effervescence.'[112] This proposition was one of those which Black 'found, upon reflection, were already seconded by experience'.[113] Hoffmann, for example, 'has observed, that quick-lime does not effervesce with spirit of vitriol; and it is well known, that the caustic spirit of urine, or of salt ammoniac, does not emit air, when mixed with acids.'[114] Black then saturated two drams of chalk with spirit of salt and compared the weight of acid used to that required to saturate a sample of quicklime made by calcining two drams of chalk. Since 'it . . . appears from these experiments, that no air is separated from quick-lime by an acid, and that chalk saturates nearly the same quantity of acid after it is converted into quick-lime, as before,' he concluded that the truth of his first proposition had been adequately proven.[115]

The second proposition asserts that quicklime regains fixed air from mild alkalis to become slaked lime once again. This, of course, was another of the predictions of his theory which had already been 'seconded by experience'. Black, however, performed the relevant experiments again, demonstrating that all three known alkalis will cause lime-water to deposit a precipitate, which upon examination turns out to be a mild calcareous earth.[116]

The third proposition asserts that quicklime is entirely soluble. 'If it be supposed,' Black argued, 'that slaked lime does not contain any parts which are more fiery, active, or subtile than others, and by which chiefly it communicates its virtues to water, but that it is an uniform compound of lime and water; it follows, that as part of it can be dissolved in water, the whole of it is also capable of being dissolved.'[117] He thought this prediction 'had less appearance of probability' than the preceding two, but he decided to test it experimentally.[118] 'I reduced eight grains of perfect quick-lime made of chalk, to an exceedingly subtile powder, by slaking it in two drams of distilled water boiling hot,' Black reported, 'and immediately threw the mixture into eighteen ounces of distilled water in a flask. After shaking it, a light sediment, which floated through the liquor, was allowed to subside; and this, when collected with the greatest care, and dried, weighed, as near as I could guess, one third of a grain.'[119] He repeated the experiment several times and then examined the insoluble sediment. 'It consisted,' he found, 'partly of an earth which effervesced violently with *aqua-fortis*, and partly of an ochry powder, which would not dissolve in that acid.'[120] The ochry powder he took to be 'an accidental or foreign admixture' in the chalk, but that part of the sediment which effervesced with acid could not be so easily ignored. 'With respect to the minute portion of alkaline earth, which composed the remainder of the sediment,' he concluded, 'it cannot be supposed to have been originally different from the rest, and incapable, from its nature, of being converted into quick-lime, or of being dissolved in water; it seems rather to have consisted of a small part of the chalk in its mild state, or saturated with air, which had either remained for want of a sufficient fire to drive it out entirely, or had been furnished by the distilled water.'[121]

Black then said that far from being surprised that some of the quicklime was neutralized by fixed air in the water, he had 'indeed expected to see a much larger quantity of sediment produced from the lime, on account of the air which water constantly contains'.[122] He had, of course, long known that there is fixed air in water. He had heard Cullen state in his lectures that 'we find there is air in water, but in an incompressible state; we find the same from many experiments of Boyle and Hales,'[123] and in his own essay Black argued that 'when slaked lime is mixed with water, the fixed air in the water is attracted by the lime, and saturates a small portion of it, which then becomes again incapable of dissolution.'[124] But it is curious that at the time when he was thinking about the third proposition, Black believed that water contained a great deal of fixed air. Why did he think water contained so large a quantity of fixed air that he should have obtained much more precipitate when he dissolved slaked lime in water than he in fact did? To answer this question we must examine certain statements contained in Cullen's paper on the cold produced by evaporating fluids, a draft of which Black examined with great interest during the first half of 1755.[125]

In his essay on cold, Cullen reported that 'when a vessel containing spirit of wine, with a thermometer immersed in it, is set under the receiver of an air-pump; upon exhausting the air, the mercury in the thermometer sinks several degrees. It becomes more especially remarkable when the air in any plenty issues out of the spirit of wine. As the spirit continues long to give out air, so it is long before the thermometer immersed in it returns to the temperature of the external air.'[126] To understand this account, and Cullen's underlying conception of air, one must recognize that Cullen was using the term 'air' to refer to both the vapour given off by a fluid boiling under reduced pressure and to the atmosphere itself. In other words, Cullen defined air as a state of matter rather than as a distinct substance or mixture of substances. According to this Newtonian view of air, the general characteristic of all airs is elasticity, which itself is merely a manifestation of the repulsive force that keeps the aerial particles apart. If air has a general chemical property, it must be derived from the acidic substance, which combines with aerial particles to fix them in solid and liquid bodies. Therefore, although Cullen encouraged his students to investigate the chemical effects of the fixation of air, his own definition of air did not attribute chemical properties to air itself. Since both the atmosphere and the vapour given off by a liquid boiling in a vacuum are elastic, Cullen saw no reason not to call them both air.

Black, however, was beginning to suspect that air itself has some distinct chemical properties. Although elastic air and all vapours share the common mechanical property of elasticity, he knew that one can distinguish between the easily condensed vapours, which have the chemical properties of the liquids from which they are produced, and the more permanent air that combines with caustic substances and neutralizes them.[127] Black was obviously thinking about this distinction and about Cullen's essay on the cold produced by evaporation when he made the following entry in his notebook: 'Nitrous aether boils *in vacuo*, and grows vastly cold. Is not the cold produced merely by the evaporation of the elastic parts; or does it emit air, which requires heat for its elastic appearance?'[128] Black later made this distinction explicit. In his chemistry lectures he told his students that Cullen imputed the violent agitation that occurs when aether is placed in an evacuated chamber 'to the extrication of a great quantity of air. It is evident, however,' he added, 'that the aether was not emitting air principally, but a great quantity of aethereal vapour, and that it was truly in a boiling state: For the elastic fluid, extricated by working the pump, had a very strong odour of aether, and the quantity of the aether in the phial, was considerably and suddenly diminished, while this boiling continued.'[129]

After having clarified the distinction between the release of air fixed in a liquid and the vaporization caused by boiling, Black began thinking about the chemical effects of the air dissolved in common water. He thought of an experiment which would answer his questions, and in August 1755 he wrote to Cullen, asking him to have the experiment performed with the new air pump in Glasgow. 'I have endeavoured to get an experiment made here with the air pump to no purpose,' Black complained.

> Dr Stewart's is not in order, and his operator is so slow and surly that I am quite tired of dunning him. Mr. Wilson told me he had sent one to Glasgow, which he was to set up immediately. If it be in order, I beg you will desire Mr torbet to try the following experiment, and to let me know the result of it as soon as he can conveniently do it.
>
> Make four or five ounces of ordinary or shell lime-water as strong as possible, and put it into a phial, which, when corked, shall be quite full, so as not to contain the smallest quantity of air. When the air-pump is ready, draw the cork of the phial, and place it immediately under the receiver, together with another phial, containing nearly the same quantity of common water, and see if the limewater upon exhausting yields any air, or if it seem to yield as much as common water.
>
> N.B. Neither of the waters must be heated. And the lime-water may be made of fresh and strong quicklime slaked with boiling water, which reduces it to a subtile powder much more perfectly and suddenly than cold water; and it may be made in a close vessel, to prevent its turning weak at the surface by throwing up a crust.[130]

By stipulating that neither sample of water was to be heated, Black was consciously insuring that the waters would not begin boiling when the pressure in the chamber was reduced. He was, in other words, using the vacuum to extract only the air dissolved in the water samples. The purpose, outcome and meaning of this experiment are made clear in the essay. 'With a view to know whether water retains its air when fully saturated with lime,' Black wrote, he had a lime-water 'made as strong as possible; four ounces of which were placed under the receiver of an air-pump, together with four ounces of common water in a phial of the same size; and, upon exhausting the receiver, without heating the phials, the air arose from each, in nearly the same quantity: From whence it is evident, that the air which quick-lime attracts, is of a different kind from that which is mixed with water.'[131]

The conclusion Black drew from this experiment appears to be so unexceptional that its true novelty and significance can be easily overlooked. He had discovered that lime-water contains nearly as much fixed air as common water but that this fixed air slakes very little of the quicklime in lime-water. Having already distinguished between air and vapour, he was now obliged to make a chemical distinction between two kinds of fixed airs. Black's interpretation of this experiment thus marks the beginning of the chemical study of pneumatic substances. Despite the confusions which continued to dog his own usage of the term 'air', he provided, in his account of this experiment, the paradigm for pneumatic chemistry.

It was then a small step from the discovery that there are two kinds of air fixed in water to the discovery that there are two kinds of elastic airs in the atmosphere, and the evidence needed to verify this inference lay ready at hand. 'That the air which quick-lime attracts . . . is also different from common elastic air,' Black argued, 'is sufficiently proved by daily experience; for lime-water, which soon attracts air, and forms a crust when exposed in open and shallow vessels, may be preserved, for any time, in bottles which are but slightly corked, or closed in such a manner as would allow free access to elastic air, were a vacuum formed in the bottle.'[132]

Black had made a discovery of great historical importance but one which had little bearing upon the truth or falsity of his theory of causticity, the sole concern of his essay on quicklime. He therefore neglected to describe the properties of this neutralizing air which he had observed in a previous experiment[133] and chose instead to bring his discussion of air to a rapid close. 'Quick-lime,' he said,

> does not attract air when in its most ordinary form, but is capable of being joined to one particular species only, which is dispersed through the atmosphere, either in the shape of an exceedingly subtle powder, or more probably in that of an elastic fluid. To this I have given the name of *fixed air*, and perhaps very improperly; but I thought it better to use a word already familiar in philosophy, than to invent a new name, before we be more fully acquainted with the nature and properties of this substance, which will probably be the subject of my further inquiry.[134]

Although compressed, Black's brief statements on the nature of fixed air initiated a new departure in the study of airs. He had determined that there are two kinds of chemically distinct airs. But since both kinds of air can be fixed, their chemical properties when fixed could no longer be attributed to the common agent, presumably an acid, which fixes them in liquid and solid substances. The specific properties of the airs therefore must be seen as properties of the aerial particles themselves, just as the different properties of the acids and alkalis were thought of as the specific qualities of their particles. His discovery of the complexity of the atmosphere demonstrated the inadequacy of the mechanistic accounts of air, it created the taxonomic subject of pneumatic chemistry and it left the received chemical theory of fixed air in a shambles. Had Black been given to speculation, he probably would have predicted that a new chemical theory of air would emerge from investigations of the specific heat properties of bodies undergoing changes of state. He seems to have regarded the other chemical properties of air as phenomena which would eventually be explained in terms of molecular textures and chemical attractions and repulsions.

In the fourth proposition Black asserted that since quicklime makes alkalis caustic by removing fixed air, pure caustic lye will contain no lime.[135] To demonstrate the truth of this assertion he prepared a clear solution of caustic potash ($Ca(OH)_2 + K_2CO_3 \rightarrow CaCO_3\downarrow + 2KOH$).[136] He then performed three tests designed to show that the alkali had lost fixed air in becoming caustic. When he saturated a sample of the lye with acid he found not 'the least effervescence or diminution of weight,'[137] such as there would have been with a mild alkali. When he dried a sample of the lye he found that, despite a considerable fixation of air, the salt in the lye 'had lost above a sixth [of its weight] in becoming caustic, and the quantity of acid saturated by two drams of it, was to the quantity of acid saturated by two drams of salt of tartar, nearly as six to five.'[138]

Finally, he observed that while the lye had little effect on lime-water, it regained its mild and effervescent properties when exposed to the atmosphere for a fortnight.[139]

Black then performed a series of experiments designed to prove that the caustic lye contains no quicklime. First he evaporated a sample of the lye, melted the salt obtained, and redissolved the salt in a small quantity of water, thereby proving that the lye contained no lime, which is only slightly soluble.[140] He then saturated a sample of lye with vitriolic acid, but he did not obtain a calcareous salt as a precipitate.[141] And as a final test he exposed a sample of the lye to the atmosphere, but it did not, he reported, 'deposit a single atom of lime'.[142]

After presenting these experiments Black added a paragraph which reveals once again that he thought of solution as the result of an elective attraction, an attraction as subject to saturation and displacement as the attraction between an acid and an alkali.[143] Since more quicklime was commonly used during the production of caustic lye than was needed to remove the fixed air from the alkali, 'we are surprised to find,' he said, that in these lyes 'little or none of the superflous quick-lime was dissolved by the water. But,' he quickly added, 'this *phaenomenon* will become less surprising, by comparing it with some similar instances in chemistry. Water may be made to deposite a salt, by the admixture of a substance which it attracts more strongly than it does that salt.'[144] Spirit of wine, for instance, will displace quicklime from lime-water but the precipitate can be immediately redissolved in fresh water. 'We may therefore,' he concluded, 'refer the above *phaenomenon* with respect to the ley, to the same cause with these; and say, that the water did not dissolve the lime, because it already contained a caustic alkali, for which it has a superior attraction.'[145] Clearly Black believed that the attraction between soluble substances and water was comparable to the attraction between caustics and fixed air or to that between alkalis and acids.

In his fifth and last proposition Black predicted that caustic lye would remove acid from magnesia and calcareous salts. He began by saying that 'we have shewn, in the former experiments, that absorbent earths lose their air when they are joined to an acid: but recover it, if separated again from that acid, by means of an ordinary alkali; the air passing from the alkali to the earth, at the same time that the acid passes from the earth to the alkali.'[146] He then argued that 'if the caustic alkali, therefore, be destitute of air, it will separate *magnesia* from an acid, under the form of a *magnesia* free of air, or which will not effervesce with acids; and the same caustic alkali will also separate a calcareous earth from acids, under the form of a calcareous earth destitute of air, but saturated with water, or under the form of slaked lime.'[147]

To prove the truth of this proposition Black looked first at the reaction between Epsom salt and lye. He reported that when 'two drams of Epsom salt were dissolved in a small quantity of water, and thrown into two ounces of the caustic ley: the mixture instantly became thick, like a decoction of starch or barley, by the *magnesia*, which was precipitated [$MgSO_4 + 2KOH \rightarrow MgO\downarrow + K_2SO_4 + H_2O$].'[148] He then 'added spirit of vitriol by degrees, until the mixture became perfectly clear, or the whole of the *magnesia* was again dissolved; which happened without any effervescence or emission of air.'[149] To show that similar reactions occur with calcareous salts he dissolved half an ounce of chalk 'in spirit of salt, the quantity of which was so adjusted, that the mixture was not acid in the least degree, and the solution was thrown into twelve ounces of caustic ley; which quantity I found, by experiment, to be sufficient for precipitating almost the whole of the chalk [$CaCO_3 + 2HCl \rightarrow CaCl_2 + CO_2\uparrow + H_2O$; $CaCl_2 + 2KOH \rightarrow Ca(OH)_2\downarrow + 2KCl$].'[150] He then filtered out the precipitate, dried it, 'then mixed it with about twenty ounces of pure water in a flask, and, after allowing the powder to subside, poured off the water, which had all the qualities of lime-water.'[151]

Would a solution of caustic alkali also remove the acid from alkali salts? Yes, Black believed, if the salts contain the volatile alkali,[152] but not if the acid is combined with a fixed alkali. It is probable, Black argued, 'that, had we . . . a method of separating the fixed alkali from an acid, without at the same time saturating it with air, we should then obtain it in a caustic form; but I am not acquainted with an instance of this separation in chemistry.'[153] There are, he noted, two reactions which seem to change alkali salts into caustic alkalis, 'but, upon examining the product of each process,' he reported, 'we find the alkali either fully or nearly saturated with air'.[154]

Having established the truth of all five propositions deduced from his theory of causticity, Black devoted the next few pages of his essay to a discussion of three curious properties of alkalis. The first of these posed a rather serious challenge to his general understanding of the relationship between heat, vaporization and elective attraction. 'When we consider that the attraction of alkalis for fixed air, is weaker than that of the calcareous earths, and reflect upon the effects of heat in chemistry,' he reasoned, 'we are led to imagine, that alkalis might be entirely deprived of their air, or rendered perfectly caustic, by a fire somewhat weaker than that which is sufficient to produce the same change upon lime; but,' as every chemist knows, 'this opinion does not seem agreeable to experience.'[155] While it is true that the alkalis 'acquire some degree of causticity in a strong fire,' they do not yield caustics which are as pure as quicklime. 'The circumstances which hinder us from rendering these salts perfectly caustic by heat, are,' Black suggested, 'their propensity to dissipation in the utmost violence of the fire, their extreme acrimony, and the imperfection of our common vessels. For, before the heat becomes very intense, the alkalis either evaporate, or dissolve a part of the crucibles in which they are contained, and often escape through their pores.'[156] Black then described his own attempt to produce a caustic alkali by calcination, and although he did not obtain a pure caustic, he was none the less 'assured by this experiment, that an alkali does really lose a part of its air, and acquire a degree of causticity, by the proper application of heat.'[157]

Since causticity is a common property of calcareous earths and alkali salts, Black was not willing to distinguish sharply between the earths and salts, as Cullen had done.[158] But certainly his predecessors had been correct when they noted that the alkalis are destroyed by the fire whereas the earths are not. Black, however, thought that specific properties merely modify the operations of general laws.

That the alkalis could not be calcined to causticity was not a sign of their essential difference from the calcareous earths, it was merely a result of their peculiar volatility and his own experimental limitations. Although he recognized the difficulties raised by the specific properties of the alkalis, his own thinking continued to be guided by an underlying conception of the general laws which link heat, elective attraction and change of state. In his determination to focus upon general causes, Black was, of course, following Cullen's advice if not his example. It was Cullen who taught him that attraction and fire are the primary causes of change and that air's 'chief properties as an elastic fluid manifestly depend on the presence of fire'.[159] Thus for Black, the faithful disciple, the progress of chemistry implied the abandonment of theories founded upon the specific properties of chemical elements such as the salts and the earths.

Although Black believed that future chemical theories would be founded upon Newtonian conceptions of force, he knew that chemists must continue to identify and classify the substances they study. Thus in his dissertation Black demonstrated that not all absorbent substances are calcareous and in his essay he established that borax does not contain a previously unknown alkali. 'Borax,' he said,

> has sometimes been referred to the class of alkalis, on account of some resemblance it bears to those salts: But it has been demonstrated by accurate experiments, that we should rather consider it as a neutral salt; that it is composed of an alkali, and of a particular saline substance called the sedative salt, which adheres to the alkali in the same manner as an acid, but can be separated by the addition of any acid whatever, the added acid joining itself to the alkali in the place of the sedative salt [borax ($Na_2B_4O_7$) + vitriolic acid (H_2SO_4) + $5H_2O$ → Glauber's salt (Na_2SO_4) + sedative salt ($4B(OH)_3$)].[160]

If borax were a distinct alkali it would contain fixed air in its natural mild state and would therefore effervesce with acids. But, Black argued, 'as this conjunction of an acid with the alkali of borax happens without the least effervescence, our principles lay us under a necessity of allowing that alkali to be perfectly free of air, which must proceed from its being incapable of union with fixed air, and with the sedative salt, at the same time.'[161] Furthermore, he contended, 'were we to mix the sedative salt with an alkali saturated with air, the air would immediately be expelled; or, the two salts, in joining, would produce an effervescence. This I found to be really the case, upon making the trial, by mixing a small quantity of the sedative salt, with an Equal quantity of each of the three alkalis, rubbing the mixtures well in a mortar, and adding a little water [$4B(OH)_3 + Na_2CO_3 \rightarrow Na_2B_4O_7 + CO_2\uparrow + 6H_2O$].'[162]

Although borax does not contain an unknown alkali, it does have a peculiar property which Black attempted to explain. He noticed that 'if a small quantity of the sedative salt be thrown into a large proportion of a dissolved fixed alkali, the sedative salt gradually disappears, and is united to the alkali, without any effervescence; but if the addition be repeated several times, it will at last be accompanied with a brisk effervescence, which will become more and more remarkable, until the alkali be entirely saturated with the sedative salt.'[163] 'This *phaenomenon* may be explained,' Black suggested, 'by considering the fixed alkalis as not perfectly satured with air: and the supposition will appear very reasonable, when we recollect that those salts are never produced without a considerable degree of heat, which may easily be imagined to dissipate a small portion of so volatile a body as air.'[164] He then confirmed this supposition by describing the properties of potassium bicarbonate. 'That the fixed alkali, in its ordinary state, is seldom entirely saturated with air,' Black said,

> seems to be confirmed by the following experiment:—I exposed a small quantity of a pure vegetable fixed alkali to the air, in a broad and shallow vessel, for the space of two months; after which I found a number of solid crystals, which resembled a neutral salt so much as to retain their form pretty well in the air, and to produce a considerable degree of cold when dissolved in water. Their taste was much milder than that of ordinary salt of tartar; and yet they seemed to be composed only of the alkali, and of a larger quantity of air than is usually contained in that salt, and which had been attracted from the atmosphere: for they still joined very readily with any acid, but with a more violent effervescence than ordinary; and they could not be mixed with the smallest portion of vinegar, or of the sedative salt, without emitting a sensible quantity of air.[165]

To understand the full meaning of Black's statement that potassium bicarbonate produces 'a considerable degree of cold when dissolved in water,' one must recall Cullen's experiments on heat and mixture.[166] Cullen argued that every instance of mixture causes a condensation of matter and a release of heat whereas every case of solution causes a rarefaction of matter and an absorption of heat. For example, Glauber's salt deprived of water generates heat when dissolved, whereas Glauber's salt which already has its water of crystallization generates cold in the same circumstances. Clearly Black viewed his fully saturated vegetable alkali as comparable to crystalline Glauber's salt. Saturation involves condensation, that is, the fixation of air or water, and therefore causes heat to flow out. But when a salt is fully saturated with water, as in the case of crystalline Glauber's salt, or with fixed air, as in the case of potassium bicarbonate, then solution in water will cause a rarefaction and an absorption of heat. Black did not linger over these points in his essay for they had little to do with the substances and properties he was examining, but as soon as he had completed his study of causticity, he turned his attention to the intriguing connections between heat and change of state.

In the final pages of his essay Black examined the elective attractions between the various alkaline substances he was studying and the acids and fixed air. By mixing magnesia with caustic lye ($MgCO_3 + 2KOH \rightarrow K_2CO_3 + MgO\downarrow + H_2O$) he determined that fixed alkali has a greater attraction than magnesia for fixed air.[167] He then combined volatile alkali with magnesia deprived of its air ($(NH_3)_2CO_3 + MgO \rightarrow MgCO_3 + 2NH_3\uparrow + H_2O$), thereby establishing that magnesia has a greater attraction than volatile alkali for fixed air.[168] He was, of course, already familiar with the relative affinities between the substances he was studying and the acids, and he therefore was able to draw attention to some very interesting differences between their elective attractions for the acids and those for fixed air. 'Having formerly shown,' Black wrote,

> that *magnesia* saturated with air separates an acid from a calcareous earth, which it is not able to do after being deprived of its air by fire [i.e., $MgCO_3 + CaCl_2 \rightarrow MgCl_2 + CaCO_3\downarrow$; but MgO and $CaCl_2$ do not react]; I now suspected that the air was the cause of this separation, because I found that it was joined to the calcareous earth, at the same time that the acid was joined to the earth of *magnesia*; and imagined, that a pure calcareous earth might possibly have a stronger attraction for acids, than a pure earth of *magnesia* [$Ca(OH)_2 + MgCl_2 \rightarrow CaCl_2 + MgO\downarrow + H_2O$].[169]

Black performed the experiment and then explained the various reactions in terms of double elective attractions. He argued that since calcareous earth without air will remove the acid from a magnesia salt, 'we must therefore acknowledge a stronger attraction between the calcareous earths and acids, than between these and *magnesia*. But, how does it then happen,' he asked,

> that if *magnesia*, saturated with air, be mixed with a compound of acid and calcareous earth, these two last, which attract one another the most strongly, do not remain united; but the acid is joined to the *magnesia*, and the calcareous earth to the air, which it attracts much more weakly than it does the acid? Is it because

> the sum of the forces which tend to join the *magnesia* to the acid, and the calcareous earth to the air, is greater than the sum of the forces which tend to join the calcareous earth to the acid, and the *magnesia* to the air; and because there is a repulsion between the acid and air, and between the two earths; or they are somehow kept asunder, in such a manner as hinders any three of them from being united together?[170]

Black then asserted that 'the first part of this supposition is favoured by our experiments, which seem to show a greater difference between the forces wherewith the calcareous earth and *magnesia* attract fixed air, than between those which dispose them to unite with the acid.'[171] But he was much more cautious about the latter part of his argument. 'The repulsions, however, hinted in the second [part], are,' he said, 'perhaps more doubtful, though they are suggested in many other instances of decomposition; but,' he quickly added, 'the bounds of my present purpose will not allow me to enter upon this subject, which is one of the most extensive in chemistry.'[172]

After discussing two more cases of double decomposition, Black set down in tabular form his own modest contribution to the collective knowledge of elective attractions. 'From some of the above experiments,' he wrote, 'it appears that a few alterations may be made in the column of acids in Mr Geoffroy's table of elective attractions; and that a new column may be added to that table, according to the following scheme, where the alkaline substances are all considered as in their pure state, and free of fixed air.

Acids	Fixed Air
Fixed Alkali	Calcareous Earth
Calcareous Earth	Fixed Alkali
Volatile Alkali and *Magnesia*	*Magnesia*
	Volatile Alkali'[173]

And then, without recapitulation or extrapolation, Black brought his essay to a close.

What became of Black's promise to make fixed air 'the subject of my further inquiry?' We know that shortly after completing his essay he performed experiments demonstrating that fixed air will not support respiration or combustion and that it is the product of respiration, fermentation and combustion.[174] But we also know that Black did not publish accounts of these experiments or carry his research on gases further forward. Pneumatic chemistry as we think of it seems to be such a natural extension of his investigation of fixed air that we are surprised to find him abandoning this subject for the study of heat. And yet, from his own point of view, the study of gases was not an attractive subject for a philosophical chemist.

We can get some idea of the relatively minor significance Black attached to his knowledge of fixed air by looking at a set of lecture notes taken by a student who attended Black's chemistry course shortly after he began teaching in Edinburgh. It is most surprising to find that in these lectures fixed air has not emerged as a topic in its own right; its distinctive properties are discussed under proposition three of his theory of causticity, just as they had been in the essay.[175] In addition to proving that quicklime is completely soluble and 'that the Air which is contain'd or Attracted by Quicklime is not the same with common Air,'[176] Black took the opportunity, while discussing the third proposition, to tell his students that fixed air will not support combustion or respiration and that it is given off by the processes of burning, breathing and fermentation. He described his typically elegant experiments, which they could perform themselves to demonstrate the truth of his statements, and then said 'I must add some very Neat Expts. in the Phylos: Trasact: by Mr. Clavendish. [sic] – Who has made observns on Mephitic Air.'[177] Cavendish, he informed his students, had shown that mephitic air is highly soluble in water and that 'the fixed air is greatly Heavier than the Common'.[178] And then, having completed his discussion of the distinctive properties of fixed air, Black moved on to the fourth proposition of his theory of causticity.

Perhaps even more revealing of Black's attitude towards the study of gases was his lack of interest in inflammable air (hydrogen), the first of the factitious airs described in Cavendish's three papers. Cavendish collected this gas as it was given off by the reaction of acids and metals, and he proved that it differs from fixed air in being insoluble in water and lighter than common air. Anyone seeking to distinguish between the gases in terms of their chemical and physical properties would have welcomed Cavendish's experiments as the second major achievement in the field of pneumatic chemistry.[179] But it would be wrong to suppose that Cavendish himself believed he had isolated one of several distinct airs. Since the quantity of gas released depends upon the weight of metal dissolved, and not upon the kind of acid used, he concluded that the inflammable air comes from the metal rather than from the acid. In more concrete terms, he asserted that when metals are mixed with acid, 'their phlogiston flies off, without having its nature changed by the acid, and forms the inflammable air'.[180] Not only did Cavendish think of inflammable air as elastic phlogiston, he was quite sure that fixed air itself, the paradigmatic factitious air, was a mixture 'of substances of different natures, part of it being more soluble in water than the rest'.[181] Brilliant and precise as Cavendish's experiments were, they were not designed to illuminate a theory which sought to explain chemical changes in terms of the elementary properties of gases.

Although Black was obviously familiar with Cavendish's inflammable air, he had good reasons for believing that it should not be considered a distinct air comparable to fixed air. At times it seems that he was in full agreement with Cavendish, as when he stated that the effervescence given off by the action of acids on metals 'depends not on a Separation of Air like the Alkalis but on a part of the principles of Inflammability, which joins the Acid & produces this elastic Vapour'.[182] But there is a fundamental difference in the weight changes caused by the release of fixed air and by the release of the inflammable vapour. 'It has been thought,' Black said, that 'fixed Air causes this Effervescence, but if we attend that vital air [i.e., atmospheric air] is convertible into Mephitic by certain Bodies, 'tis plain then the air . . . passing through the fuel is not receiving any addition or its bulk would be increas'd whereas it is diminish'd, which we can shew by Analogy the Calx of a Metal being always lessen'd when reduced to a Metallic form.'[183] In other words, the release of inflammable vapour during burning causes the weight of the common air involved to decrease as it is transformed into mephitic air. 'It would appear then,' Black concluded, that it is 'by the great quantity of the △ [i.e., phlogiston or inflammable principle] which the Mephitic Air contains that it proves noxious to the Lungs.'[184] As he then pointed out, the difference he was emphasizing could be readily perceived if one simply noted that 'the Alkaline Salts become lighter when inflam'd but the Metals heavier'.[185]

Black realized that his notion of the inflammable vapour implied the existence of a non-gravitating substance, but he believed the weight gain which occurs during inflammation justified this conclusion. Under the heading of 'Phlogiston or Principle of Inflamability—What it is,' he said that 'though many inflamable Substances seem to suffer a loss when inflamed, we see this Matter escapes so fast as scarce can be obtained; but when we do get it we find it even overballances the former principle the Substance contained. This is a fact difficult to explain & is a Strong objection to this Theory. The only answer is that tho' we have hitherto found the property of Gravitation, this does not throw out a possibility of particular kinds of Matter that may be an exception.'[186] If we think that Black was trying to found his theory of chemistry upon a set of distinct material elements, then his willingness to abandon universal gravitation will strike us as weak-minded if not downright craven. But if we recognize that he looked to heat and ultimately to the force of repulsion as the fundamental concepts in chemical theory, then we will realize that his notion of phlogiston is simply a bridge between the phenomena studied by chemists and the fundamental properties associated with his explanatory terms. A study of the properties of inflammable bodies indicates, he argued, 'that in general the property of inflamability depends on a Substance common to them all, therefore this has got the name of *Phlogiston* but I chuse rather to call it the *inflamable principel.*'[187] But, he quickly added, this name refers to a specific causal property of the substance, not to the whole substance itself, just as 'Fixed Air may be called the principle of Effervescing, but not the Air itself effervesces.'[188] In the case of phlogiston, the distinguishing properties are to be understood as manifestations of heat; 'Heat or light,' Black asserted, 'are the principles of Inflamability.'[189]

Black really should not be called a pneumatic chemist unless we wish to apply that label to all chemists who studied the properties of elastic vapours, whatever their theoretical predilections. He should, instead, be called a heat theorist and a Newtonian. Like most of the leading chemists in Great Britain before the end of the eighteenth century, he was looking for general laws which would explain changes in the states of matter. He realized that specific substances have particular properties, but these were neither to be explained nor should they be taken as a basis for explanation. Knowledge of them simply enabled one to apply the general laws of heat flow, change of state and elective attraction to particular cases. As a theorist Black had much more in common with those early nineteenth-century physicists whose studies of heat flow contributed to the science of thermodynamics than he had with the chemists who developed the new system of nomenclature and the periodic table of elements.[190]

[9]

The Theory of Latent Heat

I. ORIGINS OF THE THEORY

After the publication of Black's essay on quicklime, there could be little questioning his pre-eminence among the younger chemists in Scotland. Those who knew Black considered his theory of causticity merely the initial accomplishment of a brilliant and profoundly informed mind and they believed that if he received the kind of encouragement he deserved, Black would make other equally significant contributions to philosophical chemistry. They therefore secured for him appointments as professor and as lecturer in chemistry in the University of Glasgow. As is well known, their expectations were well founded, for during the decade he spent as a member of the faculty in the University of Glasgow, Black completed a set of experiments on heat which were every bit as brilliant as his earlier experiments on quicklime and magnesia alba.

Beneath this continuity of achievement, however, lies an apparent discontinuity of subject matter. Instead of extending his theory of causticity or undertaking the promised 'further inquiry' of fixed air, Black turned his attention to the study of heat. At first glance this shift of interest seems to signal an abandonment of chemistry in favour of physics, but anyone who has even a slight familiarity with the chemical doctrines which Black acquired while studying with Cullen will realize that heat was one of those topics which, like air, was studied by both chemists and mechanicians. The notion that there was a radical discontinuity in Black's research is further weakened by the considerable amount of evidence that indicates that in turning to the study of heat, Black was in fact returning to those general questions of chemical theory, which he had begun investigating before he took up the problem of causticity. Black thus used the freedom and security that accompanied his appointment to the faculty in Glasgow to turn his attention to those questions that he had long believed were of fundamental importance to the advancement of theoretical chemistry.

In his introduction to Black's chemistry lectures, John Robison tells us that during his student years Black 'laid down a very comprehensive and serious plan for the conduct of his studies'.[1] Robison's evidence for this statement consisted of 'a number of note-books found among his papers. There are some in which he seems to have inserted every thing as it took his fancy, in medicine, chemistry, jurisprudence, or matters of taste; and I find others into which he has transferred the same things, but has distributed them according to their scientific connections. In short, he has kept a journal and leger of his studies, and has posted his books like a merchant.' [2] In reading over these notebooks, Robison was particularly struck by 'the steadiness with which he advanced in any path of knowledge'. He found that 'things are inserted for the first time, from some present impression of their singularity for importance, but without any allusion to their connections. When a thing of the same kind is mentioned again, there is generally a reference back to its fellow; and thus the most insulated facts often acquired a connection which gave them scientific importance.' [3]

The steadiness of Black's advance towards an understanding of those subjects which interested him and the inclusion of several references to contemporary events enabled Robison to establish 'the order in which things had occurred, (for there are very few dates [in the notebooks]) and a pretty certain knowledge of the years when he made the observations'.[4] In the oldest parcels of these notes, which evidently were written while Black was studying with Cullen in Glasgow, Robison found 'queries respecting the nature of cooling mixtures, and the cold produced by liquefaction; but,' he added, 'it is not till some time after, I think not before the year 1752, that I can date any observation relative to fixed air.' [5] Clearly Black's early interest in cooling mixtures and the cold produced by liquefaction indicates that his own understanding of heat took as its point of departure Cullen's theory of heat and mixture. Thus rather than thinking that Black began his investigation of heat after presenting his theory of causticity, we should say with Robison that after getting settled in Glasgow, Black 'brought to maturity those speculations concerning the combination of heat or fire with the substance of tangible matter, which had long occupied his thoughts occasionally'.[6]

Black, like Cullen, believed that chemists who study heat phenomena should try to relate these phenomena to the specific changes in the properties of bodies which fall within their proper field of examination. To keep his own investigation within the realm of chemistry, Cullen focused his attention upon the flow of heat in chemical reactions. After examining the heat released and absorbed by different combinations of salts and solvents, he proposed his theory of heat and mixture.[7] According to this theory, when a combination of a solid and a liquid causes a release of heat, a condensation, or a more compact aggregation of atoms, has been brought about; when such a combination causes cold, a rarefaction has occurred. The experiments used to establish this law also indicated that these changes in aggregation involve changes of state. Thus when water is fixed in Glauber's salt, the reaction generates heat, whereas when crystalline Glauber's salt is dissolved in water, it generates cold. In other words, Cullen's theory implies that fluids contain a quantity of fire which is lost when they become solid and that solids absorb a quantity of fire when they become liquid. Perhaps it was this deduction from Cullen's theory which led Black to perform his earliest experiments upon cooling mixtures and the cold produced by liquefaction.

Much as he liked his theory of heat and mixture, Cullen realized that there are certain heat effects caused by changes of state which could not be explained by his formulation of the theory. He knew, for instance, that when ice at 32 degrees is mixed with a comparable quantity of water at 50 degrees, the temperature of the mixture sinks to 32 degrees during the dissolution of the ice.[8] Here was a situation involving a change of state, the liquefaction of ice, and a flow of heat from water to ice, and yet the final temperature of the mixture was the same as the initial temperature of the ice, not a temperature somewhere between the initial temperatures of the water and the ice. Although Cullen knew that one must distinguish between intensity of heat, as measured by the thermometer, and the quantity of heat in a body, he still expected to see some rise in temperature when heat flowed into a body. He therefore found the results of his ice-water experiment quite baffling.

Cullen had also hoped that his theory of heat and mixture would enable him to explain the cold produced by evaporating fluids.[9] Here was another case of change of state in which rarefaction was attended by the absorption of a great deal of heat. So long as he performed his experiments in the atmosphere, Cullen could argue that the dissolution of the liquid in the air was the cause of the drop in temperature which accompanies evaporation, but when he discovered that an even greater degree of cold could be obtained by evaporating the liquids in a vacuum, he was obliged to recognize the limitations of his theory.

What interpretation did Black give to these anomalous observations as he thought about Cullen's experiments during the period between the publication of his dissertation and his return to Glasgow? Perhaps he drew an analogy between the chemical effects of fixed air and those of fixed heat. Such an analogy suggests that heat, when fixed in a body, no longer causes the sensation of warmth, a change comparable to fixed air's loss of elasticity. Perhaps fixed heat has the effect of changing the state of aggregation, an effect comparable to the slaking power of fixed air. This analogy also suggests that heat, like air, combines with substances according to the laws of elective attraction. If these laws govern the fixation of heat, then there is a point at which each substance, at least when in a given state of aggregation, will become saturated with heat. When studying fixed air Black discovered that the reaction between acids and mild alkalis involves the elective attractions of three substances—the acid, the alkali and fixed air. Perhaps in the theory of heat and mixture one should similarly think of the heat involved as having an elective attraction which causes it to react with the solvent and the substances dissolved. Such a view of the matter would at least enable one to explain both the ice-water anomaly, in terms of the combination of heat and ice during melting, and the anomaly of evaporation in a vacuum, in terms of the combination of heat and liquid during vaporization.

There are several indications that Black's thinking actually proceeded along the lines suggested by this analogy to fixed air.[10] The very names he used to designate what came to be called specific heat reveal the direction of his thinking. Robison reports that when Black was considering various names for this property, the term '*affinity* pleased him a long while,' and that he also 'used the terms *capacity, faculty for receiving heat; appetite for heat*, &c. &c'.[11] Robison also alludes to 'the simple experiments and familiar observations by which he [Black] demonstrated the fixation (I may call it) of heat in bodies when it melts or evaporates them'.[12] And when describing the fixation of heat in bodies, 'Dr. Black said,' Robison reports, 'that it was concealed in them,—*latet*,—it was as much concealed as carbonic acid is in marble.'[13]

While Black's speculations were probably guided by the analogy between fixed air and fixed heat, his early experiments took as their point of departure the facts that Cullen had examined while formulating his theory of heat and mixture. We only know of these early experiments through the excerpts and summaries of another set of Black's notebooks published by Adam Ferguson, in his biographical memoir of Black, and by Robison, in his notes to Black's lectures.[14] Events cited in these notebooks enabled both authors to determine that they contained entries which had 'been written in 1754–5–6–7, while the Doctor was a student at Glasgow and Edinburgh, and during the first year of his Professorship in Glasgow'.[15] Although this body of evidence is slight, it is enough to reveal that after having completed his study of causticity, Black again turned his attention to his master's theory of heat and mixture.

After thinking about the temperature changes that occur when ice melts in water and those that accompany the vaporization of liquids in a vacuum, Black realized he could restate Cullen's chemical theory of heat in a way that would enable him to explain these recalcitrant facts. All he had to do was make explicit and accept the implications of Cullen's theory itself. When fire is fixed, it loses its ability to warm substances and enters into chemical combination with the body in which it is fixed. When enough fire has been fixed in a body, it forces the particles of the body into a new state of aggregation, thereby causing a change of state. Thus in studying heat, as in his investigation of the chemical attributes of fixed air, Black followed the injunctions of his teacher and discovered the specific chemical effects of a substance which his immediate predecessors had considered only in the context of the mechanical philosophy.

Black's notebooks contained many 'conjectures and queries relating to the cold produced in the liquefaction of salt and snow, and in general in the solution of salts in water.'[16] Like Cullen, he believed that in these solutions the salt, in changing from a solid to a liquid, absorbs heat, thereby causing a fall in the temperature of the solution and the sensation of cold. Solutions that produce heat are mentioned more frequently, however, in the passages quoted by Robison and Ferguson. Black asked, for instance, 'is not heat produced in solutions, when a fluid matter is converted into a solid?' As an example of a solution that yields heat he mentioned the reaction of 'calcined Glauber salt and water,' but he then noted that heat is not produced 'when solid matter is rendered fluid'.[17] The reaction between quicklime and water provided another example of a heat producing solution. Black asked 'does not the heat produced in slaking lime, and in the solution of lime and caustic alkali in acids, arise from the sudden fixing of the fluid part [i.e., water] in a firm state? taking them from the state of fluidity. Is not this the case in the setting of gypsum?'[18]

After having explored some of the implications of Cullen's theory, Black attempted to reformulate it as a conservation law. Like Newton he presented his hypothesis as a question: 'Is not the heat which is disengaged from water in becoming solid in a crystal, equal to what must be employed to disjoin it [i.e., the salt in the crystal] from the water?'[19] Black realized, of course, that such a supposition was valueless unless it could be experimentally tested, and he immediately asked 'how may this be tried?' To stay within the boundaries of Cullen's theory, the hypothesis should have been tested by examining the heat flow in the formation of deliquescent and crystalline salts, but Black soon found that the properties of these substances are too complex to provide clear answers to the questions he was asking. He determined that 'in a salt it [i.e., the water of crystallization] is not all disengaged, for the brine is still salt; and when it [i.e., the crystalline salt] effloresces [i.e., loses its water to the atmosphere], more [heat] must be employed to produce [the water] vapour.'[20]

It was probably these properties of crystalline salts that turned Black's attention to the relatively simple change of aggregation involved in the freezing and melting of water. In moving from a consideration of salt solutions to an examination of the heat flow during melting, he first attributed the 'intense cold produced by ice dissolved in an acid' to the heat needed to liquify the ice. He then argued that that 'the principle [can be] applied to crystallization'.[21] 'Is not ice crystallized water,' he asked?[22] 'Does it not always feel cold, because it melts on our handling it? This,' Black declared, 'is similar to the solution of salts in water.'[23] He then asked 'is it [i.e., the dissolving of ice in water] not similar to a solution of a saline crystal?'[24] Such a supposition, he argued, enables us to explain why 'a greater alteration of heat should be produced by mixing a neutral salt with ice or snow, than by dissolving it in water'.[25] This extension of Cullen's theory was stated in a slightly different manner in another notebook entry quoted by Robison: 'Does it not appear,' Black asked, 'on a review of cooling mixtures, that a refrigeration will always happen, when salt, in a crystalline form, (that is, combined with water) or water in a crystalline form, (that is, frozen) is made fluid by salts, or in spirit of wine. According to this supposition, salt mixed with ice should produce much more cold than salt mixed with water that is as cold as the ice.'[26] Armed with this reformulation of Cullen's theory, Black was able to explain how the temperature of mixtures of ice and warm water could fall to 32 degrees while the ice was being converted into water. According to Ferguson, it was this set of insights which brought the doctrine of latent heat into clear focus in Black's mind.[27]

Black believed that heat is an active constituent in chemical change. His attitude was clearly expressed in a notebook entry in which he asserted that 'all solution must be considered as liquefactions in consequence of heat, for there is no fluidity without heat.'[28] As an illustration of his doctrine he noted that 'salt and water, when mixed,' contain so much fixed heat as a result of the liquefaction of the salt that when frozen, they 'melt with much less heat than what will melt ice, and vastly less than what will melt salt'.[29] In other words, Black argued, 'all bodies, in becoming fluid, have occasion for more heat than in their solid state,'[30] and it is the amount of heat fixed in a body which determines what state of aggregation it will assume. Consider, for instance, Fahrenheit's experiments on the supercooling of water. If pure water is kept very still while being cooled, its temperature can be reduced far below the freezing point. But once such supercooled water has been disturbed, it freezes rapidly and in the process raises the temperature of the ice to 32 degrees. Is not this temperature rise, Black asked, the result of 'the heat that is unnecessary to ice?'[31] Clearly Black was not thinking of heat as something that simply accompanies chemical change. It is, rather, a chemical constituent that is as active in causing change of state and as specific in its properties and effects as any of the forces or principles utilized by chemists.

The active role played by heat is even more evident in vaporization, a change of aggregation that can be brought about under conditions in which the attractive forces involved in dissolution cannot act. Cullen had discovered that a great degree of cold can be produced by evaporating liquids in a vacuum. Black was impressed by these experiments, particularly by the one in which boiling nitrous aether made water freeze. To explain these rather extreme cases of heat flow, he turned to the less dramatic heat effects which accompany effervescence. Chemically, he argued, the effervescence produced by reactions that take place in the atmosphere is quite like boiling in a vacuum. 'In the effervescence of mercury in nitrous acid, whence comes the great heat which must certainly be required,' Black asked, 'for so much elastic matter must rob it, as aether boiling *in vacuo* does?'[32] When there is no effervescence to carry off the heat released in a reaction, the temperature change will be considerably greater. For instance, Black said, there is a 'much greater heat when a caustic ley is mixed with an acid, than when a mild one is; [for] none [of the heat released in the former reaction] is taken off by the elastic effervescence.'[33] He also asked, 'is not the cold produced by mixing vitriolic acid with sal ammoniac owing to the production of elastic matter?'[34] In effervescence Black found a type of vaporization resulting from a change in the state of aggregation, one which was accompanied by the absorption of a great quantity of heat.

When making entries in his notebooks Black seems to have been particularly struck by the connection between air pressure and the amount of heat needed to vaporize a substance, a connection suggested, no doubt, by simultaneously considering vaporization occurring in a vacuum and vaporization that occurs under atmospheric pressure. 'Will a boiling fluid raise the thermometer in proportion to the pressure on its surface,' he asked? 'Will this be of any use in discovering the absolute proportions of heat?'[35] And if the heat needed to vaporize a liquid is proportional to the air pressure bearing upon its surface, 'will distillation *in vacuo* save fuel?'[36] Other conjectures also suggested themselves. Ferguson mentioned one such when he noted that Black 'observed, that animal heat originates in the lungs, and that heat always accompanies the production of fixed air.'[37]

Such conjectures, however, never dominated Black's thinking. Having discovered a connection between the fixation of heat and change of state, he sought to specify and quantify his discovery, for, as is well known, Black was abstemious when it came to fabricating hypotheses. Perhaps he was thinking of Fahrenheit's experiments on supercooling when he devised his first experiment to measure the heat absorbed during the vaporization of water. In his notebook Black described 'one way of estimating the quantity of heat which is transferred and disappears in the production of vapour, or which is expended, and contributes to the formation of vapour, but cannot be discovered in the vapour by the thermometer.' His method was to 'place a phial with water, close corked, in a stove or oven, where it will acquire a degree of heat above the boiling point: Then, by some contrivance, open it suddenly, and observe how much has been converted into vapour, and what heat it retains.'[38] Although crude, this experiment clearly indicates the direction of Black's future research on heat. Having proven to his own satisfaction that heat can exist in a latent state, he set out to quantify his discovery so that others would be convinced of its truth and so that he could begin to formulate the laws that govern this newly discovered state of heat.

There can be little doubt that Black had determined the general range and meaning of his theory of latent heat before he began verifying it with quantitative and exact experiments. Ferguson quotes a statement from Black's notebooks which contains all the essential features of the theory. 'Heat,' Black observed,

> is in nature the principle of fluidity and evaporation, though, in producing these effects, it is latent in respect to the thermometer, or any sensation of ours; and as matter, otherwise quiescent, becomes voluble and volatile in liquid and in vapour, heat may be considered in nature as the great principle of chemical movement and life. If it pass through vacuity as well as through body, as it certainly does in its communication from the sun to the planets, we must consider it not as an accident in bodies, but as a separate and specific existence, not less so than light or electric matter; and though agreeing with these in some of its effects, in its nature possibly different from either.[39]

In calling heat 'the great principle of chemical movement and of life,' Black gave it an importance which Cullen had long said it should have but which it never had attained within the context of his acid-alkali conception of chemical action. But Black did not simply say that heat is important, he made it a central force in his theory of chemistry. As Robison reminds us, Black 'did not call his own discourses *Lectures on Chemistry,* but *Lectures on the Effects of Heat and Mixture.*'[40] Furthermore, Robison argues, Black's concept of capacity for heat implies chemical combination. 'Philosophers,' Robison declared,

> do not seem agreed as to the meaning affixed to this term. It is frequently used in the sense of mere *capaciousness*, or *room* for the admission of heat. But this is a very needless consideration, and unconnected with all chemical science. Heat is not merely *admitted* into bodies, but *absorbed* or *drawn* into them, in consequence of some tendency to combination, whatever the nature of this union may be; whether mere adhesion, like capillary attraction, or the mixture of metals or of watry fluids; or like the solution of saline crystals in water, and the mixture of brines with more water; or like the union of salt and water in crystallization; or like the union of acids and alkalis. In all these examples, there is some physical action, and this is continued till an equilibrium of the acting forces is effected. This is very different from mere capaciousness, and it even receives no explanation by the help of capaciousness.[41]

Black also emphasized the chemical nature of his theory when he asserted that during boiling, 'the heat . . . did not escape along with the vapour, but *in* it; probably united to every particle, as one of the ingredients of its vaporous constitution'.[42] He also believed that 'when a particle of vapour again becomes water, these atoms of heat are set at liberty by the fixed laws of chemical affinity'.[43] Whatever its limitations, Black's theory of latent heat was, in his own mind and in the minds of his students and colleagues, a theory that addressed certain chemical properties of bodies and sought to explain them in terms of the heat that is chemically fixed in the bodies.

Black began lecturing on his theory of latent heat soon after he assumed his teaching duties in Glasgow. Many years later he told James Watt that 'I began to give the Doctrine of lat[en]t Heat in My Lectures at Glasgow in the Winter 1757–58 which I beleive was the first winter of my Lecturing there . . . or if I did not give it that winter I certainly gave it in the [winter of] 1758–59.'[44] John Robison also recorded that

> in an old and rejected sheet of lectures, which some other cir-

> cumstances oblige me to refer to this year [i.e., 1758], he [Black] treats the subject [of latent heat] still as a general view, and makes no mention of the *quantity* of heat which enters into the formation of water. I think that this would not have been omitted, had the experiments been already made. I am persuaded, however, that the extension of the doctrine to the ebullition of fluids followed very soon after its establishment in the cases of liquefaction and congelation.[45]

In these first descriptive accounts Black only applied the theory to the liquefaction and congelation of substances, chiefly water, but in the following years he extended it so as to include the vaporization and condensation of water and other liquids. 'At all events,' Robison concluded, 'it is fact, that the doctrine of latent heat, precisely as it is contained in these [published] lectures, was explained and demonstrated by Dr. Black in 1761, and in every subsequent course.'[46]

2. EXPERIMENTAL CONFIRMATION

If it is assumed that heat always causes a sensation of warmth, then one can study the flow of heat which takes place in chemical reactions by using a thermometer in a straightforward manner. But if heat can become latent, that is to say, insensible, then experiments designed to reveal the quantity of heat involved in any particular change will necessarily be somewhat complicated. In experiments of the latter sort, measurements of the quantity of heat, as opposed to the intensity of heat, must be made indirectly.[47] The experimental problem is made even more difficult by the known properties of latent heat. When heat becomes fixed in bodies, it changes their state of aggregation. Does this mean that as the temperature indicated by a thermometer increases, the thermometric substance itself is undergoing a qualitative change? If this is so, then one must examine the possibility that every change in temperature recorded by the thermometer is caused by a slightly different quantity of heat being fixed in the thermometric substance. Does the thermometer, the basic instrument in all heat experiments, provide a reliable and direct measurement of the intensity of heat? 'This question,' Black told his students, 'has been overlooked, or little attended to, by some of the principal writers on thermometers. It does not appear to have occurred to Dr. Boerhaave; and Dr. Martin gives very little attention to it. I began to attend to it, and made an experiment to decide it in the year 1760, and did not then know that others had thought of it.'[48] We know, however, from Robison's summary of Black's notebooks, that by 1757 Black had already thought of 'the method for examining the thermometric scale, so as to decide whether the scale of equal expansions corresponds to equal augmentations of heat'.[49]

There are logical and physical grounds for questioning the steadiness of the relationship between temperature change and the flow of heat.[50] One can simply doubt the validity of the assertion that equal increases in temperature are caused by equal accretions of heat until it has been proven true by experiment. Black presented the problem this way in his later lectures. After describing the methods used in the construction of thermometers, he told his students that

> it remains to be considered, whether these equal increments, or diminutions of bulk, be produced by equal increments, or diminutions of heat. We can imagine the fact to be otherwise, and that in some parts of the scale, a greater addition of heat may be required to produce one degree of expansion, than in other parts. If a string be stretched, by suspending a moderate weight to it, and we add one pound to that weight, we shall make it a little longer; but, by adding a second pound, we shall not add as much more to the length of the string as the first pound added; nor will a third pound produce so much effect as the second pound. In like manner, we can imagine, that when a thermometer receives a series of equal additions to its heat, these may not produce equal increments of expansion; and, therefore, that equal increments of expansion may require for their production increments of heat very unequal among themselves.[51]

But it seems likely that when first formulating his theory of latent heat, Black was more concerned about the physical effects caused by the fixation of heat in the thermometric substance than about the logical foundations of thermometry. Robison tells us that Black had 'made experiments on some bodies which change their mode of aggregation from solid to fluid, and the contrary, not all at once, but by imperceptible degrees; and had found that, in the whole of this change, they were absorbing heat. Resin and sealing wax are very clear examples of this.'[52] He then examined the thermometric properties of these substances and found that 'a thermometer plunged into resin in very thin fusion, and compared with another standing in oil in the same vessel, shewed, from the very beginning, a very different progress of refrigeration.'[53] He concluded, therefore, that 'we are not certain that something of the same kind does not happen in water and other fluids. They may have different degrees of fluidity in different temperatures.'[54] Robison then states that 'such were the surmises which made Dr. Black think it necessary to examine the scale of the thermometer, in order to learn whether equal differences of expansion correspond to equal additions or abstractions of heat.

He made a series of experiments on this subject, and read an account of their result in the [Glasgow] literary society . . . on the 28th of March 1760.'[55]

Although Black did not know it at the time, the relationship between the thermometric scale and the flow of heat had been previously examined by several natural philosophers, including Brook Taylor.[56] As Black later told his students, the experiment used by both Taylor and himself involved

> mixing cold and hot water together, by which we can produce various heats, or temperatures of heat, the differences of which from one another we can be sure of, independently of any thermometer. If, for example, we suddenly mix a pound of hot water with a pound of cold, the excess of heat which was in the hot water, compared with the cold, being now equally divided between it and the cold water, must produce an excess or strength of heat in the mixture, just the half of that which existed immediately before in the hot water. And, by applying a thermometer to the hot water first, next to the cold, and lastly to the mixture, we learn whether the quicksilver is expanded by different heats, in proportion to the differences of these heats.[57]

Black then said that to perform this experiment well, 'attention must be given to two principal particulars'.[58] The first is that the inner diameter of the tube containing the thermometric substance be uniform, and

> the second object, of necessary attention in performing the above experiment, is to make allowance for the heating or cooling of the vessel in which the mixture is made. If the warm water is poured into the cold, the cold vessel in which the mixture is thus made, will take from the mixture some of its heat, and make it appear colder than it should otherwise do. If, on the contrary, the cold water is poured into the hot, the hot vessel will impart some of its excess of heat to the more temperate mixture, and thus raise its heat above the degree which should be produced. To avoid this deception, we must employ the two vessels of the same materials, and of the same size and weight, and then by making the experiment both the one way and the other, and taking the medium of the results, we shall learn the truth.[59]

According to Robison, the inquiries Black made in 1760 led him to conclude 'that equal additions or abstractions of heat produced equal variations of bulk in the liquor of the thermometers employed by him, and therefore that the scale of expansion was also a scale of heat.'[60] In the lectures he gave towards the end of his career, however, Black was less sure of this conclusion. He found that when the experiment had been made with all necessary precautions,

> the result has shewn, that when the thermometer is made of quicksilver, the gradual expansions of this fluid, while it is heated slowly, from the cold of melting snow to the heat of boiling water, are very nearly proportional to the additions of heat by which they are produced. There is, however, a little deviation from the exact proportion; while the heat increases, the expansions become a little greater than in proportion to the increase or addition of heat. The equable or uniform increase of heat, therefore, is attended with a small acceleration of the increase of expansion. This is much more remarkable in water, alcohol, and some other fluids, than in quicksilver. In this last fluid, this irregularity is so inconsiderable within the range of heat above mentioned, that it does not deserve any notice in common experiments, and in the ordinary use of thermometers; but if it should be necessary, on any occasion, to estimate the amount of it in thermometers of quicksilver, and of spirit of wine, it may be collected from the experiments of Mr. De Luc, and of Dr. Crawford.[61]

This lack of regularity did not upset Black because, being a chemist, he was always looking for specific changes in the properties of bodies. In attempting to explain this surprising discovery he even allowed himself one of his rare excursions into the realm of speculation. 'Here we may remark,' he said,

> that this little disproportion between the degrees of expansion and the degrees of heat, is not at all inconsistent with that supposition concerning the nature or cause of heat which I formerly mentioned as appearing to me the most probable; I mean the supposition that heat depends on the presence or abundance of an inconceivably subtile self-repelling matter, which is attracted by all other matter of the cohering or attracting kind. For the attraction of cohesion, that is, the attraction which the particles of a body have for one another, like all other attractions, is weakened by distance; and therefore when a body is expanded by heat, and the distance of its particles is a little increased, its attraction of cohesion being thus a little weakened, a further addition of heat to it can easily be imagined to produce more effect in expanding it further; the distending power of the repellent matter of heat being rather less counteracted by the attraction of cohesion, than it was at the first.[62]

But such hypothetical flights had little to do with the confirmation and quantification of the theory of latent heat, a task which remained

Black's primary concern during his early years as a professor in Glasgow.

Having determined that laboratory thermometers are reliable indicators of the sensible heat of bodies and having 'already full conviction of the truth of the doctrine' of latent heat,[63] Black had to find some means by which he could measure '*how much* heat was . . . absorbed by . . . ice, retained by water . . . while fluid, and emitted by it in the act of freezing.'[64] Perhaps while thinking about this problem he remembered the experiment in which George Martine placed two thin-walled glass vessels, one containing water and the other mercury, at an equal distance before a steady fire so that their rates of heating could be compared.[65] Since in this experiment equal quantities of heat flow into the two bodies, it may be assumed that the same amount of heat is being fixed in each body in any given time.[66] If, therefore, the temperature of one liquid rises more rapidly than that of the other, it means that the substance in the first vessel has a lower ratio of insensible to sensible heat than that in the second vessel. In the eyes of the heat theorist, of course, the properties of water and ice are as different as those of mercury and water. Therefore, Black probably thought, Martine's procedure can be used to compare the ratio of insensible to sensible heat in ice, as it goes from 32° Fahrenheit to room temperature, and water, as it warms up over the same range of temperature. 'This thought,' Robison wrote, 'occurred to him in the summer season; and as there was no ice-house then in Glasgow, he waited with impatience for the winter; and in December 1761, he made the decisive experiment, in a large hall adjoining to his college rooms, expending on the ice during its liquefaction, and obtaining from the water during its congelation, as much heat as would have raised the water somewhat more than 140 degrees in its temperature.'[67]

In his lectures Black presented a more detailed account of this experiment.

> The plan of the first [experiment to determine the quantity of latent heat contained in water] was, to take a mass of ice, and an equal quantity of water, in separate vessels, of the same size and shape, and as nearly as possible of the same heat, to suspend them in the air of a warm room, and, by observing with a thermometer the celerity with which the water is heated, or receives heat, to learn the celerity with which it enters the ice; and the time necessary for melting the ice being also attended to, to form an estimate, from these two data, of the quantity of heat which enters the ice during its liquefaction.
>
> In order to prepare for this experiment, I chose two thin

globular glasses, four inches diameter, and very nearly of the same weight: I poured into one of them five ounces of pure water, and then set it in a mixture of snow and salt, that the water might be frozen into a small mass of ice. As soon as frozen, it was carried into a large empty hall, in which the air was not disturbed nor varied in its temperature during the progress of the experiment; and in this room the glass was supported, as it were, in mid air, by being set on a ring of strong wire, which had a tail issuing from the side of it five inches long, and the end of this tail was fixed in the most projecting part of a reading desk or pulpit: And in this situation the glass remained until the ice was completely melted.

When the ice was thus placed, I set up the other globular glass precisely in the same situation, and at the distance of 18 inches to one side, and into this I poured five ounces of water, previously cooled, as near to the coldness of melting ice as possible, viz. to 33 degrees, and suspended in it a very delicate thermometer, the bulb of which being in the centre of the water, and the tube being so placed, that without touching the thermometer, I could see the degree to which it pointed. I then began to observe the ascent of this thermometer, at proper intervals, in order to learn with what celerity the water received heat, stirring the water gently with the end of a feather about a minute before each observation. The heat of the air, examined at a little distance from the glasses, was 47 degrees of Fahrenheit's scale.

The thermometer assumed the temperature of the water in less than half a minute, after which, the rise of it was observed every five or ten minutes, during half an hour. At the end of that time, the water was grown warmer than at first, by seven degrees; and the temperature of it had risen to the 40th degree of Fahrenheit's scale.

The glass with the ice was, when first taken out of the freezing mixture, four or five degrees colder than melting snow, which I learned by applying the bulb of the thermometer to the bottom of it; but after some minutes, it had gained from the air those four or five degrees, and was just beginning to melt, which point of time was then noted, and the glass left undisturbed ten hours and a half. At the end of this time, I found only a very small and spongy mass of the ice remaining unmelted, in the centre of the upper surface of the water, but this also was totally melted in a few minutes more; and, introducing the bulb of the thermometer into the water, near the sides of the glass, I found

> the water there was warmed to the 40th degree of Fahrenheit. From this it appears, that when a considerable part of the ice was melted, and the quantity of water around it was increasing, the heat was not transmitted through this water to the remaining ice altogether so fast as it was received by the water; which is easily understood, if we consider that the distance between the remaining ice and the external surface of the vessel through which the heat entered, was gradually increasing, and that heat always requires time to pass through bodies, or to be communicated from one part of them to another.
>
> It appears, therefore, from this experiment, that it was necessary that the glass with the ice receive heat from the air of the room during 21 half-hours, in order to melt the ice into water, and to heat that water to the 40th degree of Fahrenheit. During all this time, it was receiving the heat, or matter of heat, with the same celerity (very nearly) with which the water-glass received it, during the single half-hour in the first part of the experiment. For, as the water received it with a celerity which was diminishing gradually during that half-hour, in consequence of the diminution of difference between its degrees of heat and that of the air; so the glass with the ice also received heat with a diminishing celerity, which corresponded exactly with that of the water-glass, only that the progression of this diminution was much more slow, and corresponded to the whole time which the water surrounding the ice required to become warmed to the 40th degree of Fahrenheit. The whole quantity of heat, therefore, received by the ice-glass during the 21 half-hours, was 21 times the quantity received by the water-glass during the single half-hour. It was, therefore, a quantity of heat, which, had it been added to liquid water, would have made it warmer by 40–33x21, or 7x21, or 147 degrees. No part of this heat, however, appeared in the ice-water, except 8 degrees; the remaining 139 or 140 degrees had been absorbed by the melting ice, and were concealed in the water into which it was changed.[68]

Black's second experiment for measuring the heat absorbed by melting ice was probably suggested by Fahrenheit's experiments upon the distribution of heat when equal volumes of water and mercury, having different initial temperatures, are mixed together.[69] We know from Robison's summary of Black's notebooks that 'while a student of medicine in Edinburgh,' Black had set 'down queries, founded on Fahrenheit's experiments'.[70] It is also likely that Black had in mind the mixture experiments he had performed when examining the thermometer. Whatever the source of the idea, however, once

Black had thought of using mixtures to measure the quantity of latent heat in melted ice, he considered it an 'obvious method'.[71] As he told his students,

> when hot and cold water are mixed together, the excess of heat contained in the hot water is equally distributed in an instant through the whole mixture, and raises the temperature of it according to the greatness of the excess of temperature, and the proportion which hot water bore to the cold. If the quantities of hot and cold water are equal, the temperature is the middle degree between that of the hot and that of the cold. No part of the heat disappears on this occasion, so far as we can perceive, but the intensity of it only is diminished, by its being diffused through a larger quantity of matter. It was, therefore, obvious, that if a quantity of heat is absorbed, and disappears in the melting of ice, this would easily be perceived when the ice is melted with warm water.
>
> To make this experiment, I first froze a quantity of water in the neck of a broken retort, in order to have a mass of ice of an oblong form.
>
> At the same time I heated a quantity of water, nearly equal in weight to the ice, in a very thin globular glass, the mouth of which was sufficiently wide to take in the piece of ice. The water was heated by a small spirit of wine lamp applied to the bottom of the glass; it was also often stirred with the end of a feather, and a thermometer hung in it.
>
> While the water was heating, the mass of ice was taken out of the mould in which it had been formed, and was exposed to the air of a temperate room, until it was perceived to be beginning to melt over the whole of its surface.
>
> I then put a woollen glove on my left hand, and taking hold of the ice, I wiped it quite dry with a linen towel, laid it in the scale of a balance on a piece of flannel, and hastily counterpoised it with sand in the opposite scale, that I might examine the weight of it afterwards; and I immediately plunged it into the hot water, and extinguished the lamp at the same time. The lamp being small, the heat of the water had been increasing very slowly, and had almost ceased to increase; and being examined immediately before I put the ice into it, the temperature was found to be just 190 degrees. The ice was all melted in a few seconds, and produced a mixture, the temperature of which was 53 degrees.
>
> The weight of the ice, when put into the hot water, was seven ounces three drachms and a-half. The weight of the glass, with

the whole mixture in it, was sixteen ounces, seven drachms, and seven grains. The weight of the glass alone was nearly one ounce.

In considering this experiment, we may overlook quantities less than half a drachm, or thirty grains, and reckon the quantities of the different articles, by the number of half-drachms in each.

Thus the weight of the ice was 119 half-drachms.

---- Hot water	135
---- Mixture	254
---- Glass alone	16

The melting of the ice was effected, not only by the heat of the hot water, but also by that of the glass. And, by other experiments, I learned that sixteen parts of hot glass have no more power in heating cold bodies, than eight parts of equally hot water; we may therefore substitute, in place of the sixteen half-drachms of warm glass, eight half-drachms of warm water, which, added to the above quantity of warm water, make up 143 half-drachms.

The heat of this warm water was 190 degrees, that is 158 hotter than the ice; and if this heat had abated in the mixture only in consequence of the quantity and coldness of the ice, and if nothing had happened when the ice was put in, but merely a communication of this heat, and an equal distribution of it through the mixture, the temperature of the mixture should have been 158, viz. the excess of heat in the warm water, multiplied by 143, the quantity of the warm matter, and divided by 262, the quantity of the whole, which gives 86.

The mixture should have been 86 degrees warmer than melting ice; but it was found only 21 degrees warmer. Therefore, a quantity of heat has disappeared, which, if it had remained in a sensible state, would have made the whole mixture and glass warmer by 65 degrees than they were actually found to be. But this quantity of heat would be sufficient for increasing, by 143 degrees, the heat of a quantity of water, equal in weight to the ice alone. It was, however, absorbed by the ice, without in the least increasing its sensible heat.[72]

Armed with the confirming data provided by these two experiments, Black presented his doctrine before the Glasgow Literary Society. In his lectures he stated that 'these two experiments, and the reasoning which accompanies them, were read by me in the Philosophical Club, or Society of Professors and others in the University of Glasgow, in the year 1762, and have been described and explained in my lectures, there, and in Edinburgh, every year since.'[73] And

Robison, who had the registers of that society before him, was able to be even more specific and give the date of Black's presentation as 23 April, 1762.[74]

Having established the truth of his new theory, Black then used it to explain the phenomena Cullen had studied when forming his theory of heat and mixture. Black told his students that his doctrine of latent heat enabled him to explain 'some other remarkable facts relating to cold, of which no explanation had before been attempted. The facts I mean, are those experiments in which intense degrees of cold are produced by mixture.'[75] Unlike Cullen, however, Black did not concentrate his attention upon the solution of salts in water. He examined instead mixtures 'of ice and snow with different salts; and it always happens,' he emphasized, 'that while the cold is produced, the salt and snow dissolve together into a liquid'.[76] In his concluding statement, however, Black presented his theory in a way which encompassed all the facts which Cullen had relied upon in forming his chemical theory of heat and mixture. 'I think you are now sufficiently aware,' Black told his students,

> that, since fluidity is the consequence of a certain combination of calorific matter with the substance of solid bodies, it must follow, that if any substances are induced, by their chemical properties, to act on each other, so as mutually to dissolve and become fluid; and if there be no external source, from which the necessary latent or fluid-making heat can be obtained, it must be furnished by the materials themselves; that is, some of their sensible heat must disappear,—their temperature, indicated by the thermometer, must be lowered: And this must happen in a degree so much the more remarkable, as there is more of the solid matters rendered fluid, and also, in proportion as the substances, so liquefied, require a greater quantity of latent heat for this purpose.[77]

Although Black established the truth of his doctrines by studying the solidification and liquefaction of water, he did not believe that his theory was restricted to explaining the properties of water alone. Sometime in the early 1760s he therefore asked his assistant William Irvine to investigate the latent heat in several substances which are gradually transformed from a solid into a liquid as heat is added. 'The late ingenious Dr. Irvin,' Black reported in his later lectures,

> who was my pupil at Glasgow, and afterwards my successor in the chemical chair, making some experiments, at my desire, on the melting and congealing of spermaceti, and bees wax, and some other substances, found that the spermaceti in melting, absorbed a quantity of heat, which, though it did not make it

> sensibly warmer, but only fluid, would have been sufficient for making melted bees wax warmer by 175 degrees. And from experiments with tin, he found reason to conclude, that this metal, while melting, absorbs a quantity of heat equal to what might add 500 degrees of heat to it in its solid state.[78]

Reid's account of Black's early lectures reveals that Black was particularly struck by the difference between the effects heat produced in the substances Irvine studied and those it produced in water. 'Spermaceti,' Reid wrote,

> in passing from a solid to a perfectly fluid form, requires about 150° of heat, which becomes latent; bees' wax about 160°. But there is this remarkable difference between these bodies—as well as iron and some other metals on the one hand, and water on the other—that the former soften by degrees, so that there are many intermediate degrees of softness between the hardest state which the body takes by cold, and the state of perfect fluidity; whereas in water there seems to be no intermediate degree between perfect ice and perfect water. Accordingly, in spermaceti, bees' wax, and iron, the latent heat is more or less, according to the degree of softness; but in water it is always the same.[79]

Black, however, did not accept the conclusion Irvine drew from these experiments, namely that latent heat can be explained in terms of the change in specific heat which occurs as a body changes its state of aggregation.[80] Having founded his theory upon the properties of water, a common substance in which the relationship between heat and change of state is clear and dramatic, Black was not inclined to attach great significance to the curious properties of these complex substances studied by Irvine. Irvine, however, believed he had discovered a property of matter which had profound implications for the theory of heat, and although he began as Black's disciple, his interpretation of the experiments they performed together rapidly diverged from that of his master. As Robison later wrote, 'Dr. Black's speculations were anxiously directed to one favorite object,—the theory of fluidity and vapour; and Dr. Irvine's to the discovery of the absolute heats of bodies.'[81]

Robison was puzzled by the order in which Black chose to present his experiments on latent heat. Black's early notebooks reveal that he had been thinking about the heat absorbed during both melting and vaporization before he began lecturing on his doctrine in Glasgow. Yet when it came time to present the doctrine and describe the experiments that support it, Black dealt first with the latent heat of fusion. Robison was surprised that the experiments on the latent heat of vaporization, which in retrospect seem to illustrate the most striking aspects of the theory, had not been presented first. 'The fundamental phenomena [of vaporization], and the experiments for determining this part of the theory,' he wrote in his notes to Black's lectures, 'appear to me to be much more obvious and impressive than in the other case. I am so much of this opinion *now*, that I wonder, as often as I think of it, that the heat latent in steam was not detected even ages ago.'[82]

When describing the evolution of his doctrine, Black also expressed surprise at his own resistance to the notion of the latent heat of vaporization. He knew that if a pan of water is set upon a stove heated by a steady fire, 'it requires about six times the number of minutes to boil off a small quantity of water, that it takes to bring it to boil'.[83] He then said that 'I can scarcely remember the time in which I had not some confused idea of this disagreement of the fact with the common opinion [that water will be converted into a vapour as soon as it reaches the temperature at which it boils]; and I presume that it has come across the mind of almost every person who has attended to the boiling of a pot or pan. But the importance of the surmise never struck me with due force,' he admitted, 'till after I had made my experiments on the melting of ice. The regular procedure in that case, and its similarity to what appears here, encouraged me to expect a similar regularity in the boiling of water, if my conjecture was well founded.'[84]

In his lectures Black explained his reluctance to examine the phenomenon of vaporization by claiming that he had been unable to think of an experiment which would yield an accurate value for the latent heat of vaporization of water.[85] But this excuse is unconvincing, for when he decided to determine the amount of latent heat in steam, Black simply set a pan of water on a stove and noted the time it took to bring the water to a boil and the time needed to boil off all the water. It therefore seems much more likely that at the outset, Black did not consider vaporization to be solely an effect of the fixation of heat.[86] He knew that boiling and condensation, unlike freezing and melting, are strongly affected by pressure, and he may have thought that given Cullen's account of boiling in a vacuum, vaporization should not be considered a chemical process at all. Thus he could not apply his theory of latent heat to vaporization until he had determined how both latent heat and the pressure of the atmosphere affect the process of boiling.

In his lectures Black said that when philosophers first examined the phenomena of boiling, 'different opinions were formed with respect to the causes on which they depend. Some thought that this agitation was occasioned by that part of the heat, which was more than the water was capable of receiving, and which forced its way through, so as to occasion the agitation of boiling; others again imagined, that the agitation proceeded from air, which water is known to contain, and which is now expelled by the heat.'[87] But, he argued,

> neither of these accounts . . . is just or satisfactory; the first is repugnant to all our experience in regard to heat: We have never observed it in the form of an expansive fluid like air. It prevades all bodies, and cannot be confined by any vessel, or any sort of matter, whereas the elastic matter of boiling water can be confined by external pressure, as is evident in the experiments made with Papin's digester. . . . The second explanation is equally unsatisfactory, nor is it intelligible; for, if we continue to apply heat to the water, it will continue to boil to the last drop, and we should therefore conclude that it is all air. But to be satisfied that this opinion is not true, we need only apply cold, or a cooling cause, to this supposed air, and we shall immediately find it condensible into the same quantity of water from which it was produced.[88]

According to Black, airs and vapours are alike in being elastic but can be distinguished by the way in which they become fixed, vapours condensing into the liquid from which they were formed and airs being fixed by combination with solid or liquid substances. At a more detailed level of description, each of the elastic fluids has its characteristic chemical properties. Black had previously demonstrated that airs have distinctive chemical properties, and he assumed that all elastic fluids contain heat in chemical combination. He thus sought, without confusion or a falsifying reductionism, to explain their common property of elasticity in terms of the latent heat which they contain. Black's initial achievement was to separate the property of elasticity from air and attach it to the concept of heat.[89] It was this achievement that gave his theory of latent heat its range and power.

On 4 October 1762, Black began measuring the latent heat of steam. He said in his lectures that 'I procured some cylindrical tinplate vessels, about 4 or 5 inches diameter, and flat-bottomed. Putting a small quantity of water into them, of the temperature 50 degrees, I set them upon a red hot kitchen table, that is, a cast-iron plate, having a furnace of burning fuel below it, taking care that the fire should be pretty regular. After four minutes, the water began sensibly to boil, and in twenty minutes more, it was all boiled off.'[90]

After performing the experiment two more times Black

> reasoned from these experiments in the following manner: The vessels in the first experiment received 162 degrees of heat in four minutes, or 40½ degrees each minute. If we, therefore, suppose that the heat enters equally fast during the whole ebullition, we must suppose that 810 degrees of heat have been absorbed by the water, and are contained in its vapour. Since this vapour is no hotter than boiling water, the heat is contained in it in a latent state, if we consider it only as the cause of warmth. Its presence is sufficiently indicated, however, by the vapours or expansive form which the water has now acquired.[91]

It appears that even after having determined the quantity of latent heat in steam, Black continued to worry about the effects of pressure upon the formation of water vapour. We know that his friends and pupils were studying the relationship between pressure and boiling points during the period 1762–4. In 1761 or 1762 James Watt built a model engine driven by the steam produced by a Papin's digester, and he probably also then made his first experiments on the rise in the boiling point produced by an increase in pressure.[92] But it seems likely that Black was particularly encouraged by the investigation undertaken by John Robison following his return to Glasgow in 1763. Black mentioned in his lectures that 'at the time I was engaged in these experiments, Professor Robison, then one of my pupils, told me, that in some experiments which he had made with the air-pump belonging to the University of Glasgow, however hot the water was at first, the heat of it was rapidly diminished till it fell to the 90th or 91st degree of Fahrenheit.'[93] In 1764 Robison also performed experiments 'upon water, spirit of wine, mixtures of spirits and water, and sulphuric acid'. These experiments showed 'that the difference between the temperatures at which they boil, *in vacuo*, and in the open air, is nearly the same in all, namely, 120 degrees of Fahrenheit's scale.'[94] These experiments thus all indicated that while pressure has a great effect upon the boiling point of liquids, the relationship between pressure and boiling point is a general one, not one which depends upon the specific properties of the liquid being examined. Chemists could therefore continue to assume that the amount of latent heat needed to vaporize each of the chemically distinct liquids can be directly correlated to the specific properties of the substances. In other words, Robison's experiments assured Black that in formulating his chemical theory of latent heat, he could ignore the mechanical relationship between pressure and boiling point.

It also seems likely that William Irvine, who, Robison informs us, assisted Black during 'his first experiments for measuring the latent heat of steam,'[95] was thinking about the relationship between pressure and the latent heat of vaporization during this period. In his lectures Black told his students that

> almost immediately after my discovery of this remarkable absorption and concealment of heat, in the phenomena of fusion and ebullition, Dr. Irvin, whose speculations were always ingenious, conceived that there was no occasion for the distinction between sensible and latent heat, in the peculiar circumstances of these phenomena. He imagined, that in every case there is a portion of heat which does not affect the thermometer; and that the reason why there is a seeming disappearance in liquifaction and ebullition is, that when the element changes its form from that of water to that of vapour, its capacity for heat is suddenly increased; and, consequently, it requires much more heat to be thrown into it, in order to make it appear equally hot by the thermometer.[96]

Irvine focused his attention upon the total heat in bodies, that is to say, the sum of their latent and sensible heats.[97] Irvine therefore may well have argued that no matter what the pressure and temperature at which steam is formed, it always contains the same amount of total heat. This line of reasoning leads one to the prediction that steam produced at lowered pressures, and therefore at a lowered boiling point, will have a greater latent heat than steam produced at higher pressures. This conclusion, which Watt later determined to be roughly true, and thereafter came to be called Watt's law,[98] fitted well with the assumption that the laws governing the fixation of heat in each particular liquid and those governing the general relationship between pressure and boiling point can be treated independently. Encouraged by these facts and hypotheses, Black continued his research on the latent heat of vaporization.

On 9 October 1764 Black and Irvine performed an experiment designed to measure the quantity of heat released during the condensation of steam.

> Five measures (each containing 4 lb. 5 oz. and 6 dr. avoirdupois) of water, of the temperature 52°, were poured into a small still in the laboratory. The fire had been kindled about 40 minutes before, and was come to a clear and uniform state. The still was set into the furnace, and, in an hour and twenty minutes, the first drop came from the worm; and in three hours and forty-five minutes more, three measures of water were distilled, and the experiment ended. The refrigeratory contained 38 measures of water, of which the temperature, at the beginning of the experiment, was 52°. When one measure had come over, the water in the refrigeratory was at 76°. When two had come over

> it was at 100°; and when three had come over, it was at 123°.
>
> In this experiment, the heat, which emerged from three measures of water, had raised the temperature of the water in the refrigeratory from 52° to 123°, or 71°. Now 3 is to 38 as 71 to $899\frac{1}{3}$, and the heat would have raised the three measures to $899\frac{1}{3}$ degrees in its temperature, if this had been possible without converting it into vapour. The heat of the vapour from which this emerges, was 212°, or 160° more than that of the water. Taking this from 899°, there remains 739°, the heat contained in the vapour in a latent state.[99]

After describing this experiment, Black summarized his main point by saying that

> what I have related is sufficient for establishing the main principle, namely, that the heat which disappears, in the conversion of water into vapour, is not lost, but is retained by the vapour, and is indicated by its expansive form, although it does not affect the thermometer. This heat emerges again from this vapour when it becomes water, and recovers its former quality of affecting the thermometer; in short, it appears again as the cause of heat and expansion.[100]

As a final demonstration that vaporization and condensation can be explained in terms of his theory of latent heat, Black described what happens when steam is suddenly compressed. 'This,' he said,

> is a nice and troublesome experiment, and requires a very nice apparatus. A metal cylinder, nicely bored, is fitted with a piston; and the cylinder is made to terminate below, in a small glass tube, close at the end. A thermometer is so fixed to the cylinder, that its bulb is in the axis, and the tube projects outwards through the cylinder. A little water is poured into it, so as to fill the glass tube at the bottom. The piston is put into the top, and a small hole in the piston is left open. A large glass cylinder is filled with water, kept boiling hot. The water in the glass tube is made to boil briskly, the steam blowing through the hole in the piston. In this state it is immersed in the boiling water, the hole in the piston speedily stopped, and the piston thrust down with great force, at least half way. The thermometer rises immediately, and the water in the glass tube is seen to increase in bulk. The whole must be done with great dispatch.[101]

The complication raised by the variation of the boiling point with pressure had been overcome and the truth of the new theory of vaporization had been amply demonstrated.

Where does the concept of capacity for heat fit into Black's theory of latent heat? The answer is that the notions of capacity for heat and latent heat are logically quite independent of each other. In fact, Black paid almost no attention to capacity for heat while developing his theory. What he knew about capacity for heat, however, convinced him that the mechanical philosophers could not explain all the properties of heat. This knowledge was in turn important in legitimizing his chemical approach to the subject. From Cullen's discussions of Fahrenheit's mixture experiments and Martine's experiments on rates of heating, Black learned that the relative capacities which different substances have for heat cannot be correlated to their volumes or to their densities.[102] In his own lectures Black carefully demonstrated that capacity for heat is a specific property. 'It was formerly a common supposition,' he told his audience,

> that the quantities of heat required to increase the heat of different bodies by the same number of degrees, were directly in proportion to the quantity of matter in each; and therefore, when the bodies were of equal size, the quantities of heat were in proportion to their density. But very soon after I began to think on this subject, (anno 1760) I perceived that this opinion was a mistake, and that the quantities of heat which different kinds of matter must receive, to reduce them to an equilibrium with one another, or to raise their temperature by an equal number of degrees, are not in proportion to the quantity of matter in each, but in proportions widely different from this, and for which no general principle or reason can yet be assigned.[103]

As evidence of the truth of this assertion, Black gave a detailed account of Fahrenheit's mixture experiment, as described by Boerhaave, and of Martine's experiments. 'It appears, therefore, from the general results of such experiments,' he concluded,

> that if we had a thousand masses of matter, of the same size and form, but of different materials, and were to place them all in the same room, until they assumed the same temperature; were we then to introduce into that room a great mass of red hot iron, the heat of which, when communicated with all these different bodies at the same time, might be sufficient for raising the temperature of them all, by 20 degrees; the heat thus communicated from the iron, although it produced an equal effect on each of these bodies, in raising its temperature by 20 degrees, would not however be equally divided or distributed among them. Some of them would attract and retain a much greater quantity of this heat, or matter of heat, than others; and the quantity received by each would not be in proportion to their densities, but in proportions totally unconnected with it; and perhaps not any two of them would receive precisely the same quantity but

> each, according to its particular capacity, or its particular force of attraction for this matter, would attract and require its own peculiar quantity to raise its temperature by the 20 degrees, or to reduce it to an equilibrium or equality of saturation with the surrounding bodies. We must, therefore, conclude that different bodies, although they be of the same size, or even of the same weight, when they are reduced to the same temperature or degree of heat, whatever that be, may contain very different quantities of the matter of heat; which different quantities are necessary to bring them to this level, or equilibrium, with one another.[104]

Then, in a final paragraph, Black emphasized that the mechanical conception of heat as motion was incapable of accounting for the facts revealed through the careful examination of the capacities of various bodies for heat.

Clearly Black was aware that substances have different capacities for heat long before he began quantifying his theory of latent heat. Robison found that 'in some small note books, written while a student of medicine in Edinburgh, he sets down queries, founded on Fahrenheit's experiment, and projects of experiments, on a variety of substances which had exhibited puzzling appearances in this respect. He made several of these before 1757.'[105] Furthermore, as McKie and Heathcote have pointed out, Black was careful to make allowance for the thermal capacities of the vessels used in his experiments on thermometric substances and on the heat absorbed during the melting of ice.[106] The fact that Black chose not to quantify his knowledge of specific heat during his first years in Glasgow is therefore sufficient evidence that, unlike his pupil William Irvine, he did not consider knowledge of this particular property relevant to the theoretical understanding of the phenomena he was studying.

Robison wrote that on the question of specific heat, 'Mr. Watt of Birmingham was, as far as I can learn, the first who considered this subject steadily, and in a system. He was strongly incited to this examination by the great object which then occupied his attention,—the improvement of the steam engine.'[107] In his first effort towards improvement, Robison said, Watt studied

> the quantity [of steam] thus wasted by warming the cylinder and other parts of the apparatus; and some of his earliest attempts to improve the engine consisted in employing cylinders made of substances which required little heat for warming them. He made many experiments on various substances, in order to learn their properties in this respect. They were but rude, but sufficient for his purpose; and the examination would have been carried much farther, had he not contrived the separate condenser

which made this property of bodies of little moment to him.[108]

Perhaps Watt's investigation revived Black's interest in capacities for heat. Robison states that 'Mr. Watt's experiments were all communicated to his preceptor, who took the warmest interest in them; and this correspondence gave him the opportunity of knowing the uncommon genius and the great worth of his pupil,'[109] and it seems likely that given a practical reason to investigate this subject, Black decided to perform some experiments of his own. In any case, we know from Robison's notes that

> before the year 1765, Dr. Black had made many experiments on the heats communicated to water by different solid bodies, and had completely established their regular and steady differences in this respect. He mentioned these in his lectures in the University of Glasgow, and was occupied with the experiments for the same purposes when he was called to Edinburgh. He was much assisted in the experiments by Mr. Irvine, who afterwards lectured from the same chair. The registers of these experiments remain among Dr. Black's papers, most of them dated, and partly written by himself, and partly by Mr. Irvine. They are all previous to 1770.[110]

Perhaps Black decided that after verifying and quantifying his theory of latent heat, he should examine Irvine's theory of specific heat more carefully. Whatever his motivation, however, we know that Black devoted a considerable amount of time to the study of capacities for heat. But we also know that in doing so, he was simply examining more closely a phenomenon which he already knew could neither be explained in mechanical terms nor made the basis of an adequate theory of chemical change.[111]

[10]

Black's Influence on the Emerging Science of Heat

1. JAMES WATT

In June 1754, the same month that Joseph Black published his inaugural dissertation in Edinburgh, James Watt left his home town of Greenock and moved to Glasgow.[1] Eighteen years old and eager to make his own way in the world, Watt spent his first year in Glasgow living with his mother's relatives, the Muirheads, while seeking a position in which he could acquire the skills needed to become a mathematical instrument maker. His choice of trade reflected both his own mathematical proclivities, which seem to have been a characteristic of the Watts, and his intimate knowledge of his father's business. Thomas Watt, James' father, was 'a shipwright; a ship-chandler, supplying vessels with nautical apparatus, stores and instruments; a builder; and a merchant,' and he was also for nearly twenty years 'a member of the Town Council of Greenock, and, during great part of that time, its Treasurer; a magistrate; and always a zealous and enlightened promoter of the improvements of the town.'[2] Thus when James Watt set out for Glasgow he already had a rather thorough preliminary knowledge of the trade he had chosen.

In Glasgow Watt had the good fortune to be introduced to several members of the university faculty through his relative George Muirhead, who was then professor of Latin. Although James never attended a course of lectures given in the university, he did become friendly with Dr Dick, the young professor of natural philosophy who had previously been Black's closest companion in Glasgow. When Professor Dick became aware of Watt's interest in mathematical instruments, he strongly advised him to learn his trade in London and he further encouraged him with letters of introduction. Thus in June 1755 Watt set off on horseback for London. At first he encountered considerable difficulty in finding a master, for the small circle of skilled instrument makers in London were reluctant to dispense with the customary seven year apprenticeship. With Dick's letters, however, Watt gained the attention and then the friendship of James Short, a Fellow of the Royal Society and a celebrated maker of reflecting telescopes, and through Short he obtained a position with John Morgan, a maker of Hadley quadrants.[3] After a year of arduous work and frugal living Watt left London and returned to Scotland to practice his newly acquired skills.

Professor Dick was evidently glad to have Watt back in Glasgow, for shortly after his return Dick asked him to help unpack and repair a set of astronomical instruments which had been given to the university. After spending a few months on this task, Watt tried to establish himself as a craftsman in the city. He soon found, however, that not being the son of a burgess, nor having married a burgess' daughter, and not having served a regular apprenticeship, he would not be allowed to practise his trade within the city.[4] He therefore was pleased when the university decided, in the summer of 1757, to allow him to occupy an apartment and open a shop within the walls of the college and to designate him the 'Mathematical instrument maker to the University'.[5]

Shortly after moving into the university, James Watt first met John Robison.[6] Robison was a bright young man who had recently taken his degree at the university and was evidently at a loose ends. In 1757 he tried without success to obtain an appointment as assistant to Professor Dick.[7] Shortly thereafter he became interested in the astronomical observatory being built by the university and this led to his being introduced to Watt. He discovered that Watt was a man of the highest ability and an enthusiastic investigator of mechanical questions, and the two young men quickly formed a friendship which lasted until Robison's death in 1805.[8]

Late in 1756 Joseph Black returned to Glasgow to become professor of anatomy and botany and lecturer in chemistry. He was, no doubt, soon introduced to Watt by their mutual friend Professor Dick. Many years later Black recalled that after meeting Watt, 'I soon had occasion to employ him to make some things which I needed for my experiments, and found him to be a young man possessing most uncommon talents for mechanical knowledge and practice, with an originality, readiness, and copiousness of invention, which often surprised and delighted me in our frequent conversations together.'[9] Black apparently enjoyed visiting Watt's workshop and it was there, Robison later recalled, that he first met his future mentor.[10] 'My first acquaintance with Dr. Black began in your Rooms,' Robison wrote to Watt,

> when you was rubbing up McFarlanes Instruments. Dr. Black used to come in, and, standing with his back to us, amuse himself with Birds Quadrant. . . . His marked attention to me (as he told me not long ago) was owing to my saying distinctly, and giving reason for it, that Dr. Dick, my professor, had infinitely more knowledge than his Successor, who was much more

> popular. Indeed Dr. Black has often said to me that dick was one of the most sensible and manly fellows he ever knew.[11]

Throughout 1758 these three young philosophical bachelors thrived upon each others company. Black was then trying to think of ways in which his theory of latent heat could be extended and tested, Watt was struggling to make a go of it as an instrument maker, and Robison was casting about for a situation in which he could fasten his agile mind to a concrete set of problems. Serious tasks all, but evidently pursued with an enthusiasm which made light work of heavy thinking. As an example of the way in which Watt threw himself into challenges he found appealing, Robison later recalled that 'a Mason lodge in Glasgow wanted an Organ. The office bearers were acquaintances of Mr. Watt. We imagined that Mr. Watt could do any thing, and although aware *that he did not know one musical note from another*, he was asked if he could build this Organ. . . . He said Yes. But he began by building a very small one for his intimate friend Dr. Black.' Watt became so intrigued by the problems he encountered while building this model that 'he then began to study the philosophical theory of Music. Fortunately for me,' Robison wrote, 'no book was at hand, but the most refined of all, Smiths Harmonics. Before Mr. Watt had half finished his organ, he and I were compleatly masters of that most refined and beautiful theory of the Beats of imperfect consonances.'[12] In later years Black asserted that he had been so impressed by Watt's character upon first meeting him that 'I therefore contracted with him an intimate friendship, which has continued and increased ever since that time.'[13] And Watt, while anxious to correct Robison's assumption that he had attended Black's lectures in Glasgow, recalled that 'every thing I learnt from him was in conversation and by doing small mechanical jobs for him. These Conversations and those I had with you [Robison] served to give me true notions in Science and to develope the powers of my Mind.'[14]

Sometime during this buoyant year of 1758 Robison first drew Watt's attention to the steam engine. In 1765 Watt wrote in his notebook that

> about 6 or 8 years ago My Ingenious friend Mr. John Robison having conceived that a fire engine might be made without a Lever—by Inverting the Cylinder & placing it above the mouth of the pit, proposed to me to make a model of it, which was set about but never Compleated, he going abroad & I having at that time little knowledge of the machine. However, I always thought the Machine Might be applyed to other as valuable purposes as drawing Water.[15]

Robison and Watt hoped to improve the operation of the Newcomen engine, which was then widely used to pump water from deep mines. The piston of this engine was connected to one end of a great beam, the other end of which was attached to the chains that operated the pumps. The beam was so balanced that the weight of the pump chains drew the piston upwards in the cylinder. As it travelled upwards, the piston drew steam from the boiler into the cylinder. When the piston had reached the top of its stroke, cold water was sprayed into the cylinder to condense the steam. This condensation created a partial vacuum in the cylinder, thereby causing the piston to be driven downwards by the pressure of the atmosphere. Robison suggested that the traditional arrangement of the Newcomen engine could be greatly simplified by inverting the cylinder over the pumps and thereby eliminating the beam, but Watt's attempt to make a model of this modified engine evidently met with little success.[16]

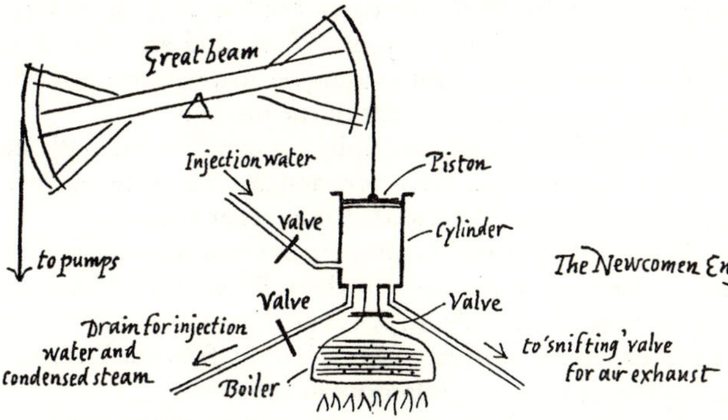

The Newcomen Engine[17]

Later in 1758 Robison left Glasgow to embark on an adventurous four years as a midshipman. The year after his departure Watt entered into a partnership designed to extend his business and he was soon obliged to devote a good deal of time to supervising the journeymen he employed to meet his steadily increasing trade.[18] Black and Watt no doubt continued to see each other often,[19] but the problems which now occupied Watt in his workshop were no longer chosen as freely or treated as philosophically as before. During the next few years he did, however, find time to construct a model of an engine driven by the pressure of steam rather than by the pressure of the atmosphere. As Watt recalled many years later,

> about the year 1761, or 1762, I tried some experiments on the force of steam, in a Papin's digester, and formed a species of steam-engine by fixing upon it a syringe one-third of an inch diameter, with a solid piston, and furnished also with a cock to admit the steam from the digester, or shut it off at pleasure, as

> well as to open a communication from the inside of the syringe to the open air, by which the steam contained in the syringe might escape. When the communication between the digester and syringe was opened, the steam entered the syringe, and by its action upon the piston raised a considerable weight (15 lb.) with which it was loaded. When this was raised as high as was thought proper, the communication with the digester was shut, and that with the atmosphere opened; the steam then made its escape, and the weight descended. The operations were repeated, and though in this experiment the cock was turned by hand, it was easy to see how it could be done by the machine itself, and to make it work with perfect regularity. But I soon relinquished the idea of constructing an engine upon this principle, from being sensible it would be liable to some of the objections against Savary's [*sic*] engine,[20] viz. the danger of bursting the boiler, and the difficulty of making the joints tight, and also that a great part of the power of the steam would be lost, because no vacuum was formed to assist the descent of the piston.[21]

Watt was not sufficiently encouraged by his experiments with steam under pressure to continue, however, and he therefore put his investigation of the steam engine aside for another year.

In 1763 Watt moved from his quarters in the university to a small apartment in the city, and in July of the following year he married his cousin Margaret Miller.[22] In leaving the university, however, Watt did not cut himself off from his scientific friends. Robison recalled that upon his return to Glasgow in 1763,

> I was happy to find Mr. Watt settled in Glasgow, as fond of science as ever. Our acquaintance was renewed; I believe, with mutual satisfaction, for I had now acquired some knowledge. . . . Our acquaintance at this time became very intimate, and I believe neither of us engaged far in any train of thought without the other sharing in it. I had had the advantage of a more regular education: this frequently enabled me to direct or confirm Mr. Watt's speculations, and put into a systematic form the random suggestions of his inquisitive and inventive mind. This kind of friendly commerce knit us more together, and each of us knew the whole extent of the other's reading and knowledge. I was not singular in this attachment. All the young lads of our little place that were any way remarkable for scientific predilection were acquaintances of Mr. Watt; and his parlour was a *rendezvous* for all of this description.[23]

Clearly the passage of time had not weakened the mutual admiration and shared ardour for science and improvement which attracted

Watt, Robison and Black to each other. Watt himself recalled an investigation which reveals that after Robison's return their joint studies were again conducted as enthusiastically as they had been in 1758. Writing in 1805, Watt remembered 'that in 1765, on seeing a perspective machine, with a double sliding motion at right angles to each other, which Dr Lind had brought from India, and had made some improvements upon, Mr R. suggested that the same thing might be better accomplished by a double parallelogram (as some parallel rulers were then made),—an idea which I pursued, and, by some improvements and additions of my own, made a useful instrument.'[24] Watt then constructed a model of this perspective machine, which, like his first organ model, he presented to his friend Joseph Black.[25]

The Newcomen engine continued to present a set of problems which interested all three philosophers. Robison wrote that upon his return to the University of Glasgow in 1763, he found Watt 'most intimately connected with Dr Black, and continually speaking of his Lectures and doctrines.'[26] It is difficult now to determine precisely what aspects of Black's heat theory Watt had mastered during Robison's absence. Shortly after Robison's return, however, Watt began an investigation which transformed the study of heat into a practical question, one which soon commanded all the concentrated power of his attentive mind. During the winter of 1763–64 John Anderson, who had succeeded Dr Dick as professor of natural philosophy, asked Watt to repair the university's model of the Newcomen engine.[27] At first, Watt later wrote, 'I set about repairing it as a mere mechanician, and when that was done and it was set to work, I was surprized to find that its boiler could not supply it with [enough] steam, though apparently quite large enough.'[28] Why wouldn't this model, having been made mechanically sound, perform as expected? This was the question that captivated Watt, and aided by his friend John Robison, who like Watt 'had not yet studied chemistry,'[29] he set about finding the answer.

In attempting to reconstruct the progress of Watt's thinking as he studied the model steam engine, we must rely on his own accounts, the earliest of which was written thirty years after the investigations took place, and on his recently published notebook, the relevant parts of which appear to date from the first three months of 1765.[30] Both these sources reveal that from the very outset Watt analyzed the steam engine in terms of its ability to turn the heat produced by the burning of fuel into a usable mechanical force. After trying to make the repaired model of the Newcomen engine operate properly, Watt wrote in his notebook that 'I was surprized at the Imense

Quantity of fuel it consumed in proportion to its Cylinder, which was only 2 In diameter.'[31] Watt never doubted that the way to improve the efficiency of the engine was to reduce the amount of heat that passed through the engine without producing any useful work. It seems likely that he concentrated upon the quantity of heat acting in the engine, rather than upon the mechanical aspects of its operation, because he was aware of Black's approach to the study of chemistry. And it was clearly because of his concentration upon the transformation of heat into 'duty' in the steam engine that Watt was able to invent the separate condenser.[32]

The task at hand was to conserve heat. 'Mr Robison being now returned,' Watt wrote in his notebook, 'it was hinted in some of our Conversations together that if the Cylinder was made of wood it would not occasion the Loss of so much heat, being less susceptible of heat or cold than Mettals.'[33] Although he doubted whether this was a practical suggestion, Watt 'was resolved to make a model of the Machine in the common way, only with a wooden Cylinder'.[34] At the same time he decided 'to begin by Making the boyler perform with as little heat as posible'.[35]

Watt first examined the way in which steam is produced by the boiler. He thought that by intermittently dropping water on a hot surface or by varying the surface area of the evaporating pan he might effect some economy of fuel, but a few experiments convinced him that 'the evaporation of water is neither in proportion to the surface nor quantity of it, but as the quantity of heat that enters it'.[36] These initial experiments brought him to the rather humbling realization that when it came to the efficient heating of water, there was 'no boyler so perfect in that Respect as the Common tea kitchen'.[37] Watt then described a few modifications he made in his boiler, such as covering the outside with wood, and concluded by saying that 'this boyler I put in practice & its effects answered my expectations [by] evaporating prodigious Quantities of Water & consuming little fuel'.[38]

Watt next turned his attention to the problem of preventing the loss of heat through the cylinder. He made a wooden cylinder out of fir, but after becoming soaked with steam it contracted in the middle, causing the piston to bind. He tried a second cylinder of fir and again encountered the same difficulty. Perhaps oak would work, Watt thought, but he knew how soft the oak sheathing on his boiler had become and he therefore doubted that success could be achieved by using a wooden cylinder.[39]

If Watt was disappointed by these early reversals, surely his next investigation was more discouraging. 'In working this [Newcomen] machine,' he wrote, 'I perceived that the water issuing from the sinking pipe was always near boyling hot & that when the Steam was weak it was a long time after opening the Steam-Cock before any Steam came out of the snifting valve.'[40] Watt was troubled by the fact that the cold water injected into the cylinder to condense the steam had itself become nearly boiling hot. It was bad enough that the injection water carried off the heat from the steam in the cylinder at the time the water was injected, but now Watt was convinced that the injection water was also absorbing a great deal of heat from the steam which was being introduced into the cylinder to start the next cycle. The water, in other words, was condensing steam which had not done any work. As Watt said, 'on consideration I found that the water in the Cylinder could not be forced out by the steam till it was near the heat of boyling water as it [the water] would till then undoubtedly Condense all the Steam Coming in Contact with it, & I have since found that the Quantity of steam Condensed would be many fulls of the Cylinder.'[41]

In the face of these apparently unavoidable heat losses, Watt temporarily gave up on the Newcomen engine. 'About this time,' he wrote, 'I conceived the Idea of making a Machine Something like Capt Savarys but to raise water to any height by suction, only by making a series of Receivers one above another so that each would raise the Water to the sucking pipe of the other, the steam to be conveyed to them thro copper pipes surrounded with Chaff to prevent the escape of heat.'[42] If the Newcomen engine could not be improved, at least when using a steam engine to pump out a mine one might be able to avoid the waste involved in heating and cooling a piston and driving heavy mechanical pumping equipment. And if a series of Savery engines were employed, then there would be no need to use steam pressure to force a column of water up a pipe. One could therefore avoid the difficult task of making steam under pressure. But the goal was heat efficiency, and as Watt immediately reminded himself in his notebook, 'before anything could be known certainly about the best manner of Constructing Fire engines, there were several facts necessary to be determined'.[43]

When he began searching for ways to improve the steam engine, Watt turned first to the writings of Belidor and Desaguliers.[44] Evidently he paid particular attention to the findings Desaguliers presented in his memoir on the vaporization of water, the same memoir that Cullen discussed at such length in his chemistry lectures.[45] Desaguliers reported that experiments made by Henry Beighton and himself on a Newcomen engine indicated that 'water in boiling is expanded 14000 times' to generate steam at atmospheric pressure.[46]

But Watt's own experiments upon boilers and cylinders convinced him 'that there was a great error in Dr Desaguliers' calculations of Mr Beighton's experiments on the bulk of steam',[47] and he therefore set out to determine the correct figure himself.

Watt took a flask and to its neck he attached a tube which extended nearly to the bottom of the flask. He then put a small amount of water in the flask and heated it at a temperature just above the boiling point of water. When all the water had been vaporized the flask was full of steam, the heavier air having been driven out through the tube first. Watt cooled the flask and weighed it. He then dried it by again heating it and blowing air through it. He then weighed the flask again. Having determined the weight of steam required to fill the flask, that is to say, the difference between the weight of the cooled flask and that of the dried flask, he then weighed the flask when full of water. The ratio of the weights of a flask full of steam and a flask full of water thus told him how much water was expanded when it was boiled at atmospheric pressure.[48] Although Black later helped Watt modify and repeat this experiment, Watt himself did 'not consider these experiments as extremely accurate'.[49] Nevertheless, his value of between 1600 and 1800 for the volume ratio of steam to water at atmospheric pressure was good enough to demonstrate the gross error of Desaguliers' figure, and it was also accurate enough to enable Watt to proceed with his analysis of the steam engine.

Watt next determined the rate at which his boiler consumed coal and the rate at which it vaporized water. Then, by using his value for the ratio of steam to water, he calculated that his boiler 'produced rather less than 7 Cubic feet of steam in a minute'.[50] He attached his boiler to a long lead pipe, which he placed in a box 'as long as itself filled with Sawdust to prevent the evaporation of heat', and found that after '16 seconds the steam came out transparent at the end of it'.[51] The end of the steam pipe was then attached to 'a Small engine of Savarys kind', and several experiments were run with various arrangements of this apparatus.[52]

Having studied both the Newcomen and Savery engines and having failed to discover any way of markedly improving either of them, Watt turned back to the study of heat for further insight into the problems which blocked his advance. By experiment he determined that equal volumes of iron, copper and water contain roughly the same quantities of heat, but, he wrote, 'tin is said by Dr Black only to hold the half of the heat of an equall bulk of water'.[53] He next turned his attention to the heat contained in steam. He knew that 'Doctor Black . . . [believed] that water during the time that it was heating to [? at] the boiling point did every minute receive Considerable additions of heat.'[54] While willing to accept the rather obvious assertion that water placed in a vessel upon a fire absorbs heat continuously until it flies off as vapour, Watt was troubled by his discovery 'that the Quantity of water used for Injection in fire engines was much greater than I thought was necessary to Cool the Quantity of water contained in the Steam down to below the boiling point'.[55] To test his suspicion that steam contains more heat than boiling water, Watt performed two experiments. In the first he 'mixed 1 part of boiling water with 30 parts of cold water' and 'found it [the mixture] only heated to the arithmetical mean betwixt the two heats & that it was scarcely sensibly heated to the finger'.[56] He found, in other words, that for every degree of heat added to the temperature of the cold water, the temperature of the boiling hot water fell 30 degrees.[57] Having determined that boiling water does not contain an extraordinary amount of heat, he examined the heat in steam. He 'took a bent Glass tube & inverted it into the nose of a tea kettle, the other end being Imersed in cold water'.[58] He then 'found on making the kettle boil that tho there was only a small increase of the water in the frigeratory, that it was become boiling hot'.[59]

Watt was surprised by the difference in the amounts of heat contained in boiling hot water and in steam, but the difference he had discovered did not remain an intellectual stumbling-block for long.[60] 'On telling it to Dr Black', Watt wrote,

> & asking him if it was possible that water under the form of steam could contain more heat than it did when [as] water it was heated to 212° or boiling hot, [at which temperature] it would receive no more heat, tho the fire &c continued the same, [and on pointing out] that the steams that go off do not appear sensibly hotter to the thermometer than 212 & that the Quantity of water that is evaporated in a minute is small in proportion to the heat that is certainly added to the water, He [Black, told Watt that he] Thought that these steams certainly carried off or contained more heat than was sensible to the thermometer & that in Certain Circumstances they might part with that heat to other bodies.[61]

Driven by his own desire to improve the steam engine, Watt had discovered a phenomenon which could be readily explained in terms of Black's theory of latent, or insensible, heat. Black, Watt wrote, 'told me that [the doctrine of latent heat] had long been a tenet of his & [he] explained to me his thoughts on the subject'.[62]

Black also demonstrated that his doctrine was true. About a week after explaining his theory to Watt, Black 'tryed the experiment with the Still'.[63] He 'took a Common still & having put a Quantity of water into it, he adapted a Refrigeratory as usual. This he filled with a measured Quantity of Water of a known heat. He then distilled till there came into his receiver a certain proportion of the Water contained in his Refrigeratory. He found the water in the Refrigeratory had acquired from this water as much heat as it would have got from an equal quantity of water heated to 1012.'[64] The experiment which Black performed for Watt was one which he and Irvine had devised just a few months earlier to determine the quantity of heat released during the condensation of steam. Watt, evidently impressed by this experiment, soon repeated it himself,[65] and he thereby convinced himself that Black's doctrine of latent heat did indeed explain the source of the heat which raised the temperature of the injection water to very near the boiling point.[66]

Far from suggesting the invention of the separate condenser, the discovery of latent heat convinced Watt that his investigation had led him to another impasse. He had searched for the imperfections in the steam engine that allowed so much heat to rush away before doing useful work and he found that the escape route was not a correctible flaw in the design or construction of the steam engine but a law of nature. To condense steam in the cylinder of a steam engine, and thereby create the vacuum that enabled the engine to convert the pressure of the atmosphere into a mechanical force, one had to transfer the latent heat of the steam to the injection water. As Robison noted, 'this great cause of loss', like the heat expended on heating and cooling the cylinder, 'seemed to be unavoidable'.[67] No matter which way he turned, Watt found that when it came to converting the heat released by fuel into useful work, the steam engine was an intrinsically wasteful device.

The steam engine remained, nonetheless, the best available engine for pumping the water out of deep mines. Therefore, after determining the amount of latent heat in steam, Watt again tried to bring his knowledge to bear on this practical problem. 'Strengthened with these facts', he wrote in his notebook, 'I could now judge of the merits of Savarys engine in Comparison with Newcomens. Savarys engine being more simple & having no pistons has in Consequence less friction than Newcomens, and the vacuum being perfect, it can suck water to 28 or 29 feet. Newcomens engines as now Constructed do not raise more than a Column of water of 16 or 18 feet high & the diameter of their Cylinders.'[68] Having concluded that Savery's engine offered the greater opportunity for improvement, Watt closed this section of his notebook by again entertaining the idea of arranging a number of these engines in series. 'If a series of Savarys engines were placed above one another so as to act in succession; if there were 10 of them Each of them would have all the time to fill itself with water that the other 9 were filling with steam. Consequently the sucking pipes might be very small. As the Cylinders could be made of wood, the expence would be much smaller than Newcomens.'[69]

The investigations of the steam engine described in the first part of Watt's notebook failed to point towards any improvement which would make that source of power a more efficient heat engine. Since he was convinced that the steam engine 'might be applyed to other as valuable purposes as drawing Water',[70] Watt must have found it particularly difficult to accept his own conclusion, namely that Savery's engine wasted less heat than Newcomen's. Haunted by the disparity between the lumbering actuality of the Newcomen engine and its boundless potential, he soon returned to the analysis of the piston driven engine. It was probably during this second examination of the Newcomen engine that Watt first recognized the problem of back pressure.[71] When cold water is injected into a cylinder full of steam, the steam is condensed and a vacuum is formed. But as the pressure in the cylinder drops, the temperature at which the water in the cylinder will boil also falls. Thus the injection water, which has been warmed by the latent heat of the steam, soon begins to boil. The steam formed by this boiling in a partial vacuum creates a back pressure, which opposes the pressure exerted by the atmosphere on the opposite face of the piston. To prevent the formation of this back pressure, the cylinder must be cooled down to a temperature at which the injection water will no longer boil, that is, to around 100°. The engine can be operated without bringing the cylinder temperature down this far at the end of each cycle, but only at the cost of reducing the load placed upon the machine. Here then was a new dilemma, one which Watt apparently became aware of between the time he performed the early experiments described in his notebook and the time he invented the separate condenser.

Watt never claimed that he discovered the phenomenon of boiling at reduced pressure and his experiments were hardly of the sort that would have brought this phenomenon to his attention. It was, rather, his close connection with Robison and Black which again enabled him to gain a deeper understanding of the problems which confronted anyone seeking to make a heat analysis of the steam engine. Black, of course, was completely familiar with Cullen's experiments on evaporation in a vacuum, and it seems that during 1764 he was particularly concerned about the relationship between the boiling points of liquids, which vary with pressure, and their latent heats of vaporization.[72] He probably suggested that John Robison, who was studying with him at the time, examine this phenomenon further. Thus while Watt was puzzling over the economy of heat in the steam engine, Robison was performing several experiments which demonstrated that hot water in an evacuated chamber boils until it has cooled down to about 90°.[73] Watt no doubt was made aware of Robison's experiments and quickly realized that they were relevant to his investigation of the steam engine. He had known all along that the operation of the Newcomen engine depended on the formation of a vacuum in the cylinder, but now he knew that 'a vacuum could not be produced in the cylinder unless that vessel were cooled by the injection below the point at which water would boil in an exhausted Vessel'.[74] Many years later Watt wrote that for this knowledge, 'I was indebted to the Experiments of Dr Cullen on the boiling of Ether *in Vacuo* and some conversations I had on the subject of that experiment with Dr B. and Mr Robison'.[75]

The problem which was solved by the invention of the separate condenser now became clear in Watt's mind. He had determined '*that to make a perfect steam-engine, it was necessary that the cylinder should be always as hot as the steam which entered it, and that the steam should be cooled down below 100° in order to exert its full powers.* The gain by such construction would be double:—first, no steam would be condensed on entering the cylinder; and secondly the power exerted would be greater as the steam was more cooled.'[76]

Having conceived of a steam engine which was perfect according to the known laws of heat, Watt had next to think of how to make such an engine a reality. The steam in the cylinder evidently posed the greatest challenge, for if it could only be removed by condensation, then the cylinder would have to be cooled. Perhaps Watt was thinking of the vacuum produced by the air pump Robison had used in his experiments on boiling under reduced pressure when it occurred to him that steam has a mechanical property which would solve his problem for him. Watt said that during his famous spring walk on the Glasgow green, 'the idea occurred to me that as steam was an elastic vapour it would expand and rush into a previously exhausted space and that if I were to produce a vacuum in a separate vessel and open a communication between the steam in the Cylinder and the exhausted vessel, that such would be the consequence.'[77] Once in the exhausted vessel the steam could be easily condensed by a spray of injection water, after which the water and whatever air had seeped in could be readily pumped out. Watt had at last found a way to eliminate some of the great waste of heat in the Newcomen engine.

Having thought of the separate condenser, Watt immediately set about making a model of his invention. As Black later recalled, 'this capital improvement flashed on his mind at once, and filled him with rapture, and he immediately made an hasty trial of it which satisfied him of its value, employing for this purpose a large brass syringe which he borrowed from a friend.'[78] Further convinced of the worth of his invention, 'his mind became now very much employed in contriving the machinery by which this improvement might be reduced to practice, and he soon planned it to such a degree, that he thought he was ready to make an experiment on a large scale.'[79] In April 1765 Watt told his friend James Lind that 'I can think of nothing else but this machine. I hope to have the decisive trial before I see you.'[80]

As Watt set about making a larger model, Black wrote, 'he was stopped by the want of funds, and he found it necessary to associate himself with some person who had money and spirit for such an undertaking, and to participate with him [in] the advantages which might be derived from this invention'.[81] Black therefore introduced him to Dr Roebuck, an entrepreneurial friend of his, who was just then badly in need of strong engines to pump out the deep coal mines he had leased from the Duke of Hamilton.[82] Roebuck and Watt soon became partners and by late summer, 1765, Watt was sending Roebuck regular reports on the progress being made on the various models he was constructing.[83]

The letters that passed between Watt and Roebuck also reveal that Watt still harboured hopes of finding additional ways to further reduce the seemingly unavoidable heat losses in the Newcomen engine. In his search for ways to 'totally prevent waste of steam,'[84] Watt began looking for an improvement which would reduce the amount of latent heat carried off by the condensing water. He performed some experiments on the latent heat of steam produced in an evacuated vessel and found that it was considerably greater than the latent heat of steam at normal pressure.[85] He also studied the 'heats of water boiling under pressure greater than that of the atmosphere'.[86] It appeared, on the basis of these investigations, that the total heat in steam is roughly constant and that the amount of latent heat in the steam used to drive an engine could therefore be reduced by operating at pressures above that of the atmosphere. Early in September 1765 Watt told Roebuck, 'that, in proportion as the sensible heat of steam increases, its latent diminishes, so, in the steam-engine, working with pressures above 15 lbs. must be more advantageous than below it; for not only the latent heat is diminished but the steam is considerably expanded by the sensible heat, which is easily added.'[87]

Roebuck, who found the doctrine of latent heat difficult to grasp,[88] was rather mystified by 'Watt's law'. 'The fact which you relate is so paradoxical,' he replied, 'that, if less accurate observers of nature than yourself and Dr. Black had been witnesses of it, I should scarce have given credit to it.'[89] Operating with steam under pressure was a grand idea, but Watt soon found that unlike the separate condenser, this innovation overtaxed the capacities of the available technology. As Black told Robison in 1767, the advantages of Watt's second radical improvement were more than offset by 'the enormous size and strength which must be given to the boiler, and the expense of fuel in supplying steam so dense and so hot'.[90] Thus the heat wasted when the steam gives up its latent heat during condensation remained, in the Watt engine, an unavoidable loss.

Professional success and new responsibilities soon led to the final breaking up of the philosophical trio in Glasgow. In 1766 Black moved to Edinburgh to succeed Cullen in the chair of medicine and chemistry, and Robison, on Black's recommendation, was appointed lecturer in chemistry in Glasgow. 'Frightened by my own good fortune,' Robison wrote, 'I was obliged to strain every nerve to do some credit to so partial a recommendation; and I was obliged to relinquish all other occupations of my thoughts.'[91] Watt, meanwhile, had to contend with the increasing obligations to his family. His improved steam engine had not yet brought in any money and he therefore tided himself over by becoming involved in other ventures. In 1767 he helped make a survey for a small canal intended to connect the Forth and Clyde rivers, and later that year he travelled to London to represent the scheme before Parliament. On his return trip he passed through Birmingham where he met Erasmus Darwin, a member of the Lunar Society, and toured Matthew Boulton's recently completed Soho works.[92] Disappointed by Roebuck's inability to provide the financial backing needed to improve and manufacture the steam engine, Watt was deeply impressed by the flourishing state of science and industry in Birmingham. He returned to Glasgow and laboured over his engine for seven more years, but in 1774, having freed himself from the partnership with Roebuck and having suffered through the death of his wife, he quit Glasgow for a well prepared and more promising future in Birmingham.[93]

Black's move to Edinburgh and Watt's later move to Birmingham did not, of course, terminate the friendship both men valued so highly. It is well known that in 1769 Black loaned Watt the money he needed to obtain his patent on the separate condenser,[94] but the common interest in chemistry shared by these two thinkers is much less often referred to. Yet as Humphry Davy insisted, 'those who consider James Watt only as a great practical mechanic, form a very erroneous idea of his character:—He was equally distinguished as a natural philosopher and a chemist, and his inventions demonstrate his profound knowledge of those sciences, and that peculiar characteristic of genius, the union of them for practical application.'[95] In Birmingham Watt continued his investigation of the steam engine by performing more refined experiments on the nature of steam; he joined his colleagues in the Lunar Society in the study of philosophical problems; he campaigned for recognition of his priority as a discoverer of the composition of water; and he actively defended Black's theory of latent heat against those who questioned its originality or its validity.[96] While Watt clearly did not invent the separate condenser by applying Joseph Black's theory of latent heat to the steam engine, his life-long interest in chemistry and his specific understanding of the subject justifies our calling him a devoted and influential disciple of his Glasgow mentor.[97]

2. WILLIAM IRVINE

William Irvine was Black's student and assistant during the latter half of the decade Black spent as a member of the university faculty in Glasgow. Robison recalled that while 'a young student of medicine', Irvine had shown himself to be a man of 'quick apprehension, sound understanding, and particularly disposed to consider every thing mathematically', and that by attending Black's lectures he had become 'greatly captivated with chemical science'.[98] Robison also recalled that Irvine

> engaged, with great pleasure and zeal, in all examinations which seemed to interest the Professor, and particularly such as would admit of mathematical consideration;—thermometrical experiments on the scale of heat;—on the connection between expansion and variation of temperature;—on the measures of heat, &c.;—all these were fields of research altogether to his mind. He supplied Dr. Black with a vast number of experiments on the equilibrium of heat, on the specific heats of different substances, and on the continued absorption and fixation of heat by glass, sealing wax, resin, and other substances, which gradually become more fluid.[99]

Although Black always spoke of Irvine's experiments with respect and appreciation, he never was particularly impressed by his pupil's criticism of the theory of latent heat. So long as Irvine was studying with Black, therefore, he was not encouraged to expand upon his objections to the theory.[100] In 1766, the year Black was called to Edinburgh, Irvine obtained his medical degree and apparently expected that he would be appointed to succeed Black as the lecturer in chemistry in Glasgow.[101] He was disappointed, however, for it was decided that Robison should receive the post, but the faculty thought so highly of Irvine that they offered him a lectureship in materia medica. Three years later, when Robison resigned to go to Russia, Irvine finally obtained the lectureship in chemistry, and he subsequently held both lectureships until his untimely death in 1787.[102]

Although Irvine seems to have defined his fundamental notions of total heat and his doctrine of capacities while studying with Black, his early views were apparently quite undeveloped and consisted of little more than private conjectures before he worked them up for his lectures on chemistry.[103] After Black and Robison had left Glasgow, Watt and Irvine, who doubtless had known each other for years, were drawn closer together by their common interest in the study of heat.[104] Encouraged by his new academic responsibilities and by his stimulating association with Watt, Irvine organized his own understanding of heat into a novel theory. Thus when Irvine began lecturing on his doctrine of capacities, his audience had the pleasure of being the first to hear a public presentation of his new theory of heat. As Watt later wrote, 'I will . . . attest that Dr. Irvine of Glasgow about the year 177½ [sic] invented a method of finding the quantity of Absolute heat in Bodies.'[105] About two years later Irvine further announced that by using his theory, he had determined the natural zero of temperature. In 1774, shortly before he left Glasgow, Watt wrote that 'my friend Dr. Irvine in winter discovered what is the lowest possible degree of heat, or the real beginning of the scale; by which means thermometers may be made, showing the real quantity of heat in bodies.'[106] Thus by 1775 one of Black's own pupils and successors had put forward a well-developed theory which posed a profound challenge to the doctrine of latent heat.

How did Irvine come to oppose his master's understanding of change of state? It seems that while studying the specific heats of substances that soften and become plastic when heated, such as wax and resin, Irvine began to wonder whether this gradual change of state should in fact be considered a chemical change. He knew that 'these substances change their capacities for heat gradually,'[107] but, except for the alteration of this property, the physical changes taking place are not accompanied by the kind of qualitative changes one sees in the reaction between an acid and an alkali. In the absence of any striking change in other specific properties, is it wise to assume that during heating 'the cause of heat', to use Irvine's phrase,[108] is being fixed in the substance being heated? Would it not be more advisable to search for some correlation between the change of state taking place and the change in the capacity for heat which accompanies it?

When he took up these questions again several years later, Irvine thought of an experiment, which enabled him to examine the effects of a change of capacity for heat upon the total heat in a body without adding heat from an external source. The experiment involved simply mixing vitriolic acid and water. After studying the heat phenomena that accompany this mixture, Irvine formulated his doctrine of capacities for heat and presented it in a paper titled 'On Heat Produced by Mixture,' which he read before the Literary Society of Glasgow.[109] In this paper Irvine argued that 'some bodies, which, on being mixed produce a very considerable degree of heat, retain . . . after their mixture so many of their original properties, and are so superficially united, that they can be separated again so as to recover all their former qualities; so that on mixing them a second time, they will produce the same degree of heat as at first. The most remarkable of this kind are the mineral acids when mixed with water.'[110] When a mineral acid and water are mixed, 'the qualities of the compound . . . are a mean between those of the acid and water before mixture. None of the principles of which either body is compounded has been lost during the mixture, or has been changed into heat or light.'[111] Therefore, Irvine concluded, 'the heat produced in this case cannot be accounted for, from any new arrangement of the principles which compose those bodies, but must depend on some other cause.'[112]

After identifying and describing the phenomenon that interested him, Irvine surveyed the mechanical and chemical theories previously advanced to explain it. Natural philosophers have generally supposed that the heat of mixture results 'from the violence with which the particles of bodies rush together, and the great friction which must thus necessarily arise.'[113] Chemists, on the other hand, have asserted that 'the bodies most disposed to produce heat in mixture, are such as are of a fiery and acrid nature, qualities which have been accounted for by supposing that a certain quantity of fire was contained in those bodies, which meeting with fluids like water, immediately produced the commotion, and the noise and smoke which water when thrown upon a fire produces.'[114] But neither the Newtonian vibrational theory of heat in fluids nor Lemery's theory of fixed fire is capable of explaining the heat given off by the mixture of an acid with water. If the attraction between the particles of acid and water is so great that they rush together and produce heat, why, Irvine asked, doesn't that attraction cause a comparable change in their other qualities? We know from experiment that 'the mixture of acid of vitriol and water seems to possess qualities exactly a mean between the acid and water before mixture, except where heat is concerned.'[115] But if 'examined as affected by heat, the qualities are so different, that it could not arise from any accidental jostling of the particles, but must be owing to some permanent change in their properties.'[116] Furthermore, Irvine argued, 'a compound exactly similar in every respect to an acid and water, may be produced by an acid, and snow or ice; yet in this case, the union of the ingredients is not attended with heat; on the contrary, they become colder than before mixture.'[117] If heat is the result of an attraction between acid and water, then cold must result from a repulsion, and we must then suppose that the attraction between acid and water can be reversed by turning water into ice. Clearly, Irvine concluded, the mechanical theory of attractive forces is incapable of explaining this heat phenomenon.

After refuting, in a similar manner, Lemery's concepts of the '*feu legerment bridé*' in acid and the '*feu pur et isole*' set loose by the mixture of acid and water,[118] Irvine presented his own explanation of the phenomenon. He said 'the reason why heat is produced by mixing fluids seems to be, that the ingredients which compose the mixed fluid are in a different state with regard to heat from what they were before. It would seem that whatever is the cause of heat, it has a greater effect in heating the bodies after the mixture than it ought to have had in proportion to its power of heating the ingredients before mixture.'[119] In other words, the specific heat of the mixture is less than the figure one would expect from a knowledge of the specific heat of each constituent. 'Two bodies, which being mixed have produced heat, are for ever after this much more easily heated and cooled, than we should have been led to expect from the properties of the ingredients before mixture. Thus a mixture of acid of vitriol and water when set before a large fire, becomes as hot as the fire can make it in a less time than in proportion to the ingredients.'[120]

Irvine performed several experiments to verify his understanding of this phenomenon. Using Martine's technique, he compared the rates at which acid, water and a mixture of the two will heat up when equal quantities are placed before a fire. He also compared their specific heats by determining how much cold crushed glass had to be added to each of the three liquids to reduce its temperature by 50°.[121] Shortly thereafter he drew the theoretical implications of these experiments. 'It seems reasonable to conclude,' he said,

> that whatever is the cause of heat, it is capable of producing a greater effect in heating the mixture than it ought in proportion to the ingredients. The acid and water before mixture contain such a quantity of the cause of heat as is sufficient to produce all the heat in them from the lowest degree of heat to the degree

> at which they are mixed. The moment that the mixture is made, the body is so changed, that the same quantity of the substance which is the cause of heat is capable of producing upon the same quantity of matter a greater effect than before, and this in the proportion of $3\frac{1}{2}$ to 4, or as 7:8. Whatever effect therefore it would produce before mixture, it will after mixture produce one-seventh more. Hence, if the two substances before they were mixed contained such a quantity of the cause of heat as is sufficient to produce when they are at 32° of Fahrenheit, an effect equal to seven times the difference between the heat of frost and that of boiling water, or seven times 180°, after mixture, it will produce eight times this difference, or eight times 180°. The two substances therefore, though as cold as frost before mixture, will be as hot as boiling water after it.[122]

Irvine's doctrine of capacities can be stated more simply. If we assume that substances differ in the amount of heat required to raise their temperatures one degree and that the cause of heat can always be detected by a thermometer, then we can determine the relative quantities of heat in bodies. Thus every substance contains a total or absolute amount of heat, which is equal to the product of its capacity for heat and its temperature, as measured on a scale which begins at the natural, or absolute, zero point of heat.[123] When acid and water are mixed, the total heat in the two constituents remains unchanged while the mixture's capacity for heat becomes less than the mean of those of the two constituents. The temperature of the mixture will therefore rise to the point at which the product of the mixture's capacity for heat and its temperature is equal to the mean capacity for heat times the initial temperature of the acid and the water. Similarly, when water becomes ice, its capacity for heat is reduced and hence it must give up a considerable amount of heat while at the freezing point before its total heat is diminished to the point at which it can become a solid.[124]

Those who have criticized Joseph Black for his supposed horror of hypothesizing might have been more tolerant had they been familiar with the thrust of Irvine's attack upon his theory of latent heat. 'I have used the term cause of heat,' Irvine said, 'rather that latent heat, or any other term in deference to the opinion of those philosophers who conceiving heat to be quality, conceive latent heat to be a latent quality, and that a latent quality is no quality at all.'[125] If one conceives of heat as the defining property of a general cause, such as a Newtonian aether, then one must assume that the cause possesses that property at all times. To assert that heat can become latent is to imply that when one cannot identify the means by which the general cause gives rise to specific phenomena, one may simply ascribe new properties to the general cause. Although Irvine did not claim that he knew what the cause of heat is, he was sure that it 'has a cause, and whatever this may be, it cannot be subject to its own effect'.[126] He knew, of course, that the cause of heat 'produces different effects in heating different bodies,' and in this respect 'is very different from the substance which is the cause of gravity, which acts equally on equal quantities of matter.'[127] The laws that govern the distribution of heat remained unknown, however, and Irvine admitted he was 'equally ignorant whether the heat produced in those mixtures, where bodies are inflamed and consumed, and never capable of exhibiting the same appearance again, be owing to the cause which I have just assigned.'[128] Although Irvine knew he could not explain reactions involving change of chemical qualities, he was confident that his doctrine of capacities would make it possible for natural philosophers to abandon the ill-conceived notion of latent heat.

Black was not very enthusiastic about Irvine's 'solution' to his 'celebrated discovery of latent heat'.[129] During his years of study under Cullen he had given considerable thought to the heat of mixture and he was convinced that his own theory provided a more comprehensive and satisfactory explanation of this phenomenon than Irvine's. Watt also, like Black, felt that Irvine's theory was 'nothing else but an explanation of Dr Black's in other words'. [130] Although Irvine's theory made possible the calculation of natural zero, there was such a high degree of disagreement between the results of the various determinations of this figure that one was entitled to question the validity of the theory itself.[131] Black therefore gently defended his own theory. 'Solidity,' he said, 'we all know depends on cohesive attraction; but on what cause does fluidity depend?' [132] The change from solid to liquid 'is such a violent change,' he argued, 'that I cannot help thinking it must depend on some powerful cause'. [133] To speak of change of capacity is to avoid naming this powerful cause, and to argue that the concept of latent heat is nonsensical is to ignore the fact that the latent heat in liquids and vapours can be recovered as sensible heat.[134]

But above all, Black must have thought, how can a chemist in the latter half of the eighteenth century doubt that certain principles, such as air and fire, can exist in two states, the fixed and the free? How, he probably asked himself, had Irvine failed to grasp this central concept during their years of joint research, or if he had once accepted it, why did he now feel compelled to reject it on such narrow grounds? Was there still some lingering unwillingness to accept the logic of chemistry within the realm of natural philosophy? In light of the popularity that Irvine's doctrines enjoyed during the three decades following their publication, it is apparent that there still was.

3. THE ASCENDANCY OF THE CHEMICAL THEORY OF HEAT

Ideas and discoveries which have been presented in lectures have in fact been made public, and in that root sense of the word we may say that Cullen's, Black's and Irvine's doctrines of heat were published soon after they were formulated. Of course such a form of publication is apt to entail the most arbitrary kind of distribution and quite often the information conveyed is either incomplete, erroneous or incomprehensible. Thus it is not too surprising to find that before 1764 Henry Cavendish apparently had a working understanding of Black's doctrine of latent heat, while that doctrine did not become known in Edinburgh before 1766.[135] In later years, when Black had secured an unchallenged reputation as Great Britain's pre-eminent chemistry teacher, more and better manuscript copies of his lectures became available. Thus long before John Robison edited Black's lectures and had them published, manuscript copies of these lectures had carried Black's ideas into the halls of Oxford and Cambridge and into the hands of the great engineer Thomas Telford.[136]

Although the printed lecture was the chief means by which the doctrines of Scottish professors became known outside their universities, [137] their ideas often first appeared in print in the dissertations of their students. When presented in this way, Black's ideas were doubly masked, first by the student author and secondly by the medical subject matter.[138] But the case of Adair Crawford reveals that interest in the new theories of heat was often aroused, and an adequate knowledge of their basic assumptions was conveyed, even when the theories were used to explain phenomena, such as animal heat, which had not played a role in the original development of the new views on heat.

The first separately printed account of Black's theory of latent heat appeared in 1770 under the title *An Enquiry into the General Effects of Heat, with Observations on the Theories of Mixture*.[139] Published anonymously, this book stands as the first, though unauthorized, printing of a portion of Black's lectures. The editor of the *Enquiry* was perfecly open in identifying Black as the author of the doctrine of latent heat, and his book contains an adequate description of the doctrine and of the experiments with which Black confirmed and quantified his theory.[140] It is now difficult to determine whether or not the *Enquiry* brought Black's ideas to the attention of any potentially interested parties who had not previously heard of them, but we do know that the scientific intelligencer J. H. de Magellan referred to the *Enquiry* as one of his sources of information in his *Essai sur la nouvelle théorie du feu élémentaire, et de la chaleur des corps*, which was published in 1780.[141]

The next printed account of Black's doctrine of latent heat was published in France in September 1772. Titled 'Experiences du Docteur Black *Sur la marche de la chaleur dans certain circonstances*,' and sent from Edinburgh by an unnamed disciple of Black's, this brief description of the doctrine and six supporting experiments was later reprinted by the Abbé Jean Rozier in his two volume *Introduction aux Observations sur la Physique*.[142] Lavoisier, who learned of this essay shortly after its arrival in Paris in 1772,[143] must have been struck by the author's bold assertion that 'L'école d'Edimbourg fera époque dans l'Histoire de la Physique, pour s'être principalement occupée du feu & de l'air fixe. Le Docteur Black est celui qui a le plus multiplié les expériences sur ces deux points importans.'[144]

We now know that well before 1772 Lavoisier had become convinced that air and fire can exist in both the fixed and the free state.[145] Having adopted a chemical approach to the study of heat, Lavoisier recognized immediately that Black's investigations paralleled his own. Since Black's research had already advanced to the stage of experimentation and quantification, Lavoisier had to act quickly to establish an independent claim to the discovery of the chemical role of fire. Therefore, on 19 August 1772, he had the permanent secretary of the Academy initial a hastily drafted set of memoirs which Lavoisier referred to as his 'Système sur les élémens'.[146] By declaring his own theory, Lavoisier evidently hoped it would be realized that on the basis of his own research he had formulated a chemical theory of heat which in no way depended upon the discoveries of the members of 'l'école d'Edimbourg'.

Several accounts of both Black's and Irvine's doctrines were published between 1778 and 1780. In 1778 and 1779 the first four volumes of the second edition of the *Encyclopedia Britannica* were issued and they contain several statements of Black's doctrines, statements that were apparently taken directly from copies of his lectures.[147] During these same years two books presenting theories of animal heat founded upon the doctrines of the Scottish chemists also appeared. In 1778 Patrick Dugud Leslie's *A Philosophical Inquiry into the Causes of Animal Heat: With Incidental Observations on Several Phisiological and Chymical Questions, Connected with the Subject* was published in London. Although this work, which is essentially a translation of Leslie's medical dissertation,[148] contains an account of Black's doctrine of latent heat, Leslie devoted most of his attention to criticizing the way in which Black used his theory to explain the production of heat in animals.[149] Despite this disagreement over Black's application of his theory, however, Leslie's book served to bring Black's ideas and experiments to the attention of other chemists.[150] In 1779, quite a different theory of animal heat was published in Adair Crawford's *Experiments and Observation on Animal Heat, and the Inflamation of Combustible Bodies; Being an Attempt to Resolve these Phaenomena into a General Law of Nature*. Crawford had visited Scotland in 1776 and had heard William Irvine lecture. Irvine's doctrine of capacities caught his attention and during the summer of 1777 Crawford began the first experiments ever performed on the specific heats of gases. Encouraged by his own research, he then applied the doctrine of capacities to the problem of animal heat. His book, which contains an account of his theory, a description of his experimental reasearch, and several statements of his indebtedness to Irvine and, to a lesser extent, Black, attracted a great deal of attention and soon became the chief source of information concerning Irvine's theory of heat.[151]

In 1779 the first attempt to bring Black's doctrine into the mainstream of natural philosophy was also published. William Cleghorn, in his dissertation *Disputatio physica inauguralis, theoriam ignis complectens*,[152] suggested that fire is a fluid composed of particles that are mutually repellent but that are attracted to ordinary matter by a force that depends on the characteristics of the particular substance involved. This theory was quite like the caloric theory already taking shape in Lavoisier's mind,[153] and it provided as good an explanation of the phenomena upon which Black had built his theory as was possible, given the conceptual assumptions of the natural philosophy of the time.[154]

By 1780 there were comprehensive accounts of the Scottish doctrines of heat available in both English and French. The author of the long article on 'Heat', which appeared in the fifth volume of the second edition of the *Encyclopedia Britannica*, presented Black's and Irvine's theories by quoting extensively from the books by Cleghorn, Leslie and Crawford.[155] A French account of these theories, with heavy emphasis upon Irvine's doctrine of capacities and Crawford's theory of animal heat, was published in 1780 in London by Magellan. Since his own success as an intelligencer depended upon publicizing new scientific ideas, Magellan made sure his *Essai* did not fall stillborn from the press. In a letter to the Royal Society of London he drew attention to the importance of Crawford's work and he wrote an introductory letter for his own *Essai* when it was republished by Rozier in 1781.[156] Magellan's efforts were not without effect. His championing of Crawford's ideas brought them to the attention of the chemists in Paris and apparently convinced them that the study of the specific heats of gases, the key to Crawford's theory of animal heat, would provide the crucial facts in the articulation of the new theory of heat. When Jean André De Luc visited Paris in 1781, he met several scientists, including Laplace, Lavoisier, Monge and Vandermonde. He later recalled that 'Dr Crawford's theory on the phenomena of heat observed when substances undergo certain changes in their nature, was then in great agitation among these philosophers, and the two last had, from the Academy, the special commission to examine and follow that new view.'[157] Thus by 1781 the French chemists realized that there were in fact two competing Scottish theories of heat and that they would have to decide for themselves which they found the more acceptable. It was a question of special importance to Lavoisier, for he had already published an explanation of the source of animal heat based upon his chemical theory of fire.[158]

In 1777 Lavoisier summarized his views on heat in two memoirs presented before the *Académie*.[159] In these memoirs he drew the implications of the ideas that he had jotted down in his sealed note of 1772 and he explained how heat is involved in the various chemical changes such as combustion, calcination and respiration, which he was then studying. After completing these memoirs Lavoisier apparently paid little attention to the theory of heat until the questions he believed had been settled were reopened by the serious consideration being given to Magellan's *Essai* and Crawford's book.[160] In the winter of 1782–3, however, he undertook, with Laplace some experiments upon the specific heats of gases and on the determination of absolute zero, and it was this research which formed the basis for their famous 'Mémoire sur la chaleur'. At the end of January, 1783, De Luc wrote to James Watt and told him that 'les chimistes de Paris, s'occupent beaucoup aujourd'hui de la chaleur, et des modifications de ces transmissions; des grands mathematiciens se joignent à eux; car la théorie de ces transmissions ou communications donnent lieu à de fort beaux problèmes. MM. Lavoisier et De La Place entr' autres sont en grand travail et publieront. Enfin il est certain, qu'on commence à fouiller vivement dans les vraies bases de la Physique.'[161] Although Lavoisier and Laplace realized that the values they obtained when they used their ice calorimeter to determine the specific heats of gases and heats of combustion were not definitive, they argued that these values indicated that no reliable figure for absolute zero could be determined and hence that Irvine's doctrine of capacities for heat remained no more than a highly dubious hypothesis. The publication of the 'Mémoire sur la chaleur' revealed that in Lavoisier and Laplace the chemical or caloric theory of heat, which appeared in Scotland as the doctrine of latent heat, had found its most powerful advocates.[162]

Joseph Black died less than a month before the beginning of the nineteenth century. During the last two decades of his life he had seen his doctrine of latent heat opposed by the disciples of Irvine and Crawford, and overshadowed by the new system of chemistry proposed by Lavoisier and his colleagues. After Black's death it took over a decade for the Irvinist doctrines to be discredited in Great Britain. In fact Black's doctrines had to face two sets of opponents during the early years of the nineteenth century. On the one hand there were the natural philosophers, such as Count Rumford, Humphry Davy and Thomas Young, who opposed the 'chemical solution' to the theoretical puzzles posed by the phenomena of heat.[163] They favoured a neo-mechanistic force or vibrational theory of heat,[164] but since they were unable to offer precise alternative explanations for the phenomena being examined, their theoretical predilections had little effect upon the actual development of the study of heat. On the other hand there were the leading British chemists of this era who continued to divide along the lines which separated Black's doctrine from Irvine's. Black's pupils and disciples, such as Thomas Thomson and William Henry, defended their master's theory of heat while John Dalton, who admitted he was 'overawed by the authority of Crawford,' became the chief proponent of Irvine's theory.[165] But even Dalton's powerful advocacy could not prevent the Irvinist doctrines from being discredited by the exact experimental research being carried on in France.

Interest in research on the nature of heat seems to have declined in France following the publication of Lavoisier's and Laplace's 'Mémoire sur la chaleur'. After 1783 Lavoisier turned his attention first to the other aspects of his new system of chemistry and, later, to the turmoil brought on by the French Revolution, and like most of his colleagues, he showed little interest in the enlarged and improved second edition of Crawford's book, which appeared in 1788.[166] Irvine's doctrine had not, however, been rendered untenable in France by the end of the first decade of the nineteenth century. In fact the contest between the caloric theory of heat and the doctrine of capacities continued largely because the available experimental techniques were not accurate enough to resolve the dispute. The French Institute therefore decided to confront this problem directly in the 1811 prize competition in physics for the First Class. The task assigned to the participants was to determine the specific heats of gases so that a solution could be found to 'the outstanding problem of whether some caloric exists in substances in a combined state or whether all the heat given out in a reaction is due to a change in the specific heat of the reacting substance'.[167] Two entries were received, and in December 1812 the panel of judges, which consisted of Berthollet, Haüy, Charles, Thenard, and Gay-Lussac, gave the prize to Delaroche and Bérard, while awarding 'special mention' to Clément and Desormes.[168] As a result of this competition the chemical theory of heat gained undisputed ascendancy in France. As winners of this competition wrote in their prize essay, 'we must reject once and for all the hypothesis whereby the release of heat which occurs when substances combine is attributed solely to a change in the capacity of these substances for caloric, and consequently we must accept, with Blake [sic], MM. Lavoisier and De Laplace and a large number of physicists, that caloric exists in a combined state in bodies.'[169]

From our vantage point, of course, this story does not end with the victory of the caloric theory of heat. We now see Black's theory of heat as a stepping stone on the way from the Newtonian synthesis of the late seventeenth century to the great nineteenth-century synthesis of physics and chemistry founded upon thermodynamics and the dynamical theory of heat. Should we therefore re-examine Black's doctrines and discoveries in the light of these later developments? There are good reasons for not doing so. Today we do not believe that the growth of science conforms to a predetermined pattern which sets the goals for each stage of development, and we therefore study its history step by step. But historians have not always been so cautious and self-effacing, nor are they likely always to remain so. Therefore, without implying uncritical acceptance, we will end this study by quoting from the powerful and suggestive appreciation of Black's doctrine of latent heat, which Henry Thomas Buckle included in his 'Examination of the Scotch Intellect During the Eighteenth Century'.

> That this remarkable theory paved the way for the doctrine of the indestructibility of force, will be obvious to whoever has examined the manner in which, in the history of the human mind, scientific conceptions are generated. The process is always so slow, that no single discovery has ever been made, except by the united labours of several successive generations. . . . Black, no doubt, fell into the error of regarding heat as a material substance, which obeys the laws of chemical composition. But this was merely an hypothesis, which was bequeathed to him, and with which the existing state of thought forced him to encumber his theory. . . . The real service which he rendered is, that, in spite of that hypothesis, which clung to him to the last, he, far

more than any of his contemporaries, contributed towards the great conception of idealizing heat, and thus enabled his successors to admit it into the class of immaterial and supersensual forces. Once admitted into that class, the list of forces became complete; and it was comparatively easy to apply to the whole body of force, the same notion of indestructibility, which had previously been applied to the whole body of matter. But it was hardly possible to effect this object, while heat stood, as it were, midway between force and matter, yielding opposite results to different senses; amenable to the touch, yet invisible to the eye. What was wanting, was to remove it altogether out of the jurisdiction of the senses, and to admit that, though we experience its effects, we can only conceive its existence. Towards accomplishing this, Black took a prodigious stride. Unconscious, perhaps, of the remote tendency of his own labours, he undermined that doctrine of material heat, which he seemed to support. For, by his advocacy of latent heat, he taught that its movements constantly baffle, not only some of our senses, but all of them; and that, while our feelings make us believe that heat is lost, our intellect makes us believe that it is not lost. Here, we have apparent destructibility, and real indestructibility. To assert that a body received heat without its temperature rising, was to make the understanding correct the touch, and defy its dictates. It was a bold and beautiful paradox, which required courage as well as insight to broach, and the reception of which marks an epoch in the human mind, because it was an immense step towards idealizing matter into force. . . . We must, therefore, ascribe to Black the signal merit that he first, in the study of heat, impeached the authority of the senses, and thereby laid the foundation of everything which was afterwards done.[170]

Notes and References

CHAPTER ONE

Scotland and the University of Glasgow

1. Unless otherwise noted, all biographical information on Cullen is taken from John Thomson, *An Account of the Life, Letters, and Writings of William Cullen, M.D.* (2 vols., vol. 1 published 1832, republished with vol. 2, Edinburgh 1859), vol. 1, passim.
2. Ibid., 1.68.
3. See ibid., 11.222–487, 710–18.
4. James Anderson, 'Cursory Hints and Anecdotes of the late Doctor William Cullen of Edinburgh,' in *The Bee, or Literary Weekly Intelligencer,* 1 (Edinburgh 1791) 45; 1–14, 45–56, 121–5, 161–6.
5. Ibid., 48–9.
6. See R. G. Cant, 'The Scottish universities and Scottish society in the eighteenth century,' in *Studies on Voltaire and the Eighteenth Century,* 58 (1967) 1956–7; 1953–66.
7. See J. D. Mackie, *A History of Scotland* (Baltimore 1964) 150–1. For additional evidence that 'the covenanting spirit had ever been keenest in the West Country,' see Henry Grey Graham, *The Social Life of Scotland in the Eighteenth Century*, 2nd. ed. (London 1928) 136.
8. The University of Edinburgh, being a municipal foundation, was much less directly involved in the struggles of the Reformation than the older bishop's colleges; see D. B. Horn, *A Short History of the University of Edinburgh* (Edinburgh 1967) 1–6.
9. Mackie, *History of Scotland*, 159.
10. J. D. Mackie, *The University of Glasgow* (Glasgow 1954) 66–7.
11. For details on the application of the religious test to the faculties of the Scottish universities, see Graham, *Social Life*, 448.
12. Mackie, *History of Scotland*, 252–3. See also *Munimenta Alma Universitatis Glasguensis, Records of the University of Glasgow from its Foundation till 1727*, 3 vols. (Glasgow 1854) 11.493–509; and Graham, *Social Life*, 420–1.
13. Graham, *Social Life*, 316.
14. J. D. Mackie, 'Glasgow University in the Eighteenth Century,' in *An Eighteenth Century Lectureship in Chemistry*, ed. Andrew Kent (Glasgow 1950) 30. For a description of the procedure used in selecting regents, see Graham, *Social Life*, 450.
15. See Graham, *Social Life*, 458–61.
16. Ibid., 339.
17. Ibid., 399 n. 2.
18. James K. Cameron, 'The Church of Scotland in the age of reason,' in *Studies on Voltaire and the Eighteenth Century*, 58 (1967) 1940; 1939–51.
19. See *Fortuna Domus, A Series of Lectures delivered in the University of Glasgow in Commemoration of the fifth centenary of its Foundation* (Glasgow 1952) 98–9.
20. *Munimenta*, 11.526–31.
21. Graham, *Social Life*, 269.
22. Ibid., 271.
23. See James Coutts, *A History of the University of Glasgow* (Glasgow 1909) 163.
24. Quoted in Graham, *Social Life*, 351 n. 1.
25. Thomas Reid, in his *Statistical Account of the University of Glasgow*, first published in 1799 but probably

written in 1794 (see editor's note in *The Works of Thomas Reid*, ed. Sir William Hamilton (Edinburgh 1846) 721) states that 'The average number of students, of all denominations, attending the different classes, is considerably above six hundred.' (*Works*, 733.) Cf. Mackie, 'Glasgow University,' who states, on page 33, that there were one thousand students in the university by the end of the century. Lawrence Stone has recently pointed out that in England, Germany, Spain, France and Italy, universities experienced 'a period of astonishing growth' from 1550 to 1650, 'a long period of decline and low enrollment' from 1650 until 1810, and then 'another period of high expansion' after 1815 and again after 1860. Surely the rapid expansion of the universities of Glasgow and Edinburgh during this middle period of decline elsewhere is one reason for the pre-eminence of Scottish university education in the eighteenth century; see Lawrence Stone, ed., *The University in Society* (2 vols., Princeton 1974) I.vii.

26. As quoted by David Murray, *Memories of the Old College of Glasgow* (Glasgow 1927) 507.

27. *Fortuna Domus*, 103.

28. Murray, *Memories*, 507. On the editions of Clarke's translations and the scope of his Newtonian commentary, see Robert E. Schofield, *Mechanism and Materialism: British Natural Philosophy in an Age of Reason* (Princeton 1970) 26–9. It should be noted that orthodox Presbyterian ministers considered Clarke's theology an Arian heresy and attacked Carmichael's colleague Professor Simson for teaching it; see Graham, *Social Life*, 354–5, and below.

29. Reid, *Works*, 30.

30. *Munimenta*, II.514.

31. Mackie, 'Glasgow University,'

31. See also chapter II of his *University of Glasgow*.

32. See W. Innes Addison, *The Matriculation Albums of the University of Glasgow* (Glasgow 1913) xi.

33. Graham, *Social Life*, 462.

34. *Munimenta*, II.578.

35. Murray, *Memories*, III.

36. Cf. W. R. Scott, *Francis Hutcheson, His Life, Teaching and Position in the History of Philosophy* (Cambridge 1900) 66 and G. E. Davie, *The Democratic Intellect: Scotland & Her Universities in the Nineteenth Century* (Edinburgh 1961), *passim*. Graham, in a rare instance of narrow-mindedness, speaks of regenting as an 'absurd' and 'preposterous' system of instruction; see *Social Life*, 449–50.

37. See Mackie, *University of Glasgow*, 170–8, and Coutts, *University of Glasgow*, 199.

38. MS. copy of Cullen's chemical lectures in the library of the Royal College of Physicians, Edinburgh, hereafter cited as RCPE MSS, C. 10, p. 27.

39. *Munimenta*, III.236.

40. Graham, *Social Life*, 269–70. The duke's comment reveals the tension that existed between the Episcopalian landed gentry and the Presbyterian kirk; see Nicholas Phillipson, 'Culture and Society in the 18th Century Province: The Case of Edinburgh and the Scottish Enlightenment,' in Stone, ed., *The University in Society*, II.424; 407–48.

CHAPTER TWO

English Natural Philosophy after Newton

1. On Newton and his circle, see Arnold Thackray, *Atoms and Powers; An Essay on Newtonian Matter-Theory and the Development of Chemistry* (Cambridge, Mass. 1970) chapter 3, and Frank E. Manuel, *A Portrait of Isaac Newton* (Cambridge, Mass. 1968) chapter 13.

2. On Desaguliers, see Schofield,

Mechanism and Materialism, 80–7.

3. *Vegetable Staticks: Or, An Account of some Statical Experiments on the Sap in Vegetables: Being an Essay towards a Natural History of Vegetation. Also, a Specimen of An Attempt to Analyze the Air, By a great Variety of Chymio-Statical Experiments* (London 1727).

4. On this topic see the unpublished Ph.D. dissertation of Arthur J. Quinn, 'Evaporation and Repulsion: A Study of English Corpuscular Philosophy From Newton to Franklin' (Princeton University 1970).

5. Isaac Newton, *Opticks or a Treatise of the Reflections, Refractions, Inflections, and Colours of Light* (London 1730). I have used the reprint of this edition (New York 1952); for the passage cited, see page 396.

6. Ibid.

7. Hales, *Vegetable Staticks*, 165–6.

8. It should be pointed out, however, that Newton emphasized the importance of the repulsive force only in his later writings. The thirty-first query first appeared as the twenty-fourth query in the first Latin edition of the *Opticks*, published in 1708, and first appeared as the thirty-first query in the second English edition of 1714; cf. Alexander Koyré, 'Les *Queries* de *l'Optique*,' *Archives internationales d'histoire des sciences*, 13 (1960) 15–29.

9. Cf. Quinn, 'Evaporation and Repulsion,' chapter 3.

10. *Vegetable Staticks*, 184–5.

11. Ibid., 203.

12. For several of Hales' experiments on calcareous substances, see ibid., experiments 53, 68, 74, 83, 99, 100 (pp. 169, 176, 178–9, 199, 222–3). A convenient table of his results can be found in J. R. Partington, *A History of Chemistry* (4 vols., London 1961–70) III (1962) 116.

13. Hales, *Vegetable Staticks*, 315.

14. Ibid.

15. Ibid., 312.

16. Ibid.

17. Ibid., 312–13.

18. Ibid., 313.

19. See, Thackray, *Atoms and Powers*, 71–3, 106–13.

20. Hales, *Vegetable Staticks*, 279.

21. Ibid., Preface, v.

22. Ibid., 316.

23. Ibid., 285. Cf. Schofield, *Mechanism and Materialism*, 78 n. 21, where he suggests, erroneously I believe, that the terms air and fire are interchangeable in Hales' theories.

24. John Desaguliers, 'An Account of a book entitul'd *Vegetable Staticks*,' *Philosophical Transactions of the Royal Society of London*, 35 (1728, nos. 398 & 399) 264–91, 323–31.

25. John Desaguliers, 'An Attempt to solve the phenomena of the Rise of Vapours, Formation of Clouds, and Descent of Rain,' *Philosophical Transactions*, 36 (1729–30, no. 407) 6–22.

26. Ibid., 6–7.

27. Ibid., 7.

28. Ibid., 10.

29. Desaguliers, 'Rise of Vapours,' 13. This entire passage is italicized in the original.

30. Ibid., 14.

31. Ibid., 15.

32. Ibid., 16.

33. Ibid., 16–17. James Watt later determined that the figure 14,000 is far too high; his own experiments indicated a figure of '1800 for the expansion of water converted into steam of the heat of boiling water'. See his notes to John Robison's *A System of Mechanical Philosophy*, ed. David Brewster (4 vols., Edinburgh 1822) II.114–15.

34. See Desaguliers, 'Rise of Vapours,' 17.

35. Ibid., 18–19.

36. For a more detailed discussion

of Cullen's comments on Desaguliers' theory of evaporation, see below, chapter 6, section 6.

37. Desaguliers, 'Rise of Vapours', 19.

38. Ibid.

39. Ibid.

40. RCPE MSS, C.15, 1,fol.60v. Since this MS is not paginated, folio numbers referring to the leaves of the MS as presently bound have been given in all references to it.

41. Desaguliers, 'Rise of Vapours,' 21.

42. RCPE MSS, C.15, 1,fol.60v.

43. Cf. Newton's own statement, *Opticks*, 405, to the effect that 'if natural Philosophy, in all its Parts, by pursuing the inductive Method, shall at length be perfected, the Bounds of Moral Philosophy will also be enlarged.'

44. Ibid., 401.

45. See Lester S. King, 'Medical Theory and Practice at the Beginning of the 18th Century,' *Bulletin of the History of Medicine*, 46 (1972) 1–15, especially pages 14–15.

46. See Ibid., 5–10 and publications cited therein.

47. For the most extensive discussion of the sources and content of Cullen's medical theories, see Thomson, *Cullen*, vols. I and II.

48. See Allen G. Debus, *The Chemical Dream of the Renaissance* (Cambridge 1968).

49. See Hélène Metzger, *Les Doctrines Chimiques en France du début du XVIIe à la fin du XVIIIe Siècle*, vol. I (Paris 1923), chapter 4.

50. On Shaw's chemistry, see F. W. Gibbs, 'Peter Shaw and the Revival of Chemistry,' *Annals of Science*, 7 (1951) 211–37; Schofield, *Mechanism and Materialism*, 154–5; Thackray, *Atoms and Powers*, 119; and I. B. Cohen, *Franklin and Newton* (Philadelphia 1956) 223.

51. George Ernest Stahl, *Philosophical Principles of Universal Chemistry*, trans. by Peter Shaw (London 1730). This translation is frequently cited in Cullen's lecture notes for his first chemistry course.

52. See, for instance, J. B. Senac, *Nouveau Cours de Chymie suivant les principes de Newton et de Stahl*, published anonymously (Paris 1723). See also, Schofield, *Mechanism and Materialism*, 46, 211–13.

53. Peter Shaw, *Chemical Lectures, Publickly read at London, In the Years 1731, and 1732; And at Scarborough, in 1733; For the Improvement of Arts, Trades, and Natural Philosophy* (London 1734), and Peter Shaw, *Scheme for a Course of Philosophical Chemistry: Or the Application of Chemical Experiments to the Improvement of Natural Knowledge*, (pp. 40–75 of *An Essay for Introducing a Portable Laboratory . . . by Peter Shaw, M.D. and Francis Hauksbee*) (London 1731). For additional information on the course of lectures offered in 1734, see Gibbs, 'Peter Shaw,' 220–4.

54. See Stahl, *Universal Chemistry, passim;* Schofield, *Mechanism and Materialism*, 211–13; Thackray, *Atoms and Powers*, 171–6; and Hélène Metzger, *Newton, Stahl, Boerhaave et la Doctrine Chimique* (Paris 1930), part II.

55. Charles Du Fay, 'A Letter from Mons. Du Fay, F.R.S. and Royal Academy of Sciences at Paris . . . Concerning Electricity,' *Philosophical Transactions*, 38 (1733–4, no. 431) 258–66. See also, John Desaguliers, 'Some Thoughts and Conjectures Concerning the Cause of Elasticity,' *Philosophical Transactions*, 41 (1739–41, no. 454) 175–85 and John Desaguliers, 'Some Thoughts and Experiments Concerning Electricity,' *Philosophical Transactions*, 41 (1739–41, no. 454) 186–93. Cf. Schofield, *Mechanism and Materialism*, 83–5, and Quinn, 'Evaporation and Repulsion,' chapters 4 & 5.

56. Thackray, *Atoms and Powers*, 119.

57. See Schofield, *Mechanism and Materialism*, 212–13. In France too the phlogiston concept came to prominence after the middle of the century; see Henry Guerlac, 'Some French Antecedents of the Chemical Revolution,' *Chymia*, 5 (1959) 73–112.

CHAPTER THREE
Medicine and Natural Philosophy in Edinburgh

1. Thomson, *Cullen*, I.67, says he practised in Shotts; Douglas Guthrie, in Kent, *Eighteenth Century Lectureship*, 50, says he practised in Auchinlee, near Shotts.
2. See Sir Alexander Grant, *The Story of the University of Edinburgh During its First Three Hundred Years* (2 vols., London 1884) I.310–15 and Horn, *University of Edinburgh*, 41–3. On medical education in Edinburgh before 1726, see John Comrie, *History of Scottish Medicine*, 2nd ed. (2 vols., London 1932) I.261–303; and on the incorporation of extra-mural teaching into the university, see R. P. Ritchies, *The Early Days of the Royal College of Phistians, Edinburgh* (Edinburgh 1899) 268–306.
3. Quoted in Grant, *University of Edinburgh*, I.300.
4. Ibid., I.304.
5. Charity balls were one popular means of raising funds for the infirmary; see Graham, *Social Life*, 99.
6. From the preamble of the February 1726 petition from the four doctors to the Town Council, quoted in Grant, *University of Edinburgh*, I.310.
7. It should be noted that all four doctors received their professional training in Leyden; see Archibald Clow, 'Hermann Boerhaave and Scottish Chemistry,' in Kent, *Eighteenth Century Lectureship*, 42–3; and J. D. Comrie, 'Boerhaave and the Early Medical School at Edinburgh,' in *Memorialia Herman Boerhaave* (Haarlem 1939) 32–9.
8. On the popularity of Moffat spring water, see Graham, *Social Life*, 52 and Alexander Carlyle, *Autobiography*, 2nd ed. (London 1860) 109–10.
9. 'Dr. Fothergill's Account of the Edinburgh School of Medicine,' reprinted in Thomson, *Cullen*, I.523–5.
10. Cf. Partington, *History of Chemistry*, III.127–8.
11. Andrew Plummer, 'Experiments on Neutral Salts, compounded of different acid Liquors, and alcaline Salts, fixt and volatile, read 7 June 1739,' in *Essays and Observations, Physical and Literary*, I (1754) 315–40.
12. Cf. Thackray, *Atoms and Powers*, 224.
13. Plummer, 'Neutral Salts,' 329.
14. Ibid., 334.
15. Ibid., 336–7.
16. 'Dr. Fothergill's account,' Thomson, *Cullen*, I.524.
17. Thomson, *Cullen*, I.40.
18. Ibid., I.58–9.
19. Ibid., I.93.
20. Quoted in Ernest C. Mossner, *The Life of David Hume* (Austin, Texas 1954) 49. On the Rankenian Club, see Mossner, 48–9.
21. See Thomson, *Cullen*, I.10–11; J. Gray, *History of the Royal Medical Society, 1737–1937* (Edinburgh 1952); Royal Medical Society, 'History of the Medical Society of Edinburgh,' in *List of Members, Laws, and Library-Catalogue, of the Medical Society of Edinburgh* (Edinburgh 1820) iii–ci; and J. D. Holmes, 'Early Years of the Medical Society of Edinburgh,' in *University of Edinburgh Journal*, Autumn, 1968; 333–40. John Christie has recently pointed out that the student medical society was modelled on the medical society founded in 1731 by the Edinburgh medical professors; see John

Christie, 'The Origins and Development of the Scottish Scientfiic Community, 1680–1760,' *History of Science*, 12 (1974) 131–6; 122–41.

22. See J. J. C. Cormack, 'The Society's Library,' in *Bulletin of the Medical Library Association*, 48 (1960) 125–39.

23. Royal Medical Society MSS, dissertations, vol. 3.

24. Ibid., vols. 4 and 5; cf. Gray, *Royal Medical Society*, 193–214.

25. Both Stevenson and Maclaurin were members of the Rankenian Club; see Mossner, *Hume*, 48. On Maclaurin, see also Christie, 'Scottish Scientific Community'.

26. Grant, *University of Edinburgh*, I.272–3, II.328–30; Davie, *Democratic Intellect*, 18.

27. Grant, *University of Edinburgh*, II.328 and William S. Howell, *Eighteenth-Century British Logic and Rhetoric* (Princeton 1971) 272–3.

28. Cf. G. E. Davie, 'Berkeley's Impact on Scottish Philosophy', *Philosophy*, 40 (1965) 222–34.

29. RCPE MSS, C.10, pp. 26–7.

30. Glasgow University Library (hereafter cited as GUL), Cullen MSS, no. 26a (number taken from the typescript list of the MSS in the first two boxes prepared by W. P. D. Wightman, dated 29 May 1956).

31. Ibid., no. 26b.

32. Ibid., no. 23.

33. John Locke, *An Essay Concerning Human Understanding*, ed. John Yolton (2 vols., London 1965) I.193 (Bk. II, ch. XXI). See also, P. M. Heimann and J. E. McGuire, 'Newtonian Forces and Lockean Powers: Concepts of Matter in Eighteenth-Century Thought,' in *Historical Studies in the Physical Sciences*, 3 (Philadelphia 1971), ed. Russell McCormmach, 233–306.

34. GUL, Cullen MSS, no. 23; Locke, *Essay*, II.83–4 (Bk. III, Ch. IX).

35. Locke, *Essay*, I.104–5 (Bk. II, Ch. VIII).

36. Ibid., I.106–7 (Bk. II, Ch. VIII).

37. Ibid., I.109 (Bk. II, ch. VIII).

38. Adam Smith apparently commented upon the increasing division of labour in philosophy, see J. G. Crowther, *Scientists of the Industrial Revolution* (Philadelphia 1963) 16.

39 Grant, *University of Edinburgh*, I.272.

40. Ibid.,

41. Carlyle, *Autobiography*, 46.

42. Grant, *University of Edinburgh*, II.299.

43. Ibid.

44. Edinburgh, 1742.

45. Colin Maclaurin, *An Account of Sir Isaac Newton's Philosophical Discoveries* (London 1748) 19.

46. See Cohen, *Franklin and Newton*, 191–2.

47. Thomson, *Cullen*, II.678.

48. *Essays and Observations, Physical and Literary*, I (1754) iii–viii. Cf. Mossner, *Hume*, 257–8.

49. See below, chapter 4.

50. Thomson, *Cullen*, I.72–3; cf. Mossner, *Hume*, 248–9.

51. Thomson, *Cullen*, II.218–20.

52. Ibid., I.607–9; Mossner, *Hume*, 590–603.

53. See Leonard Dobbin, 'A Cullen Chemical Manuscript of 1753,' in *Annals of Science*, I (1936) 138–56. See also Cullen's *éloge* of Clerk in Thomson, *Cullen*, I.525–36.

54. Thomson, *Cullen*, I.13.

55. For excerpts from their correspondence, see Thomson, *Cullen*, I.20–4, 539–65.

56. W. Innes Addison, *A Roll of the Graduates of the University of Glasgow* (Glasgow 1898) 135. On eighteenth-century Scottish medical degrees, see Lester King, *The Medical World of the Eighteenth Century* (Chicago 1958) 27–8.

CHAPTER FOUR
Moral Philosophy and Economic Improvement

1. On the date of the move, see Carlyle, *Autobiography*, 99.

2. Thomson, *Cullen*, 1.19–20, letter dated 7 May 1744.
3. Ibid., 1.21.
4. See Scott, *Hutcheson*, chapters 4 & 5.
5. On Dunlop, see Carlyle, *Autobiography*, 71.
6. Scott, *Hutcheson*, 95.
7. On the rise of the Moderates in the kirk, see Graham, *Social Life*, 353–65; on the growing movement toward Succession, which first appeared in 1737, see ibid., 371–82.
8. For evidence of Black's interest in contemporary debates in moral philosophy, see Joseph Black, *Lectures on the Elements of Chemistry*, ed. John Robison (2 vols., Edinburgh 1803), 1.xx, where Robison recalls 'the very perspicuous manner in which I have heard Dr. Black state the distinctions between the theories of ethics which had been taught at Glasgow by Dr. Francis Hutcheson, and, after him, by Dr. Adam Smith.'
9. William Leechman, 'Account of the Life, Writings, and Character of the Author,' in Hutcheson's posthumous *A System of Moral Philosophy* (2 vols., Glasgow 1755) 1.xlvii. Leechman supports this statement by citing Hutcheson's Preface to his *Essay on the Passions*, 1.18–19.
10. Scott, *Hutcheson*, 255.
11. Leechman, 'Account,' 1.iv.
12. Cf. Mossner, *Hume*, 58, 119.
13. See Hugh Trevor-Roper, 'The Scottish Enlightenment,' in *Studies on Voltaire and the Eighteenth Century*, 58 (1967) 1640; 1635–58.
14. See Scott, *Hutcheson*, chapter 11, part 2, and W. R. Scott, *Adam Smith as Student and Professor* (Glasgow 1937) 57, 112–13.
15. Norman Kemp Smith, *The Philosophy of David Hume* (London 1941) 24.
16. As quoted in Mossner, *Hume*, 76.
17. For an exact statement of Hume's impact upon subsequent philosophy, see Leszek Kolakowski, *The Alienation of Reason, a History of Positivist Thought* (Garden City, N.Y. 1968) 38–9.
18. See Mossner, *Hume*, chapter 10.
19. From Huxley's *Hume*, 11, as quoted in John Passmore, *Hume's Intentions*, revised ed. (New York 1968) 3.
20. See E. C. Mossner, *The Forgotten Hume: Le bon David* (New York 1943).
21 Passmore, *Intentions*, 133.
22. Mossner, *Hume*, 486–7.
23. Passmore, *Intentions*, 7.
24. David Hume, *History of England*, chapter 71, as quoted in Smith, *Philosophy of David Hume*, 58.
25. Passmore, *Intentions*, 50–1.
26. Alexander F. Tytler, *Memoirs of the Life and Writings of the Honourable Henry Home of Kames* (2 vols., Edinburgh 1807) 1 (appendix) 62. Cf. L. L. Laudan, 'Thomas Reid and the Newtonian Turn of British Methodological Thought,' in *The Methodological Heritage of Newton*, ed. R. E. Butts and J. W. Davis (Toronto 1970) 103–31, and G. N. Cantor, 'Henry Brougham and the Scottish Methodological Tradition,' in *Studies in the History and Philosophy of Science*, 2 (1971) 69–89.
27. Tytler, *Memoirs*, 1 (appendix) 63.
28. As quoted, without citation, in Bryson, *Man and Society*, 17. Cf. Cullen's statement to his chemistry students: 'It is greatly for our advantage to generalize as much as possible.' Quoted from a MS copy of a student's notes in J. K. Crellin, 'William Cullen & Practical Chemistry,' in *Actes du XIIe Congrès International d'Histoire des Sciences, Paris, 1968* (pub. 1971) 6. 18; 17–21.
29. RCPE MSS, C.15, 1, fol.79v.
30. Ibid., 1, foll.32v–33r.
31. RCPE MSS, C.10, pp. 336–7.
32. Hume, *Treatise of Human Nature*, ed. L. A. Selby-Bigge (Oxford 1896) 171, as quoted in

Passmore, *Intentions*, 29–30.
33. Hume, *Enquiries concerning the Human Understanding and Concerning the Principles of Morals*, ed. L. A. Selby-Bigge (2nd ed., Oxford 1902) 76, as quoted in Passmore, *Intentions*, 35.
34. GUL, Cullen MSS, no. 7.
35. On Kames, see Tytler, *Memoirs*, and Arthur E. McGuinness, *Henry Home, Lord Kames* (New York 1970). Kames' statement about Hume's ideas was recorded by Boswell and is quoted in Mossner, *Hume*, 118.
36. As quoted in Mossner, *Hume*, 134.
37. As quoted in 134.
38. Passmore, *Intentions*, 126.
39. Both Hutcheson and Leechman sided against Hume in 1745. The position was eventually offered to Hutcheson, who declined it. See Mossner, *Hume*, 157–8.
40. Scott, *Hutcheson*, 65.
41. Ibid., 62–3.
42. See ibid., 66 where Scott says this traditional emphasis upon didactic philosophy still flourished in the Scottish universities at the end of the nineteenth century. According to Davie, *Democratic Intellect*, this aspect of Scottish education has been undermined in the twentieth century by increased Anglicization and specialization. On Hutcheson's influence as a teacher, see also, Carlyle, *Autobiography*, chapter 3.
43. Quoted in Scott, *Hutcheson*, 75.
44. Ibid., 266. Cf. Graham, *Social Life*, 352, where he says 'Professor Hutcheson in Glasgow was in his class of Moral Philosophy forming a school of liberal men'.
45. Mackie, *University of Glasgow*, 228.
46. Thomson, *Cullen*, 1.30.
47. Kent, *Eighteenth Century Lectureship*, 2–3.
48. On the origins of Cullen's popularity as a teacher, see Crellin, 'Cullen and Practical Chemistry,' and J. K. Crellin, 'William Cullen: His Caliber as a Teacher and an Unpublished Introduction to his *A Treatise of the Materia Medica*, London, 1773,' in *Medical History*, 15 (1971) 79–87.
49. Thomson, *Cullen*, 1.25.
50. Cullen continued to deliver his botany lectures in Latin; see ibid., 1.28, 32.
51. Anderson, 'Cursory Hints,' 45.
52. Lyon Playfair, *A Century of Chemistry in the University of Edinburgh, being the Introductory lecture to the course of chemistry in 1858* (Edinburgh 1858) 10.
53. Thomson, *Cullen*, 1.25. Cullen, it will be recalled, had been educated by rectors, not by professors.
54. Playfair, *Century of Chemistry*, 11.
55. Anderson, 'Cursory Hints,' 48–9. For additional evidence of Cullen's openness toward his students, see the letter Benjamin Rush wrote to John Morgan, in *Letters of Benjamin Rush; Vol. I. 1761–1792*, ed. L. H. Butterfield (Princeton 1951) 28–31, and Rush's 'Eulogium upon Dr. William Cullen,' in Benjamin Rush, *Essays, Literary, Moral & Philosophical* (Philadelphia 1798) 321–43.
56. Thomson, *Cullen*, 1.97.
57. See the *Address of the Students of Medicine*, in ibid., 1. 155–9.
58. Anderson, 'Cursory Hints,' 54.
59. Letter quoted in Thomson, *Cullen*, 1.474–5. Cf. David L. Cowen, 'Liberty, laissez-faire and licensure in nineteenth-century Britain,' in *Bulletin of the History of Medicine*, 43 (1969) 30–40.
60. Thomson, *Cullen*, 1.69.
61. Ibid., 1.81–2, letter of August, 1751.
62. On Home's role as a patron, see Mossner, *Hume*, 57–62. See also ibid., 411–12 for Hume's comments on Home's character.
63. London.

64. See Schofield, *Mechanism and Materialism*, 97, 107 n., and D'Arcy W. Thompson, 'George Martine, an 18th Century Physician,' in *The Scotsman*, 27 December 1941, page 7.

65. Thomson, *Cullen*, I.60. Home's essay was eventually published many years later; Lord Kames, 'On Evaporation,' *Essays and Observations, Physical and Literary*, 3 (1771) 80–99.

66. For a brief history of the notion of improvement, see Asa Briggs, *The Age of Improvement* (London 1959) 1–3. On the improving movement in Scotland, see Henry Hamilton, *The Industrial Revolution in Scotland* (Oxford 1932), chapter 2, T. C. Smout, *A History of the Scottish People, 1560–1830* (London 1969), chapter 12; and N. T. Phillipson and Rosalind Mitchison, eds., *Scotland in the Age of Improvement* (Edinburgh 1970). On Henry Home as an improver, see Graham, *Social Life*, 203–4.

67. Thomson, *Cullen*, I.69, from a letter to his brother.

68. Ibid., I.596–7, letter dated 1749. Perhaps Cullen knew that the Honourable the Society for the Improvement in the Knowledge of Agriculture had already urged 'that Agriculture should be taught in a college way as other sciences'; quoted in Phillipson, 'Edinburgh and the Scottish Enlightenment,' 437.

69. Ibid., I.596

70. Ibid., I.62–3. Letter dated 17 Nov. 1749.

71. Ibid., I.63.

72. Ibid., I.64. In the 1770s Kames corresponded with Black on agricultural questions and he proudly claimed that Black had given his book on farming his *imprimatur*; see Tytler, *Memoirs*, I (appendix) 75–81, and Henry Home, *The Gentleman Farmer; being an attempt to improve agriculture by subjecting it to the test of rational principles* (Edinburgh 1776) xii. On Home's agricultural theories, see G. E. Fussell, *Crop Nutrition: Science and Practice before Liebig* (Lawrence, Kansas 1971), chapter 7.

73. Thomson, *Cullen*, I.70, letter dated August 1749.

74. Ibid., I.596, letter dated 17 Jan. 1750.

75. Ibid., I.71.

76. Ibid., I.82.

77. Ibid.

78. Ibid., I.83.

79. Ibid.

80. Ibid.

81. Ibid., I.545.

82. Ibid., I.83–4.

83. See the letters printed in ibid., I.84–5.

84. Ibid., I.606, letter dated Nov. 1751.

85. Ibid., I.72–3, letter dated 21 Jan. 1752. There is some evidence that Cullen's religious beliefs were only a bit more orthodox than Hume's. Writing about Cullen, Benjamin Rush stated that 'it is scarcely possible to do justice to this great man's Character either as a Scholar—a Physician—or a Man. . . . There is One thing, however, wanting in Dr. Cullen to constitute his character a complete One, viz: a Regard to Religion. . . . A Man Who has long been in a Habit of thinking for himself & of doubting the Truth of every Principle of Science Until it is proved to him by the usual Laws of Demonstration is very apt to carry the same Thirst of free Inquiry with him to religious Matters which will by no means admit of the same Demonstration that is to be found in Matters of Science. . . . I am not fully acquainted with Dr. Cullen's Principles, nor do I believe he has formed any regular System for himself. He believes in the Immateriality & Immortality of the Soul. This I have heard him

frequently declare in his Lectures. But with regard to revealed Religion he professes himself a sceptic.' (Benjamin Rush, 'Scottish Journal', a MS record of his visit to Edinburgh, 1766–68, Indiana University Library, 58, 66–8.)
86. Francis Home, *An Essay on the Contents and Virtues of Dunse-Spaw* (Edinburgh 1751).
87. Cf. Thomson, *Cullen*, 1.85.
88. Ibid., 1.86.
89. See letters in ibid., 1.87, 92–3.
90. Ibid., 1.87–8. Milton served as Argyll's 'sous-ministre' in Edinburgh.
91. Ibid., 1.88.
92. Ibid., 1.91.
93. Cf. their letters on the subject, ibid., 1.87–93.
94. Ibid., 1.92–3.
95. Ibid., 1.92, letter dated 21 Nov. 1755.
96. Ibid., 1.90.
97. Mackie, *University of Glasgow*, 229.
98. Thomson, *Cullen*, 1.98–9.

CHAPTER FIVE
Cullen's Chemical Investigations
1. Thomson, *Cullen*, 1.78.
2. Clow, Archibald and Clow, Nan, *The Chemical Revolution* (London 1952) 165.
3. Henry Hamilton, *An Economic History of Scotland in the Eighteenth Century* (Oxford 1963) 134.
4. Ibid., 139–40.
5. Clow and Clow, *Chemical Revolution*, 138.
6. James Dunbar, *Smegmatologia, or the Art of Making Potashes, Soap, and Bleaching of Linen* (Edinburgh 1736) 30, as quoted in Clow and Clow, *Chemical Revolution*, 183–4.
7. Charles Alston, *A Dissertation on Quick-Lime and Lime-Water* (Edinburgh 1752) 21.
8. Thomson, *Cullen*, 1.579.
9. Sir William Ramsay, *The Life and Letters of Joseph Black, M.D.* (London 1918) 36, letter dated 20 Sept. 1763; cf. MS. of this letter in Edinburgh University Library, hereafter cited as EUL MSS, GEN 873, vol. 1. On Sam Black, see John Black MS letter dated 19 Oct. 1752 in Public Record Office of Northern Ireland, hereafter cited as PRONI MSS, D.1950, no. 23.
10. Dublin 1771. For Black's letter to Ferguson, see Ramsay, *Life and Letters*, 52–4, and letter dated 14 Oct. 1763 in EUL MSS, GEN 873, vol. 1.
11. GUL, Cullen MSS, no. 40b.
12. Thomson, *Cullen*, 1.598.
13. Ibid., 1.74, letter dated 26 Dec. 1752.
14. Ibid., 1.76–8.
15. Ibid., 1. 78.
16. Ibid., 1.78–9.
17. Clow and Clow, *Chemical Revolution*, 177.
18. Thomson, *Cullen*, 1.93, letter dated 22 Nov. 1755. Perhaps Deskford was referring to Francis Homes' *Experiments on Bleaching* (Edinburgh 1756), which may have been circulated in manuscript to bolster Home's candidacy for the chair of chemistry in Edinburgh.
19. Hamilton, *Economic History* 112–15. On the rapid decline of the fishing industry after the Union, see Graham, *Social Life*, 512.
20. Clow and Clow, *Chemical Revolution*, 11–12, 57–8.
21. William Brownrigg, *The Art of Making Common Salt* (London 1748) 9–10. Cf. J. Russell-Wood, 'The Scientific Work of William Brownrigg, M.D., F.R.S. (1711–1800)—II,' in *Annals of Science*, 7 (1951) 77–94.
22. Hamilton, *Economic History*, 114.
23. Brownrigg, *Salt*, ix.
24. Ibid., 10.
25. See ibid., chapter II, section III, 72–93.
26. Dobbin, 'Cullen manuscript,' 155.
27. Thomson, *Cullen*, 1.75.
28. Ibid., 1.598.

29. Ibid., 1.599.
30. Ibid., 1.74–5.
31. Dobbin, 'Cullen Manuscript.' For a discussion of this paper, see below, chapter 6, section 3.
32. Thomson, *Cullen*, 1.597.
33. Ibid., 1.592.
34. Ibid., 1.593.
35. Ibid., 1.593–4.
36. Ibid., 1.594.
37. Clow and Clow, *Chemical Revolution*, 472.
38. Thomson, *Cullen*, 1.593.
39. Ibid., 1.594–5.
40. Charles Alston, 'Extracts of Two Letters from Dr. Alston, Botany Professor at Edinburgh: to Dr. Mortimer, Secr. R. S. The first dated 17 March 1749; the second August, 9 1750,' *Philosophical Transactions of the Royal Society of London*, 47 (1751–2, published 1753) 266; 265–6.
41. Alston, *Dissertation on Quick-Lime*, 4.
42. Ibid., 47.
43. Robert Whytt, *An Essay on the Virtues of Lime-Water in the Cure of the Stone* (Edinburgh 1752) 42–3.
44. For further discussion of the chemical aspects of this debate, see Henry Guerlac, 'Joseph Black and Fixed Air,' *Isis*, 48 (1957) 124–51, 433–56.
45. See Charles Alston, *Dissertation on Quick-Lime*, 2nd ed. (Edinburgh 1754) iv.
46. From a MS copy of Cullen's chemical lectures, quoted in E. J. Holmyard, *Makers of Chemistry* (Oxford 1931) 165–6.
47. Thomson, *Cullen*, 1.67, 600.
48. Ibid., 1.85.
49. See ibid., 1.66–7, 602–4.
50. Ibid., 1.539.
51. Ibid., 1.38–9.
52. Ibid., 1.590.
53. See ibid., 1.586, 610.
54. Ibid., 1.30.
55. Cf. RCPE MSS, C.15, vol. 1; and below, chapter 6, sections 5 and 7.
56. Thomson, *Cullen*, 1.588–9; on Melville, see Schofield, *Mechanism and Materialism*, 243–4.
57. Thomson, *Cullen*, 1.60; and GUL, Cullen MSS, no. 46.
58. Thomson, *Cullen*, 1.537, 576.
59. Ibid., 1.603.
60. Ibid., 1.57–8; cf. Dobbin, 'Chemical Manuscript'.
61. Thomson, *Cullen*, 1.52; GUL, Cullen MSS, no. 20.
62. Thomson, *Cullen*, 1. 583; cf. ibid., 1.580–3, 53–4. For an extended discussion of Cullen's theory of heat and mixture, see below, chapter 6, section 7. On the Literary Society of Glasgow, which was founded in 1752 with Adam Smith as a charter member and which David Hume joined in 1753, see Mossner, *Hume*, 273.
63. Thomson, *Cullen*, 1.85.
64. Ibid., 1.591–2.
65. *Essays and Observations, Physical and Literary*, 2 (1756) 145–56. For a discussion of the contents of this paper, see below, chapter 6, section 7.
66. Cullen's brief article was widely circulated and frequently cited in the subsequent literature on heat: On its influence among German physicists, on Saussure and on Dalton, see Thomas S. Kuhn, 'The Caloric Theory of Adiabatic Compression,' *Isis*, 49 (1958) 133–5; 132–40; and Robert Fox, *The Caloric Theory of Gases from Lavoisier to Regnault* (Oxford 1971) 39–60; on its influence on Lavoisier, see J. B. Gough, 'Nouvelle contribution à l'étude de l'évolution des idées de Lavoisier sur la nature de l'air et sur la calcination des métaux,' *Archives internationales d'histoire des sciences*, Number 88–9 (July–December, 1969) 270–1; 267–75, and Douglas McKie and Niels H. de V. Heathcote, *The Discovery of Specific and Latent Heats* (London 1935) 48; on Benjamin Franklin's interest in Cullen's experiments, see Thomson, *Cullen*, 1.39; on Henry

Cavendish's knowledge of Cullen's paper, see Partington, *History of Chemistry*, III.307; and on James Watt's familiarity with Cullen's discovery, see D. S. L. Cardwell, *From Watt to Clausius—The Rise of Thermodynamics in the Early Industrial Age* (Ithaca, New York 1971) 46.

CHAPTER SIX

Cullen's Chemical Doctrines

1. RCPE MSS, C. 15, vol. I. The notes for the first lecture in this set are dated 2 Nov. 1748 (I,fol. 157r.). For a more extensive description of the Cullen chemical MSS in the RCPE and GUL, see W. P. D. Wightman, 'William Cullen and the Teaching of Chemistry,' *Annals of Science*, 11 (1955) 154–65; and his 'William Cullen and the Teaching of Chemistry—II,' *Annals of Science*, 12 (1956) 192–205. A new catalogue of the MSS in the RCPE has been prepared since the publication of Wightman's articles. Therefore, to go from the citations given in the present notes to the list of MSS prepared by Wightman (*Ann.Sci.*, 11.157), one must have both the old and the new catalogue numbers:

New Catalogue Numbers	*Old Catalogue Numbers*
C.10	Ch.5.16 & Ch.5.17
C.14	Ch.5.18
C.12	Ck.11.2
C.15, vol. I.	Ck.11.3

2. Cf. GUL, Cullen MSS, no. 18, which is a smooth copy of the latter part of Cullen's first lecture and is dated 2 Nov. 1748.
3. GUL, Cullen MSS, no. 26b.
4. Ibid., no. 26a.
5. Ibid., no. 26b.
6. Ibid.
7. Ibid.
8. Ibid. For a discussion of this definition in the context of Homberg's writings, see Partington, *History of Chemistry*, III.46.
9. GUL, Cullen MSS, no. 26b.
10. Ibid.
11. Ibid.
12. Ibid. Although he cites Shaw on this point, Cullen knew that Stahl had asserted that chemists could employ whatever means of analysis they found useful; see RCPE MSS, C.15, I,fol. 2r, where Cullen says that limiting the means used by chemists to fire is 'an oversight in Homberg corrected by Stahl'.
13. GUL, Cullen MSS, no. 26b.
14. Ibid., no. 4.
15. Ibid.
16. Ibid.
17. Ibid.
18. Ibid.
19. Ibid., no. 26a.
20. Ibid., no. 4.
21. Ibid. Newton also considered these very specific physical properties the subject matter of chemistry and of the utmost importance in natural philosophy. See Henry Guerlac, 'The Background to Dalton's Atomic Theory,' in *John Dalton & the Progress of Science*, ed. D. S. L. Cardwell (Manchester 1968) 56–91, especially 70–1.
22. RCPE MSS, C.15, I,fol. 3r & v.
23. GUL, Cullen MSS, no. 5.
24. Ibid.
25. RCPE MSS, C.15, I,fol. 157r.
26. RCPE MSS, C.15. I,fol. 5r. See also, 'William Cullen's History of Chemistry,' in Kent, *Eighteenth Century Lectureship*, 15–27.
27. GUL, Cullen MSS, no. 5.
28. Kent, *Eighteenth Century Lectureship*, 15.
29. Ibid., 16.
30. GUL, Cullen MSS, no. 5.
31. Kent, *Eighteenth Century Lectureship*, 17.
32. Ibid., 18.
33. GUL, Cullen MSS, no. 5.
34. Kent, *Eighteenth Century Lectureship*, 21.
35. GUL, Cullen MSS, no. 5.
36. Kent, *Eighteenth Century Lectureship*, 24.

37. Ibid., 24–5.
38. Ibid., 25.
39. GUL, Cullen MSS, no. 5.
40. Kent, *Eighteenth Century Lectureship*, 25.
41. GUL, Cullen MSS, no. 5.
42. Kent, *Eighteenth Century Lectureship*, 26.
43. GUL, Cullen MSS, no. 5.
44. Kent, *Eighteenth Century Lectureship*, 27.
45. GUL, Cullen MSS, no. 5.
46. Ibid.
47. GUL, Cullen MSS, no. 7.
48. Ibid.
49. RCPE MSS, C.10, p. 22. Perhaps it was the Scottish doubt about the connection between reason and practical judgment which made Cullen less confident concerning the usefulness of chemistry than many of his contemporaries; see Gibbs, 'Peter Shaw,' 224, n. 39, where he quotes the following passage from a student copy of Cullen's chemical lectures: 'As to the Arts, Manufactures, Trade, Commerce &c in Life to which Chemistry may be subservient, the Drs. Boerhaave & Shaw may be consulted, but with Caution, least a Person shou'd Acquire too high an Opinion of it.'
50. GUL, Cullen MSS, no. 7.
51. Ibid.
52. RCPE MSS, C.10, p. 26.
53. RCPE MSS, C.15, 1,fol. 86r.
54. GUL, Cullen MSS, no. 7.
55. Cf. Charles Gillispie, 'The Natural History of Industry,' *Isis*, 48 (1957) 398–407.
56. GUL, Cullen MSS, no. 7.
57. Ibid.
58. Ibid.
59. Ibid.
60. Ibid.
61. Ibid.
62. Ibid., no. 40b, p. 4.
63. Ibid., no. 7.
64. Ibid.
65. Ibid.
66. Ibid.
67. Ibid.
68. Ibid. Cf. Thackray, *Atoms and Powers*, 14–15, 22, on transmutation and the Newtonian theory of matter.
69. GUL, Cullen MSS, no. 7. It must be remembered that during the eighteenth century good families in Scotland did not consider it below their dignity to place younger sons as 'apprentices to shopkeepers, who under the vaguely comprehensive title of "merchant" might deal in anything from tallow-candles to brocade, from tobacco to Tay pearls'. Graham, *Social Life*, 33. Cf. the social origins of Scottish university students during the eighteenth century, as described in ibid., 455–8.
70. GUL, Cullen MSS, no. 7.
71. GUL, Cullen MSS, no. 23.
72. Ibid.
73. Ibid.
74. Ibid.
75. Ibid.
76. Ibid.
77. Ibid.
78. Ibid. Cf. RCPE MSS, C.15, 1,fol. 10v, where Cullen cites *Opticks*, page 375 as the source of this quotation.
79. RCPE MSS, C.10, p. 61.
80. RCPE MSS, C.15, 1,fol. 11r.
81. Ibid.
82. For discussions of Alston's approach to botany, see the article on Alston in the *Dictionary of National Biography* and Thomson, *Cullen*, 11.577–8. For Cullen's appreciation of Linnaeus and evidence that the two men corresponded, see Thomson, *Cullen*, 1.33–5.
83. RCPE MSS, C.15, 1,fol. 27r.
84. Ibid., 1,fol. 27v.
85. Dobbin, 'Cullen Manuscript,' 146.
86. Ibid., 144. G. F. Rouelle, who was Cullen's contemporary and Lavoisier's chemistry teacher, organized the contents of his lectures according to the three

realms of nature; see Rhoda Rappaport, 'G. F. Rouelle: An Eighteenth-Century Chemist and Teacher,' in *Chymia*, 6 (1960) 93; 68–101.

87. RCPE MSS, C.15, 1,fol. 11r & v.

88. See Marie Boas Hall, 'The History of the Concept of Element,' in Cardwell, *John Dalton*, 21–39.

89. From a book by Tachenius, as quoted in Partington, *History of Chemistry*, II.293.

90. Partington, *History of Chemistry*, II.143.

91. 'Experiments in Consort upon the Solution of Salts in Water. Read before the Royal Society, January 18, 1676/7,' published in Nehemiah Grew's *Anatomy of Plants* (London 1682) 296–304. For a brief discussion of Grew's paper and Boerhaave's citation of it, see Guerlac, 'Black and Fixed Air,' 146 n. 97.

92. Hermann Boerhaave, *A New Method of Chemistry . . . Translated from the original Latin of Dr. Boerhaave's Elementa Chemia, as Published by Himself . . . By Peter Shaw, M.D.*, 3rd ed. (2 vols., London 1753) I.104, 120.

93. RCPE MSS, C.15, 1,fol. 20r.

94. Stahl, *Universal Chemistry*, 150.

95. RCPE MSS, C.15, 1,fol. 13r & v.

96. Ibid., 1,fol. 11r & v.

97. Ibid., 1,fol. 11v.

98. Dobbin, 'Cullen Manuscript,' 144–5. Cullen's classification of salts clearly owes a good deal to the system of G. F. Rouelle, published in his 'Sur les sels neutres,' *Mémoires de l'Académie royale des Sciences* (1744) 353–64.

99. Boerhaave, *New Method*, I.542. On Boyle and Stahl, see Cullen's brief history of the analysis of sea salt, Dobbin, 'Cullen Manuscript,' 150.

100. *Mémoires de l'Académie royale des Sciences*, 1736, 215–32.

101. Ibid., 216.

102. Partington, *History of Chemistry*, III.69–70 and Robert P. Multhauf, *The Origins of Chemistry* (New York 1967) 347–8.

103. Cf. Partington, *History of Chemistry*, II.720.

104. For Pott's system, see ibid., II.718. For Cullen's use of Pott's categories, see GUL, Cullen MSS, no. 24, and RCPE MSS, C.10, pp. 45–8. See also Thomson, *Cullen*, I.580 for a letter in which Black thanks Cullen for having Pott's tables transcribed for him.

105. Dobbin, 'Cullen Manuscript,' 150.

106. Ibid.

107. Ibid., 150–1.

108. Ibid., 153.

109. Ibid. When preparing the magnesia alba he needed for his dissertation research, Joseph Black used 'the common method' of precipitating it from a solution of impure Glauber's salt. See Joseph Black, *Dissertatio medica inauguralis, de humore acido a cibis orto, et magnesia alba* (Edinburgh 1754). This work was reprinted in vol. 2 of the *Thesaurus medicus Edinburgensis novus* (4 vols., Edinburgh 1785) and a translation was published under the title 'Joseph Black's Inaugural Dissertation,' trans. A.Crum Brown, in the *Journal of Chemical Education*, 12 (1935) 225–8, 268–73. All citations of this work refer to the translated version. Black's description of the way he prepared magnesia alba can be found on pages 268–9.

110. Dobbin, 'Cullen Manuscript,' 154–5.

111. Ibid., 155.

112. Brownrigg, *Salt*, 75.

113. Ibid., 76.

114. Ibid.

115. Ibid., 78–9. On scratch, see also Clow and Clow, *Chemical Revolution*, 56.

116. Brownrigg, *Salt*, 79.

117. Ibid., 80.

118. Ibid.

119. Maurice Crosland, *Historical*

Studies in the Language of Chemistry (London 1962) 108. See also 'Calx' in the *Oxford English Dictionary*.

120. RCPE MSS, C.15, 1,foll. 115v & 118r.

121. Ibid., 1,fol. 118r.

122. Ibid., 1,fol. 118v.

123. Ibid., 1,fol. 119r & v.

124. J. B. Van Helmont, *Van Helmont's Works* (London 1664) 841. The work cited is a reissue of the 1662 English translation of Van Helmont's writings first published under the title *Oriatrike*.

125. John Mayow, *Medico-Physical Works*, trans. A. Crum Brown and Leonard Dobbin (Edinburgh 1907) 169.

126. Stahl, *Universal Chemistry*, 144.

127. Charles DuFay, 'Sur le sel de Chaux,' *Mémoires de l'Académie royale des Sciences* (1724) 89; 88–93.

128. Crosland, *Historical Studies*, 97. For examples of the identification of alkalinity and effervescence, see the definitions of alkalinity of Sylvius de le Boë and Boerhaave quoted in Partington, *History of Chemistry*, II.285 and II.755 respectively.

129. RCPE MSS, C.15, 1,fol. 28v.

130. GUL, Cullen MSS, no. 24a, p. 5.

131. 'Sur le Sel de la Chaux,' *Mémoires de l'Académie royale des Sciences* (1745) 93–106.

132. Ibid., 105. On the *sel sélénitique*, see G. F. Boulduc, 'Analyse des eaux de forges, et principalement de la source apellée la Royale,' *Mémoires de l'Académie royale des Sciences* (1735) 443–53. It is obvious that at some point Malouin's quicklime or lime-water became contaminated with sulphate salts.

133. See his 'Observations sur la chaux et sur le plastre,' *Mémoires de l'Académie royale des Sciences* (1747; published 1752) 678–96.

134. See his 'Diverses Experiences sur la Chaux,' *Mémoires de l'Académie royale des Sciences* (1747; published 1752) 59–81.

135. See Duhamel, 'Chaux,' 64–6.

136. Ibid., 69–70.

137. Ibid., 70.

138. Ibid., 80.

139. Brownrigg, *Salt*, 80.

140. Ibid., 81.

141. See Frederick Hoffmann, 'Animadversiones & experimenta circa magnesiam albam, tutem & gratium insipidum pulverem laxantem,' *Opera Omnia Physico-Medica* (6 vols., Geneva 1740) IV.497–8; and 'De lixivio a coctione salis communis relictio & ex eo prodeunte terra laxante sive magnesia, & sale sic dicto Ebsoniensi,' *Opera Omnia*, IV.500–1.

142. Epsom salt (magnesium sulphate) is soluble whereas gypsum (calcium sulphate) is not.

143. For Brownrigg's discussion of this problem, see *Salt*, 81; for Black's see his 'Inaugural Dissertation,' 268, 270.

144. Dobbin, 'Cullen Manuscript,' 152.

145. Ibid., 154.

146. Ibid., 149.

147. Thomson, *Cullen*, I.578.

148. RCPE MSS, C.15, 1,fol. 13v.

149. Ibid., 1,foll. 13v & 14r.

150. RCPE MSS, C.10, pp. 182, 198.

151. RCPE MSS, C.15, 1, foll. 15r-16r.

152. Ibid., 1,fol. 16r & v.

153. Ibid., 1,foll. 17v-18r.

154. Ibid., 1,fol. 22v.

155. Ibid.

156. Ibid.

157. Ibid., 1,fol. 23r.

158. RCPE MSS, C.12, I.35.

159. GUL, Cullen MSS, no. 34.

160. RCPE MSS, C.15, 1,fol. 23r.

161. Ibid., 1,fol. 32v.

162. See above, chapter 4.

163. On the French Newtonians, see Thackray, *Atoms and Powers*, sections 5.5 and 7.2.

164. RCPE MSS, C.15, 1,fol. 33r.

165. Ibid.

166. Ibid.

167. Ibid.

168. Ibid., 1,fol. 33r & v. Cf. E. F. Geoffroy, 'Table des différents rapports observé en chimie entre différentes substances,' *Mémoires de l'Académie royale des Sciences* (1718) 202–12.
169. See Cullen's letter to George Fordyce, Thomson, *Cullen*, 1.126–7, and RCPE MSS, C.15, 1,fol. 33v.
170. RCPE MSS, C.15, 1,foll. 33v-34r.
171. See Maurice Crosland, 'The Use of Diagrams as Chemical "Equations" in the Lecture Notes of William Cullen and Joseph Black,' *Annals of Science*, 15 (1959) 75–90.
172. Cf. the new diagrams Cullen developed in 1759; Thomson, *Cullen*, 1.570–1, and Crosland, 'Diagrams,' 80–1.
173. For a description of the Dutch doctrines, see Schofield, *Mechanism and Materialism*, chapter 7.
174. RCPE MSS, C.15, 1,fol. 130r.
175. Ibid., 1,fol. 130r & v.
176. Ibid., 1,fol. 40r.
177. Ibid., 1,fol. 40v.
178. Ibid.
179. Ibid., 1,fol. 41r.
180. Ibid., 1,fol. 41r & v.
181. Ibid., 1,fol. 41v.
182. Ibid., 1,fol. 45r.
183. Ibid., 1,fol. 45v.
184. Ibid., 1,fol. 43r.
185. Ibid.
186. Ibid.
187. Quoted from Boerhaave's *A New Method of Chemistry*, trans. by Peter Shaw (London 1741) vol. 1, 290–1 in McKie and Heathcote, *Specific and Latent Heats*, 13.
188. McKie and Heathcote, *Specific and Latent Heats*, 14. Cf. Cardwell, *Watt to Clausius*, 28–9, for a brief discussion of this experiment. Apparently Boerhaave's account of this experiment did not accurately represent Fahrenheit's understanding of it; see Partington, *History of Chemistry*, III. 155 n.5.
189. RCPE MSS, C.15, 1,fol. 43r.
190. Ibid., 1,fol. 43r & v.
191. RCPE MSS, C.15, 1,fol. 43v.
192. Ibid.
193. Ibid.
194. On Martine, see above, chapter 4.
195. Martine, *Essays*, 257–8. For a discussion of Martine's experiments, see McKie and Heathcote, *Specific and Latent Heats*, 14–15.
196. RCPE MSS, C.15, 1,fol. 44v.
197. RCPE MSS, C.15, 1,fol. 45v.
198. Ibid. For Musschenbroek's own account of these experiments I have consulted his *Essai de Physique*, translated from the Dutch by P. Massuet (2 vols., Leyden 1739) 1.452–4.
199. Although Musschenbroek measured only the degree of expansion caused by a specific intensity of fire, Cullen interpreted his experiments as a confirmation of Martine's demonstration that the rate at which bodies heat up is not proportional to their densities; see RCPE MSS, C.15, 1,fol. 43v. Black, whose theories of specific and latent heats arose from considerations of the different rates at which bodies heat up, avoided this confusion and made no mention of Musschenbroek's findings; see the quotation from John Robison's account of Black's discoveries in McKie and Heathcote, *Specific and Latent Heats*, 36.
200. RCPE MSS, C.15, 1,fol. 109r & v.
201. Ibid., 1,fol. 110r.
202. Ibid., 1,fol. 111r.
203. Ibid., 1,fol. 111v.
204. Ibid., 1,fol. 112r.
205. Ibid., 1,fol. 113r.
206. Fahrenheit's observations were reported in the *Philosophical Transactions*, 33 (1724) 78; cf. McKie and Heathcote, *Specific and Latent Heats*, 30.
207. RCPE MSS, C.15, 1,fol. 113r.
208. Ibid., 1,fol. 113v.
209. Ibid., 1,fol. 114r.
210. See above, chapter 2.

211. RCPE MSS, C.15, 1,fol. 114v.
212. Ibid.
213. Ibid., 1,fol. 115r & v.
214. Ibid., 1,foll. 115v & 118r.
215. Ibid., 1,foll. 118r–119r. For a brief history of eighteenth-century British attempts to weigh fire, see McKie and Heathcote, *Specific and Latent Heats*, 138–48.
216. RCPE MSS, C.15, 1,fol. 119r & v.
217. Ibid., 1,fol. 120r.
218. Ibid., 1,fol. 120r & v.
219. Ibid., 1,fol. 121r & v.
220. Ibid., 1,fol. 122r.
221. Ibid., 1,fol. 123r.
222. Ibid., 1,fol. 123v.
223. Ibid., 1,fol. 124r.
224. Ibid., 1,fol. 124v.
225. Ibid., 1,foll. 126–127r.
226. Ibid., 1,fol. 127r & v.
227. Ibid., 1,fol. 130v.
228. Ibid.
229. Ibid., 1,foll. 130v–131v.
230. Ibid., 1,foll. 131v–132r.
231. Ibid., 1,fol. 132r & v. As a physical model, Cullen's account of the relationship between attraction, repulsion and distance is similar to the ideas of Boscovich, whose works were known and discussed by the natural philosophers in Scotland from the middle of the century onwards; see Thomas Melville's letter to Cullen, dated 21 Feb. 1752, in Thomson, *Cullen*, 1.588–9, and Richard Olson, 'The Reception of Boscovich's Ideas in Scotland,' *Isis*, 60 (1969) 91–103.
232. RCPE MSS, C.15, 1,foll. 132v–133r.
233. Ibid., 1,fol. 133r.
234. Ibid., 1,fol. 133r & v.
235. Ibid., 1,fol. 133v.
236. Ibid., 1,fol. 134v.
237. RCPE MSS, C.10, p. 87. Cf. Bryan Robinson, *Dissertation on the Aether of Sir Isaac Newton* (Dublin 1743 and London 1747). For discussions of Robinson's ideas, see Thackray, *Atoms and Powers*, 135–40 and Schofield, *Mechanism and Materialism*, 107–14.
238. RCPE MSS, C.15, 1,fol. 57v.
239. RCPE MSS, C.10, p. 87.
240. RCPE MSS, C.15, 1,fol. 57v.
241. RCPE MSS, C.15, 1,fol. 20v. During the latter part of his career as a chemistry teacher Cullen added a sixth heading, the aerial bodies, to the list of principles he used to classify chemical substances; see RCPE MSS, C.10, p. 34.
242. RCPE MSS, C.15, 1,fol. 58r.
243. Ibid.
244. Ibid.
245. Black, *Elements of Chemistry*, 1.286–7.
246. RCPE MSS, C.15, 1, fol. 58v.
247. Ibid., 1,foll. 58v–59r.
248. Ibid., 1,fol. 59r.
249. Ibid., 1,foll. 59v–60r.
250. Ibid., 1,fol. 60v.
251. Ibid., 1,foll. 60v–61r. Also, see above, chapter 2.
252. RCPE MSS, C.15, 1, fol. 61r.
253. Ibid., 1,fol. 61r & v.
254. Ibid., 1,fol. 62r.
255. Ibid., 1,fol. 63r & v.
256. Ibid., 1,fol. 63v.
257. Ibid., 1,fol. 66r & v. Also, see above, chapter 4.
258. RCPE MSS, C.15, 1, fol. 66v.
259. Ibid., 1,fol. 67r.
260. GUL, Cullen MSS, no. 46, p. k. On Nieuwentyt, see Schofield, *Mechanism and Materialism*, 137–40.
261. RCPE MSS, C.15, 1, fol. 69r.
262. Ibid., 1,fol. 69r & v.
263. Ibid., 1,foll. 69v–70r.
264. Ibid., 1,fol. 70r.
265. Ibid., 1,fol. 70r & v. As Andrew Kent has pointed out in his 'Gas-producing Solids, part one,' *M & B Laboratory Bulletin*, 6 (no. 1, 1964) 10; 8–10. Hoffmann also 'noted the acid character of the effervescence from mild alkalies'.
266. RCPE MSS, C.15, 1, fol. 70v. The similarities between Hales' theory, in which acidic vapours fix amphibious air, and Cullen's theory, in which acid fumes fix both aether (fire) and air, are unmistakable.
267. Ibid., 1,fol. 71r. Again notice

Cullen's acceptance of the Newtonian assertion that all substances can be transformed into truly elastic vapours.

268. Ibid., I,fol. 71r & v.

269. Ibid., I,fol. 72r.

270. Ibid., I,fol. 72r & v.

271. Joseph Black, 'Experiments on *Magnesia alba,* Quicklime, and some other Alcaline Substances,' *Essays and Observations, Physical and Literary*, 2 (1756) 157–225; reprinted separately with original title as Alembic Club Reprint no. 1 (Edinburgh 1898), which version is quoted here and in all other references to Black's essay; page 19. In his earlier dissertation Black argued further that 'certainly air can lie hid in metallic calces: seeing that it has been expelled, by the distinguished Hales, from minium itself.' See Black, 'Dissertation,' experiment xxiv, p. 272, and Hales, *Vegetable Staticks*, 286–7, experiment cxix.

272. RCPE MSS, C.12, I.65.

273. Ibid., II.51. For earlier formulations of the acid theory of calcination, see the statements of Hoffmann and Tachenius in Partington, *History of Chemistry*, II.199, 699.

274. RCPE MSS, C.12, II.51.

275. Ibid.

276. Ibid., I.24–6.

277. Ibid., I.25.

278. Ibid., I.26.

279. Ibid.

280. Ibid.

281. See Cantor, 'Scottish Methodological Tradition,' 80–2.

282. RCPE MSS, C.10, p. 238.

283. Ibid., 339.

284. Ibid., 272.

285. Cf. the theories advanced in several dissertations submitted by medical students in Edinburgh during the latter part of the eighteenth century, as summarized in J. R. Partington and Douglas McKie, 'Historical Studies on the Phlogiston Theory.—III. Light and Heat in Combustion,' in *Annals of Science*, 3 (1938) 337–71.

286. RCPE MSS, C.12, I.66–7. For a brief account of the experiments referred to, see Hall, 'Concept of Element,' 25–6. Cullen also asserted that the medicinal qualities of the chemical substances first introduced into the pharmacopeoeia by the Paracelsians could only be determined by clinical observation; see Thomson, *Cullen*, I.613 and II.629–30.

287. From a MS copy of a student's notes of Cullen's medical lectures, quoted in W. P. D. Wightman, 'Wars of Ideas in Neurological Science—From Willis to Bichet, and from Locke to Condillac,' in *History and Philosophy of Knowledge of the Brain and its Functions*, ed. F. Poynter (London 1958) 141.

288. Quoted in Thomson, *Cullen*, I.216.

289. GUL, Cullen MSS, no. 26a.

290. See above, chapter 5.

291. RCPE MSS, C.15, I,fol. 80r & v.

292. See the three letters from Black to Kames in Tytler, *Memoirs*, I (appendix) 75–81; and Home, *Gentleman Farmer*, 292–303.

293. RCPE MSS, C.15, I,fol. 81r.

294. Thomson, *Cullen*, I.52. Some MS remains of these lists can be found in GUL, Cullen MSS, no. 20.

295. From Cullen's 'Some Further Remarks on the Generation of Heat and Cold by Mixture,' published in Thomson, *Cullen*, I.581; 580–3, cf. Ibid., I.53.

296. Ibid., I.581.

297. Ibid., I.53.

298. Ibid., I.580.

299. Ibid., I.580–1.

300. Ibid., I.581.

301. Ibid.

302. Ibid.

303. Ibid.

304. Ibid., I.582.

305. Ibid., I.583.

306. Cullen, 'Cold produced by

Evaporating Fluids,' 145–6.
307. Ibid., 146.
308. Ibid., 147.
309. Ibid., 149–50.
310. Ibid., 150.
311. Ibid.
312. Ibid., 151.
313. Ibid.
314. Ibid., 152.
315. Ibid., 152–3. Thomson, in *Cullen*, 1.60, mentions Wallerius' experiments as the source of Cullen's information on evaporation in vacuo. See also Black, *Elements of Chemistry*, 1.209.
316. Cullen, 'Cold produced by Evaporating Fluids,' 153. On the origin and strict definition of the term adiabatic, see Fox, *Caloric Theory of Gases*, 39.
317. Cullen, 'Cold produced by Evaporating Fluids,' 153–4. Note that Cullen uses the term air when referring to both vaporized alcohol and the ambient atmosphere, i.e., like Desagulier he believed that all vapours are identical in so far as they all manifest the action of the repulsive force which makes them elastic.
318. Ibid., 154. For a more complete description of the apparatus used in these experiments, see Black, *Elements of Chemistry*, 1.162–3.
319. Cullen, 'Cold produced by Evaporating Fluids,' 154–5.
320. Ibid., 155–6.
321. Ibid., 156.
322. Quoted in Thomson, *Cullen*, 1.52.
323. Ibid., 1.57.

CHAPTER SEVEN
Education and Apprenticeship

1. See extracts from the Black family Bible, PRONI MSS, T.1073, no. 1 and D.1950, no. 4A. See also the original draft of Joseph Black's autobiographical note, EUL MSS, GEN 874. Published versions of this note have been taken from a smooth copy not in Black's hand; see Ramsay, *Life and Letters*, 3–5, and Guerlac, 'Black and Fixed Air,' 127–8.
2. See letter from John Black in Thomson, *Cullen*, 1.47–8.
3. PRONI MSS, D.1401, no. 4. This document, a power of attorney issued by John Black, is quoted in Henry Riddell, 'The Great Chemist, Joseph Black, his Belfast Friends and Family Connections,' *Proceedings of the Belfast Natural History and Philosophical Society*, Session 1919–1920, 3 (1920) 61–2; 49–88.
4. See John Black's diary where, in the mid-1750s, he recorded that his was one of the three Protestant households in Bordeaux; PRONI MSS, T.1073, no. 8, fol. 7.
5. A thorough search of the relevant French archives failed to unearth any record of Joseph's Black's birth or baptism; see Guerlac, 'Black and Fixed Air,' 130–1.
6. PRONI MSS, T.1073, no. 1; D.1950, no. 4A.
7. Ibid., D.719, no. 6, letter dated February 1741.
8. Ibid., D.719, no. 8.
9. Ibid., D.1401, no. 4.
10. Ibid., T.1073, no. 7, fol. 1. See also Guerlac, 'Black and Fixed Air,' 129–30.
11. Thomson, *Cullen*, 1.47–8.
12. Quoted in Adam Ferguson, 'Minutes of the Life and Character of Joseph Black, M.D.,' *Transactions of the Royal Society of Edinburgh*, 5 (1805) 102; 101–17.
13. See L. M. Cullen, *Anglo-Irish Trade, 1660–1800* (New York 1968) 148–9.
14. Ferguson, 'Life and Character,' 101–2.
15. EUL MSS, GEN 874, autobiographical note.
16. PRONI MSS, D.1401, no. 4.
17. EUL MSS, GEN 874, autobiographical note.
18. PRONI MSS, T.1073, no. 5.

19. Ibid., D.719, no. 4.
20. Ibid., D.719, no. 60.
21. Ibid., D.719, no. 5.
22. EUL MSS, GEN 874, autobiographical note.
23. See Riddell, 'Joseph Black,' 82–3.
24. EUL MSS, GEN 874, autobiographical note. We know from a letter that Black was still in Belfast in May 1743; PRONI MSS, D.719, no. 11.
25. Ibid., D.719, no. 10A.
26. Ibid., D.719, no. 16.
27. EUL MSS, GEN 874, autobiographical note. Ramsay, *Life and Letters*, 6, says Black 'attended George Ross's classes in "Humanity" [i.e., Latin], and James Moore's in Greek,' but he does not indicate what source provided him with this information.
28. Addison, *Matriculation Albums*, 34. The complete entry is quoted by Ramsay, *Life and Letters*, 6–7, and by Guerlac, 'Black and Fixed Air,' 132.
29. Addison, *Matriculation Albums*, xii.
30. See Guerlac, 'Black and Fixed Air,' 132–3.
31. Addison, *Matriculation Albums*, xii.
32. See ibid., 36, 38.
33. Black, *Elements of Chemistry*, I.xx.
34. The chronology suggested here would have Black beginning his medical studies a year later than that suggested by Guerlac, see 'Black and Fixed Air,' 133.
35. EUL MSS, GEN 874, autobiographical note.
36. Black, *Elements of Chemistry*, I.xxi.
37. PRONI MSS, D.719, no. 18, letter dated 1 June 1748.
38. EUL MSS, GEN 874, autobiographical note. Although Cullen was lecturing on medicine when Black began his professional studies, he was not yet a professor.
39. It is possible that Black attended the chemistry course given by Carrick and Cullen in 1747–8 and had already decided to continue his chemical studies before beginning his medical education.
40. EUL MSS, GEN 874, autobiographical note.
41. Ibid.
42. Thomson, *Cullen*, I.583. Cf. Guerlac, 'Black and Fixed Air,' 134, 453 n. 224.
43. PRONI MSS, D.719, no. 34.
44. Ibid., D.719, no. 36A.
45. Ibid., D.719, no. 43.
46. Ibid., D.719, no. 44.
47. EUL MSS, GEN 874, 64 family letters, no. 3.
48. PRONI MSS, T.1073, no. 7, fol. 8.
49. Thomson, *Cullen*, I.48, letter dated 9 Dec., 1751. For the date of Black's departure, see PRONI MSS, T.1073, no. 7, fol. 8.
50. Ibid., T.1073, no. 7, fol. 9.
51. Ibid., D.1950, no. 24.
52. Ibid., T.1073, no. 10.
53. Black, *Elements of Chemistry*, I.xxiii.
54. For a discussion of these experiments, see below, chapter 8, section 1.
55. Black, *Elements of Chemistry*, I.xxvi. Guerlac, in 'Black and Fixed Air,' 148, argues that 'Robison's inferences are suspect. His statement that he found queries concerning cooling mixtures etc. in the "oldest parcels" of notes is not confirmed by Adam Ferguson, who gives excerpts of these queries in his "Life of Dr. Black," placing them in 1756–7, which is more likely.' But the error is Guerlac's, not Robison's. The notes Ferguson referred to on pages 106–7 of his memoir are quoted by Robison in his *notes* to Black's lectures, not in the introduction; see Black, *Elements of Chemistry*, I.525–6. Guerlac's claim that a later dating for Black's earliest experiments on quicklime is 'more likely' rests

upon his reconstruction of the sequence of Black's research. Unlike Guerlac, I believe Black understood the relationship between causticity and fixed air *before* he began his investigations on magnesia alba, and Robison's dating of Black's early experiments supports this chronology.

56. EUL MSS, GEN 874, 64 family letters, no. 5.

57. Ferguson, 'Life and Character,' 102–3.

58. EUL MSS, GEN 874, 64 family letters, no. 5.

59. Ibid.

60. Ibid.

61. On the lime-water debate, see above, chapter 5.

62. Black, *Elements of Chemistry*, I.xxvi.

63. Thomson, *Cullen*, I.573.

64. Ibid.

65. Ibid., I.574.

66. Ibid., I.577.

67. Ibid.

68. EUL MSS, GEN 874, 64 family letters, no. 6, letter dated 10 Mar. 1753.

69. PRONI MSS, D.1950, no. 25.

70. Ibid., D.1950, no. 19.

71. Ibid., T.1073, no. 7, fol. 14.

72. Thomson, *Cullen*, I.573, letter dated 10 Feb. 1753.

73. Ibid., I.574.

74. Ibid., I.576.

75. Ibid., I.576–7.

76. EUL MSS, Dc.2.76[8], letter dated 8 Mar. 1754; printed in full in Guerlac, 'Black and Fixed Air,' 149–50.

77. PRONI MSS, T.1073, no. 7, fol. 18.

78. EUL MSS, GEN 873, vol. I. Published version in Ramsay, *Life and Letters*, 17–20.

79. PRONI MSS, T.1073, no. 7, fol. 19.

80. EUL MSS, GEN 874, 64 family letters, no. 7.

81. See above, chapter 5.

82. EUL MSS, GEN 874, 64 family letters, no. 7.

83. See PRONI MSS, T.1073, no. 7, foll. 19 & 20.

84. Black, *Elements of Chemistry*, I.xxviii–xxix.

85. PRONI MSS, T.1073, no. 8, fol. 2.

86. EUL MSS, GEN 874, 64 family letters, no. 8, letter dated 2 Sept. 1755.

87. Thomson, *Cullen*, I.50.

88. Ibid., I.51, letter dated 18 June 1754.

89. See Joseph's letter to his father dated 1 June 1754; EUL MSS, GEN 873, vol. I; published in Ramsay, *Life and Letters*, 18–20.

90. Thomson, *Cullen*, I.51.

91. See EUL MSS, GEN 874, 64 family letters, no. 7, letter dated 23 Nov. 1754.

92. See the footnote on the first page of Black's essay, as printed in *Essays and Observations, Physical and Literary*. In early August 1755 Black wrote to Cullen and told him that 'my paper on Lime has been read to the Society; and Lord Kames, who was not present, desiring afterwards to see it, I sent him the only legible copy I had, which is the reason I have not shewn it to you before now.'—Thomson, *Cullen*, I.580.

93. See below, chapter 8, section 3. Guerlac, in 'Black and Fixed Air,' 450–1, points out that the discussion of this experiment was probably added to Black's essay after it had been read before the Philosophical Society.

94. EUL MSS, GEN 874, 64 family letters, no. 8.

95. Ibid.

96. PRONI MSS, T.1073, no. 8, fol. 5.

97. See above, chapter 4.

98. This of course was Cullen's essay on the cold produced by evaporation.

99. See above, chapter 4.

100. See Comrie, *Scottish Medicine*, I.359, 362. See also EUL MSS, GEN 873, vol. I, letter dated 4 April

1757; published in Ramsay, *Life and Letters*, 32–3. The Thomas Hamilton who succeeded to the chair of anatomy was no doubt the same Thomas Hamilton who in 1741 replaced William Hunter as Cullen's surgical partner in Hamilton; see Thomson, *Cullen*, I.15.

101. See below, chapter 9.

102. See his letters to his father, published in Ramsay, *Life and Letters*, 32–6.

103. See Ramsay, *Life and Letters*, 35; EUL MSS, GEN 874, 64 family letters, no. 9, letter dated 4 Feb. 1761; 1764 letter from Black to Adam Smith published in Graham, *Social Life*, 461 n.

104. See Comrie, *Scottish Medicine*, I.361.

105. On the use of lime in bleaching and the publication of Black's memoir, see above, chapter 5. See also EUL MSS, GEN 873, vol. I, foll. 9r–15v, 20r, and Ramsay, *Life and Letters*, 36–7, 52–4.

106. See Ramsay, *Life and Letters*, 36–7, 54. On the Carron Iron Works, see Hamilton, *Economic History*, 193–7.

107. EUL MSS, GEN 874, 64 family letters, no. 12, letter dated 30 June 1766.

108. Two of these were Thomas Reid, professor of moral philosophy (see his letters to Drs Andrew and David Skene in Reid, *Works*, 39–48) and Alexander Wilson, professor of astronomy (see his letter of 21 January 1768 to Black in which he asks if the rise of temperature which accompanies the appearance of clouds or fog 'can be accounted for by your theory of Latent Heat'. EUL MSS, GEN 873, vol. I, fol. 26).

109. On the petition, see above, epigraph to Part II.

110. EUL MSS, GEN 874, 64 family letters, no. 13, letter dated 26 Sept. 1766.

111. For information on Black's activities after 1766, see Henry Guerlac, 'Joseph Black,' in *Dictionary of Scientific Biography*, vol. 2 (New York 1970) 173–83, and Partington, *History of Chemistry*, vol. 3, ch. 4.

CHAPTER EIGHT

Black's Study of Causticity and Fixed Air

1. On the dating of these notes, see above, chapter 7.

2. Black, *Elements of Chemistry*, xxvi.

3. Ibid.

4. Guerlac, 'Black and Fixed Air,' 147.

5. Cf. Hales, *Vegetable Staticks*, 285, and Whytt, *Essay on Lime-Water*, 78. For evidence of Hales' influence upon Whytt's approach to the study of lime-water, see Robert Whytt, 'Essay towards the Discovery of a safe Medicine for dissolving the Stone,' in *Medical Essays and Observations*, 5, part 2 (1744) 668; 667–750; and Whytt, *Essay on Lime-Water*, 38.

6. Black, *Elements of Chemistry*, I.xxvi.

7. William Homberg, 'Observation sur la quantité d'acides absorbées par les alcalis terreux,' *Mémoires de l'Académie royale des Sciences* (1700) 64–71.

8. Black, 'Dissertation,' 272.

9. Ibid., cf. Black, *Experiments*, 18.

10. Black, *Elements of Chemistry*, I.xxvi, cf. Guerlac, 'Black and Fixed Air,' 437. Although it is helpful to use modern chemical terms and equations to describe the fundamental reaction being examined, I am, of course, fully aware that the actual constituents and reactions being studied were not as simple as my formulas imply.

11. See above, chapter 5.

12. Alston, *Dissertation*, 7–10.

13. Ibid., 18.

14. Ibid., 19.

15. Guerlac, 'Black and Fixed Air,' 146.
16. Black, *Elements of Chemistry*, I.xxvi.
17. See above, chapter 6, section 6.
18. Black, *Elements of Chemistry*, I.xxvi. For additional evidence of Black's early interest in and knowledge of magnesia, see his letter to Cullen, Thomson, *Cullen*, I.578.
19. Black, 'Dissertation,' 269.
20. Thomson, *Cullen*, I.576.
21. For Alston's rather casual attempt to control the heat of his solutions, see his *Dissertation*, 35–6; for Whytt's efforts, see his *Essay on Lime-water*, 27–32.
22. Thomson, *Cullen*, I.576. Cf. Whytt, *Essay on Lime-water*, 39–40: 'In order to find the specifick Gravities of these different Lime-waters . . . I weighed, with the assistance of my colleague Dr. *John Stuart*, Professor of Natural Philosophy [and Russell's predecessor in that chair], a large Glass-phial (filled with sand, and hermetically sealed) first in Air, . . . then immersed in the Fountain-water of this City used also in making the several Lime-waters,' and then in the lime-waters themselves, the ratio of the various weights being taken as the inverse ratio of the specific gravities of the various liquids.
23. Black, 'Dissertation,' 271, experiment xix.
24. Ibid. Cf. Guerlac, 'Black, and Fixed Air,' 434, where he totally misconstrues Black's statement. According to Guerlac, Black is saying that 'the explanation of this astonishing difference between magnesia alba and the calcined magnesia usta—that the first gave a pronounced effervescence, and the latter none at all—"for a long time seemed to me very obscure".'
25. On the need to distinguish between the caustifiable and calcinable substances, see above, chapter 6, section 3.
26. Black, 'Dissertation,' 270. The term 'absorbent' seems to have referred primarily to the pharmaceutical properties of the alkaline earths; see Ibid., 228, where he says that the excess acid in the stomach 'is changed by alkaline salts and absorbent powders'.
27. Ibid., 270. Note that while Hoffmann had previously distinguished magnesia from calcareous earth by contrasting the solubilities of their sulphate salts, he had not distinguished between magnesia and those substances that can be made caustic.
28. Ibid., 226.
29. Cf. Guerlac, 'Black and Fixed Air,' 438; but see Henry M. Leicester, *Development of Biochemical Concepts from Ancient to Modern Times* (Cambridge, Mass. 1974) 128–9.
30. Thomson, *Cullen*, I.51.
31. Black, 'Dissertation,' 226.
32. Ibid., 227.
33. Ibid., 227–8.
34. Ibid., 228.
35. Ibid., 268. For full references to the two papers by Hoffmann which Black cites, see above, chapter 6, section 3.
36. Black, 'Dissertation,' 268.
37. Ibid., 270.
38. Ibid.
39. Ibid., 269.
40. Ibid.
41. Ibid.
42. Ibid.
43. Ibid.
44. Ibid.
45. See above, chapter 8, section 1, for quotations of Black's statements concerning his expectations and disappointments.
46. Black, 'Dissertation,' 270.
47. Ibid.
48. Ibid.
49. Ibid.
50. Ibid. Cf. the analogous and frequently discussed problem posed

by the effect of lime-water on volatile alkali and sal ammoniac; see Stahl, *Universal Chemistry*, 150–2, and E. F. Geoffroy, 'Eclaircissements sur la Table inserée dans les Mémoires de 1718. Concernant les Rapports observés entre différentes Substances,' *Mémoires de l'Académie royale des Sciences* (1720) 21–2; 20–34.

51. Black, 'Dissertation,' 270.

52. Ibid. Black cites Hales' experiment lii, in which it is reported that 'a dram of volatile Salt of Sal armoniack, soon distilled over with a gentle heat; but tho' the expansion in the receiver was double that of heated Air alone, yet no Air was generated, but two and an half cubick inches were absorbed [$NH_4Cl \rightarrow HCl + NH_3 \uparrow$; $2NH_3 + H_2O + CO_2 \rightarrow (NH_4)_2CO_3$].' (Hales, *Vegetable Staticks*, 168.)

53. Black, 'Dissertation,' 270.

54. Ibid.

55. Ibid., 271.

56. Ibid.

57. Ibid.

58. Ibid.

59. Ibid.

60. Ibid.

61. If Black was misled by thinking that the reaction between calcareous salts and magnesia alba indicates that magnesia has a stronger attraction for acids than the calcareous earths have (e.g., $CaCl_2 + MgCO_3 \rightarrow CaCO_3 \downarrow + MgCl_2$; experiment viii), then he was saved from this error in his later essay by his use of the theory of double elective attractions; see below, chapter 8, section 3. Alston, while trying to use Black's experiments to reinforce his own theory of quicklime, added a confusion of his own. 'There is,' he reported in the second edition of his *Dissertation* (1754), 'a substance, more nearly allied to [quicklime] than any Salt, which seems to [precipitate the calcareous matter in lime-water] effectually; I mean the *Magnesia alba*, or that alcaline Earth which is the *Basis* of Sea-salt. This I owe to Jos. Black, M.D.' (p. iv) Black must have winced upon hearing magnesia alba identified as the base of sea salt, which he knew from Duhamel's analysis to be the alkali soda.

62. Black, 'Dissertation,' 271.

63. Ibid.

64. Ibid.

65. See above, chapter 8, section 1.

66. Black, 'Dissertation,' 271.

67. Ibid. The difficulties Black encountered in his attempts completely to calcine magnesia may account for his intense interest in the construction of furnaces. See, for instance, Eric Robinson and Douglas McKie, eds. *Partners in Science: Letters of James Watt and Joseph Black* (London 1970) 9–10 n. 2.

68. Black, 'Dissertation,' 271–2. For Guerlac's interpretation of Black's understanding of this experiment, see his 'Black and Fixed Air,' 434–5, where he has Black 'brooding' over the lack of effervescence when magnesia usta is treated with acids and 'suddenly' recollecting the experiments of Stephen Hales.

69. Black, 'Dissertation,' 272.

70. Ibid.

71. Ibid.

72. Ibid.

73. Ibid.

74. Ibid.

75. Ibid.

76. See above, chapter 8, section 1.

77. Black, 'Dissertation,' 272.

78. Ibid.

79. Ibid. In the final paragraph of this experiment Black suggested that the explosion of fulminate of gold is caused by the rapid release of its fixed air; see above, chapter 6, section 6.

80. Black, 'Dissertation,' 272.

81. Ibid., 272–3.

82. Ibid., 273.

83. Ibid.

84. Thomson, *Cullen*, I.50; cf. Guerlac, 'Black and Fixed Air,' 437.
85. See Thomson, *Cullen*, I.577–9.
86. Cf. Robison in Black, *Elements of Chemistry*, I.xxviii; and Guerlac, 'Black and Fixed Air,' 443.
87. Black, *Experiments*, 11. For full citation and brief description of Marggras's memoir, see Partington, *History of Chemistry*, II.726; for the French investigations of the selenitic salt, see above, chapter 6, section 3.
88. Black, *Experiments*, 11.
89. Ibid., 14.
90. Ibid., 25.
91. Ibid., 14.
92. In the preface to the second edition of his *Dissertation*, Alston reported that Black 'found that 200 Grains of his *magnesia alba* reduced 126 Ounces . . . of Lime-water, to pure insipid Water; while the Weight of the well-dried *magnesia* was increased to four Drams and thirty-five Grains [i.e., 275 grains].' (p. iv.)
93. See above, chapter 5.
94. Cf. Black, *Experiments*, 12, where he says that 'having thus found *magnesia* to differ from the common alkaline earths, the object of my next inquiry was its peculiar degree of attraction for acids, or what was the place due to it in Mr Geoffroy's table of elective attractions,' a statement which would have been out of place in a medical thesis.
95. Ibid.
96. Ibid.
97. Ibid.
98. Ibid., 12–13.
99. Ibid., 26.
100. Ibid., 36–7.
101. Ibid., 37.
102. Ibid., 22.
103. Quoted in Guerlac, 'Black and Fixed Air,' 451.
104. Cf. Guerlac's assertion that Black presented his experiments in the same order in which he had performed them; 'Black and Fixed Air,' 433, 436, 444.
105. Black, *Experiments*, 22. Cf. the second set of experiments described by Robison; see above, chapter 8, section 1. See also Black, *Experiments*, 23–4, where he describes the experiment by which Marggras, 'an exceedingly accurate and judicious chemist,' demonstrated 'that the calcareous earths really lose a large quantity of air when they are burnt to quick-lime'.
106. Ibid., 22. Cf. the fourth experiment described by Robison.
107. Ibid., 22–3. Cf. the first experiment described by Robison.
108. Ibid., 23. Cf. the third experiment described by Robison.
109. Ibid., 24.
110. Ibid., 24–5.
111. Ibid., 26.
112. Ibid.
113. Ibid., 27.
114. Ibid., 27–8.
115. Ibid., 28.
116. Ibid., 28–9.
117. Ibid. ,26–7.
118. This prediction directly contradicted Stahl's and Alston's assertion that quicklime contains two earths, one soluble and the other insoluble; on Stahl, see above, chapter 6, section 3; for Alston's view, see his *Dissertation*, 11, where he says that 'Quick-lime consists of a dissolvable and an undissolvable Part in Water, scarcely one third Part being dissolvable'. Ironically Black presented this third proposition in his later lectures as a consequence of Alston's own experiments; see Thomas Cochrane, *Notes from Doctor Black's Lectures on Chemistry 1767/8*, ed. Douglas McKie (published privately by Imperial Chemical Industries Limited, Cheshire, 1966) 70: 'In experiment III. Dr. Alston found he could obtain 4 or 500d. times the weight of [lime] water by adding always

fresh [water to a sample of quick-lime], hence it [i.e., the quicklime] can be dissolv'd totaly in time.' For a brief account of Alston's experiment, see above, chapter 5.
119. Black, *Experiments*, 29.
120. Ibid., 29–30.
121. Ibid., 30.
122. Ibid.
123. RCPE MSS, C.15, I,fol. 72r.
124. Black, *Experiments*, 24.
125. See above, chapter 6, section 7.
126. Cullen, 'Cold produced by Evaporating Fluids,' 153–4.
127. On Newton's and Hales' distinction between easily condensable vapours and true permanent air, see above, chapter 2.
128. Black, *Elements of Chemistry*, I.525. Robison says that these notes were written between 1754 and 1757; ibid. Nitrous aether was one of the liquids which Cullen studied; see above, chapter, 6 section 7.
129. Black, *Elements of Chemistry*, I.163.
130. Thomson, *Cullen*, I.579–80.
131. Black, *Experiments*, 30.
132. Ibid.
133. See above, chapter 8, section 2.
134. Black, *Experiments*, 30–1.
135. Ibid., 27. This proposition directly contradicted Alston's theory of the caustification of alkalis; see his *Dissertation*, 19, where he says that 'quick-lime rectifies or acts on alkaline fixed Salts, by separating from them some Part of the coarsest of their Earth, and substituting in place thereof its most soluble and subtile Part, whereby their Strength, as a Menstruum, is considerably increased.'
136. Black, *Experiments*, 31.
137. Ibid., 32.
138. Ibid.
139. Ibid., 32–3.
140. Ibid., 33.
141. Ibid., 33–4.
142. Ibid., 34.
143. Cf. Cullen's similar view of solution; see above, chapter 6, section 7.
144. Black, *Experiments*, 34.
145. Ibid., 35.
146. Ibid., 27.
147. Ibid.
148. Ibid., 35.
149. Ibid.
150. Ibid.
151. Ibid., 35–6.
152. See above, this section.
153. Black, *Experiments*, 37.
154. Ibid.
155. Ibid., 38.
156. Ibid.
157. Ibid., 39.
158. See above, chapter 6, section 3. Cf. the title of Black's essay, where he speaks of alkaline substances rather than earths and salts.
159. RCPE MSS, C.15, I,fol. 58r.
160. Black, *Experiments*, 40–1. In his lectures Cullen presented borax, together with quicklime, alkaline metallic calces, alum and essential salts, in an 'appendix to the doctrine of salts'; see GUL, Cullen MSS, no. 34. The reaction Black is referring to was described by C. J. Geoffroy in his memoir 'Nouvelles expériences sur le borax,' published in the *Mémoires de l'Académie royale des Sciences*, 1732; see also Partington, *History of Chemistry*, III.56. On the identification of boric acid and sedative salt, see ibid., III.71–2.
161. Black, *Experiments*, 41.
162. Ibid.
163. Ibid.
164. Ibid.
165. Ibid., 42–3. On the identity of the salt being examined, see Partington, *History of Chemistry*, III.139.
166. See above, chapter 6, section 7.
167. Black, *Experiments*, 43.
168. Ibid., 43–4.
169. Ibid., 44.
170. Ibid., 44–5. When lecturing, Black, like Cullen, used diagrams to illustrate his ideas about double

decomposition. For instance, although fixed alkali has a greater attraction than calcareous earth for vitriolic acid, if one mixes vitriolated tartar with a calcareous earth saturated with marine acid, the calcareous earth will take the vitriolic acid away from the fixed alkali ($K_2SO_4 + CaCl_2 \rightarrow CaSO_4 \downarrow + 2KCl$). Black used the following diagram when explaining this reaction.

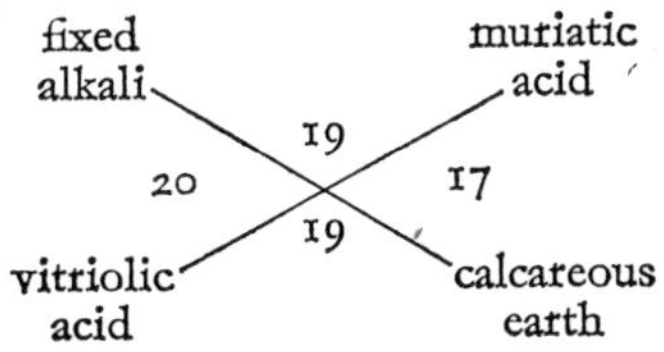

The numbers assigned to the various attractive forces are arbitrary but their relative magnitude reveals that although the attraction between fixed alkali and vitriolic acid (20) is larger than that between calcareous earth and vitriolic acid (19), the total force attracting fixed alkali to muriatic acid and vitriolic acid to calcareous earth (38) is larger than that attracting fixed alkali to vitriolic acid and muriatic acid to calcareous earth (37); see Cochrane, *Notes from Black's Lectures*, 59–60. By substituting the constituents of the reaction being discussed in the essay, we obtain the following diagram:

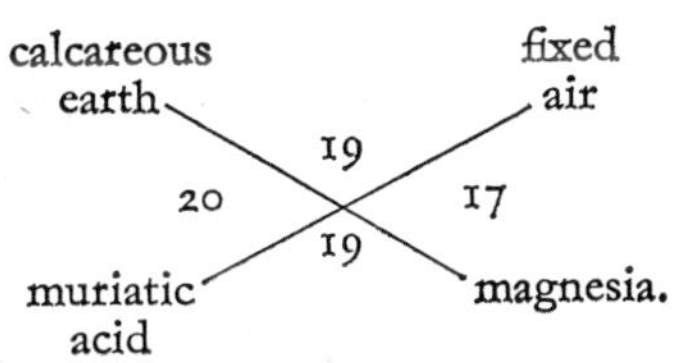

On Cullen's and Black's use of diagrams, see Crosland, 'Use of Diagrams as Chemical 'Equations,' especially 81n.14.

171. Black, *Experiments*, 45.

172. Ibid.

173. Ibid., 46.

174. For a description of these experiments see Black, *Elements of Chemistry*, II. 87–8, and Guerlac, 'Black and Fixed Air,' 451–3.

175. Cf. Cullen who treated the aerial bodies as a separate class of substances in his later lectures; see above, chapter 6, section 6.

176. Cochrane, *Notes from Black's Lectures*, 71.

177. Ibid., 72. Cf. Henry Cavendish, 'Three Papers, containing Experiments on factitious Airs,' *Philosophical Transactions of the Royal Society of London*, 56 (1767) 141–84. These papers were read to the Society on 29 May, 6 November, and 13 November 1766, and it is remarkable how rapidly Black incorporated Cavendish's research into his lectures.

178. Cochrane, *Notes from Black's Lectures*, 72.

179. Cf. Guerlac, 'Black and Fixed Air,' 454–6.

180. Quoted in Partington, *History of Chemistry*, III.314.

181. Quoted in ibid., III.316, where he points out that Black later located the source of Cavendish's error concerning the homogeneity of fixed air.

182. Cochrane, *Notes from Black's Lectures*, 111.

183. Ibid.

184. Ibid.

185. Ibid. For Cullen's views on the two kinds of calcination, see above, chapter 6, section 6.

186. Cochrane, *Notes from Black's Lectures*, 84. Cf. Cullen's statement that 'there are only two elements, one of them the gravitating matter, the other a subtile aether'; see above, chapter 6, section 6.

187. Cochrane, *Notes from Black's Lectures*, 83.
188. Ibid.
189. Ibid. Cf. Cullen's reduction of phlogiston to acidity rather than to heat; see above, chapter 6, section 6. For a general survey of the idea that phlogiston has negative weight, see J. R. Partington and Douglas McKie, 'Historical Studies on the Phlogiston Theory,' parts I and II, *Annals of Science*, 2 (1937) 361–404, and 3 (1938) 1–58.
190. On the relationship between the various formulations of the phlogiston theory and the development of pneumatic chemistry, see Bernard Langer, 'Pneumatic Chemistry, 1772–1789: A Resolution of Conflict,' unpublished Ph.D. dissertation (University of Wisconsin 1965). On heat theory in the early nineteenth century see Richard Olson, 'Count Rumford, Sir John Leslie, and the Study of the Nature and Propagation of Heat at the Beginning of the Nineteenth Century,' *Annals of Science*, 26 (1970) 273–304, Fox, *Caloric Theory of Gases*, and Ernst Mach, *Die Principien der Wärmelehre* (Leipzig 1896).

CHAPTER NINE
The Theory of Latent Heat

1. Black, *Elements of Chemistry*, I.xxii.
2. Ibid., I.xxii–xxiii.
3. Ibid., I.xxiii.
4. Ibid.
5. Ibid.
6. Ibid., I.xxxiv.
7. See above, chapter 6, section 7.
8. See above, ibid.
9. See above, ibid.
10. Cf. Fox, *Caloric Theory of Gases*, 25.
11. Black, *Elements of Chemistry*, I.505. Cf. the discussion of heat capacity in I.82–3, where the term saturation is used. For information concerning the first use of the term specific heat, see McKie and Heathcote, *Specific and Latent Heats*, 130.
12. Black, *Elements of Chemistry*, I.xxxiv–xxxv.
13. Ibid., I.513. Black's concept of latent heat, like Newton's concept of gravitational attraction, can be seen as incorporating, in a modified and slightly ambigious form, a type of explanation characteristic of the philosophical chemistry of the sixteenth and early seventeenth centuries; see, e.g., Francis Bacon's use of the phrase 'latent heat' to explain the heat given off by decaying vegetable matter and by quicklime when slaked (*The New Organon and Related Writings*, ed. Fulton H. Anderson (New York 1960) 139–43).
14. Ferguson, 'Life and Character,' 106–8; Black, *Elements of Chemistry*, I.525–7.
15. Ibid., I.525.
16. Ferguson, 'Life and Character,' 106.
17. Black, *Elements of Chemistry*, I.526. Black is here referring to the difference between the solution of deliquescent Glauber's salt, which produces heat because it fixes the water of crystallization in the salt, and the solution of crystalline Glauber's salt, which produces cold. For Cullen's discussion of this difference, see above, chapter 6, section 7.
18. Black, *Elements of Chemistry*, I.525; Ferguson, 'Life and Character,' 107.
19. Black, *Elements of Chemistry*, I.526.
20. Ibid.
21. Ibid., I.525.
22. Ibid.
23. Ferguson, 'Life and Character,' 107.
24. Black, *Elements of Chemistry*, I.525.
25. Ibid.
26. Ibid., I.526. The entry continues

as follows: 'Will not the progress of cooling be retarded while a solution of salt crystallizes? Try if this happens while fats become solid. Do not these salts produce the greatest cold which are dissolved by the smallest quantity of water?'

27. Ferguson, 'Life and Character,' 106.

28. Black, *Elements of Chemistry*, I.526.

29. Ibid.

30. Ferguson, 'Life and Character,' 106.

31. Ibid., 107. See above, chapter 6, section 5 for Cullen's failure to discuss Fahrenheit's discovery. For other discussions of this phenomenon, which Black often mentioned in his lectures, see Black, *Elements of Chemistry*, I.127–30, 138 n, and Reid, *Works*, 44–5.

32. Black, *Elements of Chemistry*, I.525–6.

33. Ibid., I.526.

34. Ibid.

35. Ibid. On Black's and Irvine's views on absolute heat, see chapter 10, below.

36. Black, *Elements of Chemistry*, I.526. See also, I.188–9. For Watt's experiments on the efficiency of generating steam under various pressures, see chapter 10, section 1, below.

37. Ferguson, 'Life and Character,' 107. Cf. Black, *Elements of Chemistry*, I.xxxviii, where Robison says that Black conceived the accretion of solid matter as a source of a part at least of the warmth of animals'. On this topic see also Everett Mendelsohn, *Heat and Life; The Development of the Theory of Animal Heat* (Cambridge, Mass. 1964) chapter 5.

38. Black, *Elements of Chemistry*, I.525. For a more complete description of this experiment, as performed by Black and by James Watt, see ibid., I.159–61.

39. Ferguson, 'Life and Character,' 108. Cf. Hales' notion of the role of air in nature; see above, chapter 2. It is interesting to note that in this early statement Black took it as self-evident that heat can be transmitted through a vacuum. For clear evidence that he considered such transmission a specific form of radiation, see his letter to J. Watt, written in 1768 and published in Robinson and McKie, *Partners in Science*, 10–11. For a discussion of this question at a later time, see the Leslie-Rumford debate as described in Olson, 'Rumford and Leslie,' 278–80.

40. Black, *Elements of Chemistry*, I.lxiv.

41. Ibid., I.506–7.

42. Ibid. I.161.

43. Ibid., I.165.

44. Robinson and McKie, *Partners in Science*, 83–4; letter dated 15 March 1780.

45. Black, *Elements of Chemistry*, I.527.

46. Ibid., I.527; cf. I.xl–xli.

47. Cf. Black's indirect study of the properties of fixed air.

48. Black, *Elements of Chemistry*, I.57.

49. Ibid., I.525.

50. See ibid., I.xl.

51. Ibid., I.56–7.

52. Ibid., I.xxxix.

53. Ibid.

54. Ibid.

55. Ibid.

56. Brook Taylor, 'An Experiment Proving the Expansion of the Liquid in the Thermometer to be as the Degree of Heat,' *Philosophical Transactions of the Royal Society*, 32 (1723) 291. Cf. Cardwell, *Watt to Clausius*, 22–3.

57. Black, *Elements of Chemistry*, I.57–8.

58. Ibid., I.58.

59. Ibid. This second caution again indicates that Black's examination of the thermometer was prompted by his interest in the physical effects of latent heat, for as McKie and Heathcote have pointed out, the attention he devoted to the heat lost to the vessels reveals an awareness of the differences in the specific heats of the substances involved in the experiment; see McKie and Heathcote, *Specific and Latent Heats*, 125.

60. Black, *Elements of Chemistry*, I.xxxix.

61. Ibid., I.59; see also, Robison's note, I.505–6.

62. Ibid., I.59. This hypothesis, which clearly has a great deal in common with the aether density hypothesis advanced by Cullen (see above, chapter 6, section 5), had been most fully developed by William Cleghorn, and it was Cleghorn's formulation which Black presented in his lectures (see William Cleghorn, *Disputatio Physica Inauguralis, Theoriam Ignis Complectens* (Edinburgh 1779), translated by Douglas McKie and Niels H. de V. Heathcote, with notes, as William Cleghorn's *De Igne* (1779),' *Annals of Science*, 14 (1958, published 1960) 1–82).

63. Black, *Elements of Chemistry*, I.xxxvi.

64. Ibid.

65. See above, chapter 6, section 5.

66. But see also the comments on this experiment in D. S. L. Cardwell, *Steam Power in the Eighteenth Century* (London 1963) 39.

67. Black, *Elements of Chemistry*, I.xxxvi. Cf. Cleghorn, '*De Igne*,' 17, where he says that 'although neither the absolute nor the relative quantity of fire is to be inferred by means of the thermometer, yet Black with the insight of genius has discovered a method by which the relative quantity of fire in bodies may be known.'

68. Black, *Elements of Chemistry*, I.120–2. In March, 1766 Thomas Reid sent a friend an account of Black's lectures. His letter contains a description of the above experiment which indicates that the account published in the lectures edited by Robison is essentially the same as that which Black presented in his earliest lectures on latent heat. Heat, see Reid, *Works*, 45.

69. See above, chapter 6, section 5.

70. Black, *Elements of Chemistry*, I.503–4.

71. Ibid., I.122.

72. Ibid., I.122–5. Cf. Reid, *Works*, 45.

73. Black, *Elements of Chemistry*, I.125 n. Cf. Robinson and McKie, *Partners in Science*, 84.

74. Black, *Elements of Chemistry*, I.xxxviii.

75. Ibid., I.130–1.

76. Ibid., I.131.

77. Ibid., I.133. Cf. I.135 where Black illustrates his conclusion by pointing out that 'if three parts of water are added to one part of [sal ammoniac], the liquefaction lowers the temperature fully 20 degrees of Fahrenheit'. For evidence that Black treated the heat of mixture this way when lecturing in Glasgow, see Reid's letter of December 1765, where he says: 'I have attended Dr Black's lectures hitherto. His doctrine of latent heat is the only thing I have yet heard that is altogether new. And, indeed, I look upon it as a very important discovery. . . . It gives a great deal of light to the phenomena of heat that appear in mixture, solution, and evaporation.' (Reid, *Works*, 42.)

78. Black, *Elements of Chemistry*, I.137. On Irvine, see Andrew Kent, 'William Irvine, M.D.,' in *An Eighteenth Century Lectureship in Chemistry*, ed. Andrew Kent (Glasgow 1950) 140–50. Irvine first attended Black's lectures in 1761–2; see Robinson and McKie, *Partners in Science*, 84. On Irvine's doctrine of heat, see chapter 10, section 2, below.

79. Reid, *Works*, 44, letter dated Mar. 1766.

80. See William Irvine, *Essays Chiefly on Chemical Subjects* (London 1805) 88, 92; see also below. Black, it will be remembered, had previously studied these slowly melting substances while establishing the reliability of the thermometer; see above, this section.
81. Black, *Elements of Chemistry*, I.505.
82. Ibid., I.527.
83. Ibid., I.156.
84. Ibid.
85. See ibid., I.156–7.
86. Cf. the note in which Robison speaks of: 'Another circumstance in the vaporous combination of heat, which I think of immense importance to the philosophy of chemistry, namely, its being counteracted by mechanical force. Not only can the elasticity or mutual repellency be counteracted by pressure, but *the chemical union of heat and the matter of a body can be prevented by it.* The instant that the pressure on the surface of boiling water is removed, the union takes place; for steam is produced, and the temperature of the water is diminished.' (Ibid., I.520.)
87. Ibid., I.152–3.
88. Ibid., I.153.
89. For the background to Lavoisier's understanding of this problem, see Martin Fichman, 'French Stahlism and Chemical Studies of Air, 1750–1770,' *Ambix*, 18 (1971) 94–122, especially page 117.
90. Black, *Elements of Chemistry*, I.157. The date on which the experiment was made is mentioned immediately after the passage quoted. For a more detailed account of this experiment, as recorded in a set of student's notes on the lectures Black delivered in 1770, see note 25 in Cleghorn, '*De Igne*,' 68–9.
91. Black, *Elements of Chemistry*, I.158. Similar values were extracted from the data obtained in the second and third experiments.
92. See Watt's notes in Robison, *System of Mechanical Philosophy*, II.113; Watt's notebook in Robinson and McKie, *Partners in Science*, 440–1; and below, chapter 10, section 2.
93. Black, *Elements of Chemistry*, I.161–2.
94. Ibid., I.151.
95. Ibid., I.xliv.
96. Ibid., I.194.
97. Cf. Black's earlier query about 'the absolute proportions of heat' in bodies; see above, chapter 9, section 1.
98. See Cardwell, *Watt to Clausius*, 53–4, and W. A. Smeaton, 'Some comments on James Watt's published account of his work on steam and steam engines,' in *Notes and Records of the Royal Society of London*, 26 (1971) 38–40; 35–42. For Watt's earliest and rather offhanded statement of his law, see James Patrick Muirhead, *The Life of James Watt, with selections from his correspondence*, 2nd ed. revised (London 1859) 161; also see below, chapter 10, section 1.
99. Black, *Elements of Chemistry*, I.171–2; see also Robinson and McKie, *Partners in Science*, 92. In the next paragraph Black discusses the heat lost during the experiment and concludes that the true value of the latent heat of steam must be considerably greater than the value obtained from this experiment.
100. Black, *Elements of Chemistry*, I.173. Cf. I.174–5, where Black says he had thought of using ice to measure the amount of heat contained in steam, but since he never made the experiment, full credit for the invention of the ice calorimeter belongs to Lavoisier. Among themselves, however, Black and his disciples were decidedly critical of the results Laplace and Lavoisier obtained

with their ice calorimeter. See Robinson and McKie, *Partners in Science*, 128, 139–40; Robert Schofield, *The Lunar Society of Birmingham*, (Oxford 1963) 266–7; and Irvine, *Essays*, 130. As Partington and Talbot have emphasized, the Scots who engaged in the experimental study of heat during the eighteenth century had no reason to be embarrassed about their methods or their results,—'Crawford's experiments on the specific heat of gases, though perhaps deplorable, were better than those of Lavoisier and Laplace for example and were remarkably good for the time.' (G. R. Talbot, comments on Robert Fox, 'Dalton's Caloric Theory,' in *John Dalton & the Progress of Science*, ed. D. S. L. Cardwell (Manchester 1968) 201; 187–202.

101. Black, *Elements of Chemistry*, I.178. Clearly Black believed his theory could explain what came to be called adiabatic compression; cf. Kuhn, 'Caloric Theory of Adiabatic Compression' and Fox, *Caloric Theory of Gases*. See also the close relationship between Black's theory and Erasmus Darwin's study of the adiabatic expansion of air; see Schofield, *Lunar Society*, 169, 267–9, and Fox, *Caloric Theory of Gases*, 54–60.

102. See above, chapter 6, section 5.

103. Black, *Elements of Chemistry*, I.79.

104. Ibid., I.82–3.

105. Ibid., I.504.

106. McKie and Heathcote, *Specific and Latent Heats*, 125–6; cf. above, this section.

107. Black, *Elements of Chemistry*, I.504.

108. Ibid. For Watt's account of these experiments, see Robison, *System of Mechanical Philosophy*, II.113–14. Also see the letter, written in March 1780, in which Watt said that 'I was the first that tryed experiments to find the quantities of heat absorbed by some metals by wood and by some other bodies in the year 1763 but invented no theories on the subject —it being done in consequence of Dr. Blacks discoveries.' (Robinson and McKie, *Partners in Science*, 77.)

109. Black, *Elements of Chemistry*, I.504.

110. Ibid.

111. Cf. Guerlac, 'Joseph Black,' in *Dictionary of Scientific Biography*, II.178, col. 2, where he says that 'it is commonly assumed that Black discovered *specific heats* before his work on latent heat. This is not the case.' Guerlac appears to be arguing that Black should not be credited with 'discovering' specific heats until he began quantifying them around 1764. See also Cardwell, *Watt to Clausius*, 37, where he admits he is unable to account for Black's neglect of his 'discovery' of specific heat.

CHAPTER TEN

Black's Influence on the Science of Heat

1. Unless otherwise indicated, all biographical information is taken from Muirhead, *Life of Watt*.

2. Ibid., 12–13.

3. See R. J. Law, *James Watt and the Separate Condenser, An Account of the Invention* (A Science Museum Monograph, London 1969) 7.

4. On Watt's exclusion from the practice of his trade within the city, see H. W. Dickinson and Rhys Jenkins, *James Watt and the Steam Engine* (Oxford 1927) 17.

5. Cf. Black's 'History of Mr. Watt's Improvement of the Steam Engine,' in Muirhead, *Life of Watt*, 58, and in Robinson and McKie, *Partners in Science*, 253.

6. Although there is some uncertainty concerning the date of their first meeting, I believe Muirhead is correct in arguing that they probably met in 1757; see his

Life of Watt, 60 n., and cf. Robinson and McKie, *Partners in Science*, 410.

7. See Samuel Smiles, *Lives of Boulton and Watt; principally from the original Soho MSS* (London 1865) 114–15, and Sir Edmund Whittaker, 'John Robison, M.A.,' in *An Eighteenth Century Lectureship in Chemistry*, ed. Andrew Kent (Glasgow 1950) 126–32.

8. See Robinson and McKie, *Partners in Science*, 256–9, 388, 410–13, and Muirhead, *Life of Watt*, 60–73, 96–100.

9. Ibid., 58; cf. Robinson and McKie, *Partners in Science*, 253.

10. For a description of the rooms in which Watt had his shop, see Smiles, *Boulton and Watt*, 107.

11. Robinson and McKie, *Partners in Science*, 321–2; cf. Muirhead, *Life of Watt*, 61. Dr Dick died in 1757 while still young and he was succeeded by Dr John Anderson. Anderson later founded Anderson's Institution, which was the forerunner of the Royal Technical College, now the University of Strathclyde, in Glasgow; the model for the Royal Institution in London; and the progenitor of the Mechanics' Institutes. See A. E. Musson and Eric Robinson, *Science and Technology in the Industrial Revolution* (Manchester 1969) 179–82. In his introduction to Black's lectures (Black, *Elements of Chemistry*, I.vii–viii), Robison gives a slightly different account of his first meeting with Black.

12. Robinson and McKie, *Partners in Science*, 259–60. These investigations evidently took place before 1759 for we know Watt was building organs during the years Robison was away from Glasgow; see Muirhead, *Life of Watt*, 46–7.

13. Robinson and McKie, *Partners in Science*, 253; cf. Muirhead, *Life of Watt*, 58.

14. Robinson and McKie, *Partners in Science*, 416.

15. Ibid., 434. In quoting from Watt's notebook I have simplified and punctuated the exact transcription provided by Robinson and McKie.

16. For a more detailed diagram of the Newcomen engine, see Cardwell, *Watt to Clausius*, 16.

17. In an account of his first experiments with the steam engine, which he wrote in 1805, Watt said that Robison also suggested that this source of power could be used to drive wheeled vehicles; see Robinson and McKie, *Partners in Science*, 410–11; cf. 416–19, and Robison, *System of Mechanical Philosophy*, II.113.

18. Muirhead, *Life of Watt*, 44–5.

19. See Smiles, *Boulton and Watt*, 112, on Watt's and Black's membership in the Anderston Club in Glasgow.

20. 'In Savery's steam-pump, steam . . . was conducted through a pipe, fitted with a valve, into an oval vessel full of water, expelling the water upwards through a second pipe. When the oval vessel was full of steam, the steam was condensed by pouring cold water on the outside. This created a partial vacuum, so that when the vessel was connected, through another pipe, with water at a lower level, atmospheric pressure forced this water upwards into the oval vessel. There were two vessels, filled and emptied alternately, the whole cycle of operations being controlled by suitably located valves. In an attempt to increase the height to which his engine would pump water, Savery used high-pressure steam, but this presented constructional problems which could not at that time be satisfactorily solved.' T. K. Derry and T. I. Williams, *A Short History of Technology* (Oxford 1961) 316.

21. Robison, *System of Mechanical Philosophy*, II.113. Cf. the brief history in Watt's notebook, Robinson and McKie, *Partners in Science*, 434. On the apparatus employed in this experiment, see Law, *Watt and the Separate Condenser*, 10, 16. Perhaps it was this attempt to build a model of an engine driven by steam that led Watt to make his first experiments on the relationship between pressure and the boiling point of water at pressures greater than that of the atmosphere. See Robinson and McKie, *Partners in Science*, 440–1. See also Black's query about the relationship between boiling point and pressure; above, chapter 9, section 1.

22. Muirhead, *Life of Watt*, 49.

23. Ibid., 62–3; cf. Robinson and McKie, *Partners in Science*, 257–8.

24. Ibid., 411; cf. Muirhead, *Life of Watt*, 51.

25. Ibid., 50; cf. Robinson and McKie, *Partners in Science*, 433. For further information on Watt's perspective machine, see Muirhead, *Life of Watt*, 50–6, and H. W. Dickinson, *James Watt, Craftsman & Engineer* (Cambridge 1936).

26. Robinson and McKie, *Partners in Science*, 379.

27. Ibid., 254, 418; see also Muirhead, *Life of Watt*, 92–3.

28. Robison, *System of Mechanical Philosophy*, II.113–14 n.

29. Muirhead, *Life of Watt*, 66.

30. On Watt's accounts, see Dickinson and Jenkins, *Watt and the Steam Engine*, 93, and Smeaton, 'Watt's published account'. Although the surviving evidence does not enable us to establish firmly the dates on which the experiments described in foll. 7r–20r of the notebook were performed (see Robinson and McKie, *Partners in Science*, 434–40), it is fairly certain that this portion of the notebook was written during the first quarter of 1765. (1) The notebook itself is clearly dated 1765 (see ibid., 434, 440; cf. Black's letter, ibid., 92). (2) The experiments described in the notebook obviously preceded Watt's invention of the separate condenser, which occurred in the spring of 1765. (3) It was during the course of the experiments described in the notebook that Watt first turned to Black for an explanation of his doctrine of latent heat (ibid., 439). We know, from one of Watt's letters to Roebuck (Muirhead, *Life of Watt*, 161), that by September 1765 Watt had a clear understanding of the latent heat of steam. It also appears that several decades after the event Watt had forgotten precisely when he first learned of Black's doctrine. In 1780 he wrote that 'I myself became instructed in the doctrine about the year 1763' (Robinson and McKie, *Partners in Science*, 85), and in his notes to Robison's article on the steam engine he stated that Black explained his doctrine to him in the summer of 1764 (Robison, *System of Mechanical Philosophy*, II.116). It may be that this latter date was taken from the reference to Watt's earliest experiments which Robison included in his edition of Black's lectures (Black, *Elements of Chemistry*, I.173).

31. Robinson and McKie, *Partners in Science*, 434.

32. See Muirhead, *Life of Watt*, 66, and Cardwell, *Watt to Clausius*, 54. It is worth noting that Watt's 1769 patent application bore the title 'A New Method of Lessening the Consumption of Steam and Fuel in Fire Engines.' See also Robison's statement (Muirhead, *Life of Watt*, 90) that Watt's idea 'was to apply his invention to the steam-engines as they existed'.

33. Robinson and McKie, *Partners in Science*, 434. I agree with Cardwell when he says that it is anachronistic to ask whether Watt and Robison were thinking about the specific heat or the conductivity of metal when they thought of using a wooden cylinder; see Cardwell, *Steam Power*, 39; cf. Smeaton, 'Watt's published account,' 37–8.

34. Robinson and McKie, *Partners in Science*, 435.

35. Ibid.

36. Ibid.

37. Ibid.

38. Ibid. Cf. Robison's account in Muirhead, *Life of Watt*, 66–7.

39. Robinson and McKie, *Partners in Science*, 436.

40. Ibid.

41. Ibid.

42. Ibid.

43. Ibid.

44. Robison, *System of Mechanical Philosophy*, II.113; cf. Robinson and McKie, *Partners in Science*, 434.

45. See above, chapter 2 and chapter 6, section 6.

46. Desaguliers, 'Rise of Vapours,' 16–17.

47. Robison, *System of Mechanical Philosophy*, II.115.

48. See Robinson and McKie, *Partners in Science*, 436–7 and photograph C; see also Robison, *System of Mechanical Philosophy*, II.115–16. For Watt's characterization of his experimental apparatus, see Smiles, *Boulton and Watt*, 119 n.1.

49. Robison, *System of Mechanical Philosophy*, II.115.

50. Robinson and McKie, *Partners in Science*, 437. It was after making this calculation that Watt became convinced that 'the quantity of steam used in every stroke by the [Newcomen] engine' is 'several times the full of the cylinder'. See Robison, *System of Mechanical Philosophy*, II.116.

51. Robinson and McKie, *Partners in Science*, 437.

52. Ibid., 438.

53. Ibid. On Watt's investigation of specific heat and on Black's and Irvine's examination of the specific heat of tin, see above, chapter 9, section 2. See also, Black, *Elements of Chemistry*, I.137, and Irvine, *Essays*, 450–1.

54. Robinson and McKie, *Partners in Science*, 438; cf. 426–7.

55. Ibid., 438–9.

56. Ibid., 439.

57. Cf. ibid., 427, where the editors erroneously assert that Watt's record of this experiment is inaccurate.

58. Ibid., 439 and photograph E.

59. Ibid., 439.

60. See Muirhead, *Life of Watt*, 66 n.

61. Robinson and McKie, *Partners in Science*, 439.

62. Ibid.

63. Ibid.

64. Ibid. As the editors point out in their introduction to this notebook (ibid., 427), Watt always added 212 to the latent heat of vaporization to get the total heat of the vapour. Thus to compare his value to that determined by Black and Irvine (see above, chapter 9, section 2, and Black, *Elements of Chemistry*, I.171–2), one must subtract 212 from 1012.

65. Robinson and McKie, *Partners in Science*, 439.

66. If Watt in fact discovered the phenomenon of latent heat in the summer of 1764 and learned of Black's doctrine shortly thereafter (cf. Robison, *System of Mechanical Philosophy*, II.116), then he may have been much more intimately involved in the experimental confirmation of the theory of latent heat than has previously been supposed. See Black, *Elements of Chemistry*, I.173, where, after describing the experiment on the condensation of steam which he and Irvine made in October 1764, Black says that 'a few weeks later, Mr. Watt made similar experiments with a smaller still, better fitted for a trial of this question, and much more manageable.' But see also Black's letter to Watt, written in June 1780, where he says 'I have notes of an experiment of yours upon my principles made at Edinburgh with a small Still J. Linds which makes the latent heat of ordinary Steam about 800°—This experiment was made in the year 1765.' (Robinson and McKie, *Partners in Science*, 92.)

67. Muirhead, *Life of Watt*, 67.

68. Robinson and McKie, *Partners in Science*, 440.

69. Ibid.

70. Ibid., 434.

71. My attention was first directed to the problem of back pressure by Mr David Channell in a paper he read before the Midwest Junto for the History of Science in April 1972. On back pressure, see also Donald Fleming, 'Latent Heat and the Invention of the Watt Engine,' in *Isis*, 43 (1952) 5; 3–5.

72. See above, chapter 9, section 2.

73. Black, *Elements of Chemistry*, I.161–2.

74. Robinson and McKie, *Partners in Science*, 418.

75. Ibid. See also Watt's 'Plain Story,' in Muirhead, *Life of Watt*, 84, and cf. E. Robinson and A. E. Musson, *James Watt and the Steam Revolution; a documentary history* (New York 1969) 192.

76. From Watt's 'Plain Story,' Muirhead, *Life of Watt*, 87.

77. Robinson and Musson, *Watt and Steam Revolution*, 42. The interview from which this account is taken was first published in the *Transactions of the Glasgow Archaeological Society*, I (1859). Cf. the similar account in Robison, *System of Mechanical Philosophy*, II.116. On the date of this insight, see Muirhead, *Life of Watt*, 59, 67–9, 73, 94–5.

78. Robinson and McKie, *Partners in Science*, 254.

79. Ibid. For Watt's own description of these earliest ancillary improvements, see Robison, *System of Mechanical Philosophy*, II.117.

80. Muirhead, *Life of Watt*, 94–5.

81. Robinson and McKie, *Partners in Science*, 254.

82. See ibid., 255, and Muirhead, *Life of Watt*, 160.

83. See ibid., 161–6, and Law, *Watt and the Separate Condenser*. For an account of some of Watt's boiler improvements and an indication of the widespread interest in his research on the steam engine, see the letter dated 20 December 1756, in Reid, *Works*, 42.

84. Muirhead, *Life of Watt*, 161; letter to Lind, 4 September, 1765.

85. See Robinson and Musson, *Watt and Steam Revolution*, 46–7, and Robison, *System of Mechanical Philosophy*, II.8, 10 n. See also Robinson and McKie, *Partners in Science*, 92. Watt's 1765 figure for the latent heat of steam in vacuo was somewhat over 1200°. Perhaps these experiments were suggested by Black's query, 'will distillation *in vacuo* save fuel?' (see above, chapter 9, section 2.)

86. Robinson and McKie, *Partners in Science*, 420.

87. Muirhead, *Life of Watt*, 161; cf. Robinson and Musson, *Watt and Steam Revolution*, 46–7.

88. See Black's comment in Robinson and McKie, *Partners in Science*, 255.

89. See James Patrick Muirhead, *The Origin and Progress of the Mechanical Inventions of James Watt—Illustrated by his Correspondence with his Friends and the Specifications of his Patents* (3 vols., London 1854) I.6, letter dated 13 September 1765.

90. Muirhead, *Life of Watt*, 71.

91. Ibid., 70.

92. Ibid., 165–8.

93. Ibid., 238–9. Cf. Musson and Robinson, *Science and Technology*, 143: 'Even before Watt came to

Birmingham, the members of the Lunar Society were interested in the design of steam engines. Boulton had invented a steam engine of his own, which he discussed with Erasmus Darwin and Benjamin Franklin.'

94. Muirhead, *Life of Watt*, 176.

95. From Davy's remark published in the *Proceedings of the Public Meeting held at Freemason's Hall, on the 18th June, 1824, for Erecting a Monument to the Late James Watt* (London 1824); reprinted in D. F. G. Arago, *Historical Eloge of James Watt*, trans. and ed. by J. P. Muirhead (London 1839) 191.

96. On Watt's later experiments on steam, see his notes to Robison, *System of Mechanical Philosophy*, II; his notebook in Robinson and McKie, *Partners in Science*, 441–90; Muirhead, *Life of Watt* and *Mechanical Inventions of Watt*; Dickinson and Jenkins, *Watt and Steam Engines*; H. W. Dickinson, *The Garret Workshop of James Watt* (Science Museum Technical Pamphlet No. 1, London 1929); Cardwell, *Watt to Clausius*; and Robinson and Musson, *Watt and Steam Revolution*. On Watt's activities in the Lunar Society, see Schofield, *Lunar Society*, and David F. Larder, 'An Unpublished Chemical Essay of James Watt,' in *Notes and Records of the Royal Society of London*, 25 (1970) 193–210. On Watt's discovery of the composition of water, see J. Watt, *Correspondence of the late James Watt on his Discovery of the Theory of the Composition of Water*, ed. J. P. Muirhead (London 1846). On Watt's attitude towards Black's chemical discoveries, see Robinson and McKie, *Partners in Science*, 66, 76–89, 128; and Watt, *Correspondence on Water*, 40–1.

97. On the legend of the dependence of Watt's invention upon Black's theory, see Cardwell, *Watt to Clausius*, 41–55, and Fleming, 'Latent Heat and Watt Engine'. For Watt's own statements on the relationship between the instruction he received from Black and his invention of the separate condenser, see Robinson and McKie, *Partners in Science*, 416, and Robison, *System of Mechanical Philosophy*, II.ix.

98. Black, *Elements of Chemistry*, I.xliv; see also, Robinson and McKie, *Partners in Science*, 359–60.

99. Black, *Elements of Chemistry*, I.vliv; see also above, chapter 9, section 2.

100. Cf. Black, *Elements of Chemistry*, I.139, where Black states that Irvine was puzzled by the results of his experiments on bees wax until Black explained them in terms of his theory of latent heat.

101. See Kent, 'William Irvine,' 141. On the appointment of Black's successors, see also the letters in Reid, *Works*, 45–9.

102. On Irvine's botanical and chemical research, see Kent, 'William Irvine,' 141–3, 146–7, and Irvine, *Essays*.

103. Cf. Black, *Elements of Chemistry*, I.141–2, where Black says he only knew of Irvine's theory through the accounts of it provided by Adair Crawford and others.

104. See Robinson and McKie, *Partners in Science*, 420, where Watt says that 'in 1772 I made other experiments, in which I was assisted by Dr Irvine Senior, both upon the boiling of water under pressures greater and lesser than that of the atmosphere'.

105. Ibid., 87, letter dated March 1780. On Irvine's use of the term absolute heat, see Adair Crawford, *Experiments and Observations on Animal Heat, and the Inflammation of Combustible Bodies. Being an Attempt to Resolve these Phaenomena into a General Law of Nature* (London 1779) 2. For the phrase 'doctrine of capacities', see Irvine, *Essays*, x.

106. Muirhead, *Mechanical Inventions*

of Watt, II.78. The phrase 'natural zero' was used by William Irvine Sr., see *Essays*, 164, and by his son; see William Irvine Jr., 'A Letter from Mr. Irvine Concerning the Late Dr. Irvine, of Glasgow, His Doctrine, which Ascribes the Disappearance of Heat, Without Increase of Temperature, to a Change of Capacity in Bodies, and that of Dr. Black, which Supposes Caloric to Become Latent by Chemical Combination with Bodies; with Particular Remarks on the Mistakes of Dr. Thompson, in His Accounts of these Doctrines,' in (Nicholson's) *Journal of Natural Philosophy, Chemistry, and the Arts*, 6 (1803) 28; 25–31.

107. Irvine, *Essays*, 92, cf. 182–3.

108. Ibid., 187.

109. See ibid., xiv, 163–89. See also, Robinson and McKie, *Partners in Science*, 87, and Black, *Elements of Chemistry*, I.142 n. and I.505. Cullen, it will be recalled, presented his own 'Remarks on the Generation of Heat and Cold by Mixture' before this society nearly two decades earlier; see above, chapter 6, section 7.

110. Irvine, *Essays*, 167–8.

111. Ibid., 168.

112. Ibid.

113. Ibid., 169.

114. Ibid., 169–70.

115. Ibid., 174.

116. Ibid.

117. Ibid.

118. Ibid., 175–6.

119. Ibid., 177.

120. Ibid.

121. Ibid., 178. For a discussion of Irvine's contribution to calorimetry, see McKie and Heathcote, *Specific and Latent Heats*, 122–30.

122. Irvine, *Essays*, 180–1.

123. Cf. ibid., 164.

124. Ibid., 183, cf. 51–3. Cf. also Robinson and McKie, *Partners in Science*, 87, and Black, *Elements of Chemistry*, I.142. For an account of Irvine's theory as presented with diagrams, see Crawford, *Animal Heat*, 2nd ed. with very large additions (London 1788) 379, and John Dalton, *A New System of Chemical Philosophy* (2 vols., Manchester 1808–27), vol. I, Part I. See also Robert Fox, 'Dalton's Caloric Theory,' in *John Dalton & the Progress of Science*, ed. D. S. L. Cardwell (Manchester 1968) 187–202; Fox, *Caloric Theory of Gases*, 25–7; and Cardwell, *Watt to Clausius*, 55–6.

125. Irvine, *Essays*, 187.

126. Ibid.

127. Ibid., 187–8.

128. Ibid., 188.

129. The phrase is Crawford's; see his *Animal Heat*, 1st ed., p. 17 n. See also *Encyclopedia Britannica*, 2nd ed., vol. 5 (Edinburgh 1780) article on 'Heat,' 3541, col. I.

130. Robinson and McKie, *Partners in Science*, 77; Black, *Elements of Chemistry*, I.195.

131. The first published calculation of absolute zero can be found in J. H. Magellan, *Essai sur la nouvelle théorie du feu élémentaire, et de la chaleur des corps* (London 1780) 176, but Crawford later said that the idea was suggested by Irvine (see his *Animal Heat*, 2nd ed., 453–4). Magellan's *Essai* was published as the fourth *traité* in his *Collection de différens traités sur des instruments d'astronomie, physique*, etc. (165–92), and was republished in Rozier's *Observations sur la Physique*, 17 (1781) 375–86, 411–22. On Magellan, see Henry Guerlac, *Lavoisier—The Crucial Year; The background and origins of his first experiments on combustion in 1772* (Ithaca, New York 1961) chapter 2; on the idea of absolute zero, see Irvine, *Essays*, 164, and McKie and Heathcote, *Specific and Latent Heats*, 130–7; for an example of the computation of natural zero according to Irvine's theory, see Irvine, *Essays*, 122–3, and Fox, *Caloric Theory of Gases*, 27–8; for Black's and Robison's reservations about the figures obtained, see Black, *Elements of Chemistry*, 1.64, 1.507–8.

132. Ibid., 1.142.

133. Ibid.

134. Ibid., 1.142–4.

135. On Cavendish, see McKie and Heathcote, *Specific and Latent Heats*, 52–3. On Black's doctrine in Edinburgh, see Reid, *Works*, 45, letter dated 23 March 1766, in which Reid concludes his description of Black's doctrine of latent heat by saying 'I cannot find that Cullen or the Edinburgh people know anything of this matter'. See also the absence of any mention of Black or his doctrine in the articles on fire and heat in the *Encyclopedia Britannica; or a Dictionary of Arts and Sciences, Compiled upon a New Plan . . .* by a Society of Gentlemen in Scotland, 1st ed. (3 vols., Edinburgh 1771) vol. II.

136. See Musson and Robinson, *Science and Technology*, 75, 168, 175.

137. Cf. the many published versions of Cullen's medical lectures.

138. Cf. the list of medical dissertations on animal heat in Mendelsohn, *Heat and Life*, 111 n. 12.

139. London.

140. For an account of the *Enquiry*, see McKie and Heathcote, *Specific and Latent Heats*, 50–1.

141. See Magellan, *Essai*, 173–4. It appears that J. A. De Luc did not know of the *Enquiry* when he wrote his *Recherches sur les modifications de l'atmosphère* (2 vols., Geneva 1772) and his *Idées sur la météroIogie* (2 vols., London 1786–7); see McKie and Heathcote, *Specific and Latent Heats*, 51–2. On De Luc, see Paul A. Tunbridge, 'Jean André De Luc, F.R.S.,' *Notes and Records of the Royal Society of London*, 26 (1971) 15–33.

142. See Douglas McKie, 'The 'Observations' of the Abbé Francois Rozier (1734–93)—1,' *Annals of Science*, 13 (1957, published 1958) 73–89, in which the text of this paper is reprinted (86–9). See also, McKie and Heathcote, *Specific and Latent Heats*, 45–8. For the letter in which James Watt referred Magellan to this paper, see Robinson and McKie, *Partners in Science*, 86.

143. See Henry Guerlac, 'A Lost Memoir of Lavoisier,' *Isis*, 50 (1959) 128–9; 125–9, and Guerlac, *Crucial Year*, 92–7.

144. McKie, 'Observations,' 86.

145. See Robert Siegfried, 'Lavoisier's View of the Gaseous State and Its Early Application to Pneumatic Chemistry,' *Isis*, 63 (1972) 59–66; 59–78; J. B. Gough, 'Lavoisier's Early Career in Science: an Examination of some New Evidence,' *British Journal for the History of Science*, 4 (1968) 52–7; and Fichman, 'French Stahlism,' 115–22.

146. See Guerlac, *Crucial Year*, 92–7. The documents Lavoisier had initialed were titled 'Essay sur la nature de lair' and consist of two incomplete memoirs—'Refflexions sur lair et sur sa combinaison dans les mineraux' and 'Refflexions sur la combinaison de la matierre du feu dans les corps'—and two brief concluding notes. These documents have been published in ibid., 215–23.

147. See, for instance, the articles on 'Congelation', in vol. III (1778), especially page 2189, and 'Evaporation', in vol. IV (1779), especially pages 2846–8. Cf. Robinson and McKie, *Partners in Science*, 80, for letter dated 9 March 1780, in which Watt refers Magellan to the articles in the first four volumes of the second edition of the *Encyclopedia*.

148. *De caloris Animalium causa* (Edinburgh 1775). The dissertation was published under Leslie's

Christian names, Patrick Dugud.

149. For an account of Leslie's theory of animal heat, see Mendelsohn, *Heat and Life*, 114–23.

150. For Magellan's reference to Leslie's work, see his *Essai* 173–4.

151. On Crawford's theory of animal heat and on the translations of his book, see Mendelsohn, *Heat and Life*, 123–33. On Crawford's book, see also McKie and Heathcote, *Specific and Latent Heats*, 38–42. See also Robinson and McKie, *Partners in Science*, 76, for a letter dated 1 March 1780, in which Watt told Magellan he could learn about 'Dr. Black's ingenious doctrines of heat' from the books by Leslie and Crawford.

152. See above, chapter 9, note 62.

153. See his 'De la combinaison de la matière du feu avec les fluids évaporables, et de la formation des fluides élastiques aëriformes,' dated 5 September 1777, *Mémoires de l'Académie royal des Sciences* (1777, published 1780) 420–32; and his 'Mémoire sur la combustion en général,' dated 5 September 1777, ibid., 592–600. Cullen, but not Black, is mentioned in the first of the above memoirs; for an early draft of this memoir, see Gough, 'Nouvelle contribution'. On the development of Lavoisier's view of heat and caloric, see Fox, *Caloric Theory of Gases*, 9–11.

154. Cleghorn's dissertation also contains the first published values for capacities for heat; see Cleghorn, '*De Igne*,' 21 and note 14.

155. Vol. v was published in 1780; for the article on heat, see 3539–55.

156. See J. H. Magellan, 'Extrait d'une Lettre de M. Magellan, de la Société Royale de Londres, sur les Montres nouvelles qui n'ont pas besoin d'être montées, sur celle de M. Mudge & sur l'Ouvrage de M. Crawford,' *Observations sur la Physique*, 16 (1780) 60–3, and his 'Lettre de M. Magellan, à l'Auteur de ce Journal, sur le Mémoire suivant,' ibid., 17 (1781) 369–75.

157. J. A. De Luc, 'To the conductors of the Edinburgh Review,' a letter dated 18 April 1805, in *Edinburgh Review*, 6 (1805) 511; 502–15.

158. See his 'Expériences sur la Respiration des Animaux, et sur les Changements qui Arrivent à l'Air en Passant par leur Poumon,' dated 3 May 1777, *Mémoires de l'Académie royale des Sciences*, (1777, published 1780) 185–94; and his 'Mémoire sur la combustion'. On Lavoisier's theory of animal heat, see Mendelsohn, *Heat and Life*, 134–9.

159. See his 'De la combinaison de la matière du feu' and his 'Mémoire sur la combustion'.

160. On the paucity of heat experiments in Lavoisier's laboratory notebooks between 1777 and 1780, see Marcellin Berthelot, *La Révolution Chimique—Lavoisier* (Paris 1890) 279–95, and Fox, *Caloric Theory of Gases*, 35. It is clear from the citations in his 1783 memoir on heat (A. L. Lavoisier and P. S. Laplace, 'Mémoire sur la chaleur,' dated 18 June 1783, *Mémoires de l'Académie royale des Sciences* (1780, published 1784) 355–408) that Lavoisier was familiar with Crawford's book, which is cited, and with Magellan's *Essai*. His familiarity with Magellan's work is revealed by his use of Kirwan's figures for the specific heat of ice, Kirwan's table of specific heats having been first published in Magellan's *Essai*.

161. Watt, *Correspondence on Water*, 12.

162. Cf. Fox, *Caloric Theory of Gases*, 10, 35–6, 138, 147–8. See also ibid., 6 n., where Fox says that the term *calorique* came into use in Lavoisier's circle around 1783–4.

163. The phrase is Davy's; see ibid., 119.

164. Cf. Schofield, *Mechanism and Materialism*, Part III, 'Neo-Mechanism, 1760–1815'.
165. On Thomson and Henry, see Fox, *Caloric Theory of Gases*, 105–6; on Dalton, see ibid., 76, 109–15, and Fox,'Dalton's Caloric Theory'.
166. Cf. Fox, *Caloric Theory of Gases*, 36–9, 50.
167. From the announcement of the competition, as quoted in ibid., 134.
168. The two essays were later published, see F. Delaroche and J. E. Bérard, 'Mémoire sur la détermination de la chaleur spécifique des différens gaz,' *Annales de Chimie et de Physique*, 85 (1813) 72–110, 113–82, and N. Clément and C. B. Desormes, 'Détermination expérimentale du zéro absolu de la chaleur et du calorique spécifique des gaz,' *Journal de Physique, de Chimie, d'Histoire Naturelle et des Arts*, 89 (1819) 321–46, 428–55. For a complete account of this competition, see Fox, *Caloric Theory of Gases*, 134–50.
169. Delaroche and Bérard, 'La chaleur spécifique,' 174–5, as quoted in Fox, *Caloric Theory of Gases*, 142.
170. Henry Thomas Buckle, *History of Civilization in England* (2 vols., London 1861) II.492–4.

Bibliography of Sources Cited

MANUSCRIPTS

Edinburgh University Library (EUL)

GEN 873 Three volumes of Joseph Black's correspondence.
vol. 1, 1754–83
vol. 2, 1781–6
vol. 3, 1785–93

GEN 874 Assorted bundles of Black's papers.
a) 64 family letters, 1748–99, mainly to his brother George.
b) 26 letters, mainly to Joseph Black.
c) Technical papers, autobiographical note, sketch.
d) Black's account books, ledger of his estate, settlement.
e) 9 letters from John Robison to George Black, 1799–1801, about the publication of Joseph Black's lectures.

GEN 875F Large folded letters removed from GEN 873.

Glasgow University Library (GUL)

For a brief survey of the Cullen MSS in the Glasgow University Library, see W. P. D. Wightman, 'William Cullen and the Teaching of Chemistry', *Annals of Science*, 11 (1955) 154–65; 157–8, 161–2, 164–5. Wightman has also provided the Library with a typed list of the contents of the first two boxes of these MSS. When citing these MSS in this book I have used the MSS numbers given on this list.

Indiana University Library

Rush, Benjamin 'Scottish Journal', a MSS record of his visit to Edinburgh, 1766–8.

Public Record Office of Northern Ireland (PRONI)

D.719 80 original Black family letters, 1739–66.

D.1401 Varied collection of original documents containing a number of items relevant to the Black family and to the Stewarts to whom they were connected by marriage.

D.1950 41 Black family letters, 1668–1858.

T.1073 52 Black family letters and papers, 1716–1807.

Royal College of Physicians, Edinburgh (RCPE)

For citation of Wightman's description of these MSS and discussion of their subsequent renumbering, see above, chapter 6, note 1.

C.10 Two volume set of student lecture notes on Cullen's chemistry course, probably from the 1760s.

C.12 Student notes on Cullen's chemistry lectures, 1757–8.

C.15, vol. 1 Cullen's own lecture notes for chemistry course, 1748–9.

Royal Medical Society

MSS Dissertations.

UNPUBLISHED DISSERTATIONS

Langer, Bernard 'Pneumatic Chemistry, 1772–1789: A Resolution of Conflict'. University of Wisconsin 1965.

Quinn, Arthur J. 'Evaporation and Repulsion: A Study of English Corpuscular Philosophy From Newton to Franklin'. Princeton University 1970.

PRINTED WORKS

Addison, W. Innes *The Matriculation Albums of the University of Glasgow*. Glasgow 1913.

— *A Roll of the Graduates of the University of Glasgow*. Glasgow 1898.

Alston, Charles *A Dissertation on Quick-Lime and Lime-Water*. Edinburgh 1752.

— *Dissertation on Quick-Lime*. 2nd ed. Edinburgh 1754.

— 'Extracts of Two Letters from Dr. Alston, Botany Professor at Edinburgh: to Dr. Mortimer, Secr. R. S. The first dated 17 March 1749; the second August 9, 1750', *Philosophical Transactions of the Royal Society of London*, 47 (1751–2, published 1753) 265–6.

Anderson, James 'Cursory Hints and Anecdotes of the late Doctor William Cullen of Edinburgh', *The Bee, or Literary Weekly Intelligencer*, 1 (Edinburgh 1791) 1–14, 45–56, 121–5, 161–6.

[Anon.]. 'Experiences du Docteur Black, *Sur la marche de la chaleur dans certain circonstances*', *Introduction aux Observations sur la Physique*, (September 1772) 156–66. Reprinted in Douglas McKie 'The "Observations" of the Abbé François Rozier (1734–93)—1', *Annals of Science*, 13 (1957, published 1958) 73–89; 86–9.

Arago, D.F.G. *Historical Eloge of James Watt*, trans. and ed. J.P. Muirhead. London 1839.

Bacon, Francis *The New Organon and Related Writings*, ed. Fulton H. Anderson. New York 1960.

Berthelot, Marcellin *La Révolution Chimique—Lavoisier*. Paris 1890.

Black, Joseph *Dissertatio medica inauguralis, de humore acido a cibis orto*. Edinburgh 1754. Reprinted in vol. 2 of *Thesaurus medicus Edinburgensis novus* (4 vols., Edinburgh 1785). English translation by A. Crum Brown in 'Joseph Black's Inaugural Dissertation', *Journal of Chemical Education*, 12 (1935) 225–8, 268–73.

— *Lectures on the Elements of Chemistry*, ed. John Robison. 2 vols. Edinburgh 1803.

— *An Enquiry into the General Effects of Heat, with Observations on the Theories of Mixture*, London 1770.

— 'Experiments on *Magnesia alba*, Quicklime, and some other Alcaline Substances', *Essays and Observations, Physical and Literary*, 2 (1756) 157–225. Published separately with original title as Alembic Club Reprint No. 1 (Edinburgh 1898).

Boerhaave, Hermann *A New Method of Chemistry . . . Translated from the original Latin of Dr. Boerhaave's Elementa Chemia, as Published by Himself . . . By Peter Shaw, M.D.* 3rd ed. 2 vols. London 1753.

Boulduc, G.F. 'Analyse des eaux de forges, et principalement de la source apellée la Royale', *Mémoires de l'Académie royale des Sciences*, 1735, 443–53.

Briggs, Asa *The Age of Improvement*. London 1959.

Brownrigg, William *The Art of Making Common Salt*. London 1748.

Bryson, Gladys *Man and Society: The Scottish Inquiry of the Eighteenth Century*. Princeton 1945.

Buckle, Henry Thomas *History of Civilization in England*. 2 vols. London 1861.

Cameron, James K. 'The Church of Scotland in the age of reason', *Studies on Voltaire and the Eighteenth Century*, 58 (1967) 1939–51.

Cant, R. G. 'The Scottish universities and Scottish society in the eighteenth century', *Studies on Voltaire and the Eighteenth Century*, 58 (1967) 1953–66.

Cantor, G. N. 'Henry Brougham and the Scottish Methodological Tradition', *Studies in the History and Philosophy of Science*, 2 (1971) 69–89.

Cardwell, D. S. L. *From Watt to Clausius—The Rise of Thermodynamics in the Early Industrial Age*. Ithaca, New York 1971.

—ed. *John Dalton & the Progress of Science*. Manchester 1968.

—*Steam Power in the Eighteenth Century*. London 1963.

Carlyle, Alexander *Autobiography*. 2nd ed. London 1860.

Cavendish, Henry 'Three Papers, containing Experiments on factitious Airs', *Philosophical Transactions of the Royal Society of London*, 56 (1767) 141–84.

Christie, John 'The Origins and Development of the Scottish Scientific Community, 1680–1760', *History of Science*, 12 (1974) 122–41.

Cleghorn, William *Disputatio Physica Inauguralis, Theoriam Ignis Complectens*. Edinburgh 1779. Translated and annotated by Douglas McKie and Niels H. de V. Heathcote, 'William Cleghorn's *De Igne* (1779)', *Annals of Science*, 14 (1958, published 1960) 1–82.

Clément, N. and Desormes, C. B. 'Détermination expérimentale du zéro absolu de la chaleur et du calorique spécifique des gaz', *Journal de Physique, de chimie, d'Histoire Naturelle et des Arts*, 89 (1819) 321–46, 428–55.

Clow, Archibald 'Hermann Boerhaave and Scottish Chemistry', in *An Eighteenth Century Lectureship In Chemistry*, ed. Andrew Kent. Glasgow 1950.

Clow, Archibald and Clow, Nan *The Chemical Revolution*. London 1952.

Cochrane, Thomas *Notes from Doctor Black's Lectures on Chemistry 1767/8*, ed. Douglas McKie. Cheshire 1966.

Cohen I.B. *Franklin and Newton*. Philadelphia 1956.

Comrie, J. D. 'Boerhaave and the Early Medical School at Edinburgh', in *Memorialia Herman Boerhaave*. Haarlem 1939.

— *History of Scottish Medicine*. 2nd ed. 2 vols. London 1932.

Cormack, J. J. C. 'The [Royal Medical] Society's Library', *Bulletin of the Medical Library Association*, 48 (1960) 125–39.

Coutts, James *A History of the University of Glasgow*, Glasgow 1909.

Cowan, David L. 'Liberty, laissez-faire and licensure in nineteenth-century Britain', *Bulletin of the History of Medicine*, 43 (1969) 30–40.

Crawford, Adair *Experiments and Observations on Animal Heat, and the Inflammation of Combustible Bodies. Being an Attempt to Resolve these Phaenomena into a General Law of Nature*. London 1779. 2nd ed., with very large additions, London 1788.

Crellin, J. K. 'William Cullen & Practical Chemistry', *Actes du XIIe Congrès International d'Histoire des Sciences, Paris 1968* (published 1971) 6. 17–21.

— 'William Cullen: His Caliber as a Teacher and an Unpublished Introduction to his *A Treatise of the Materia Medica*, London 1773', *Medical History*, 15 (1971) 79–87.

Crosland, Maurice *Historical Studies in the Language of Chemistry*. London 1962

— 'The Use of Diagrams as Chemical "Equations" in the Lecture Notes of William Cullen and Joseph Black', *Annals of Science*, 15 (1959) 75–90.

Crowther, J. G. *Scientists of the Industrial Revolution*. Philadelphia 1963.

Cullen, L. M. *Anglo-Irish Trade, 1660–1800*. New York 1968.

Cullen, William 'Of the Cold produced by Evaporating Fluids, and of some other Means of producing Cold', *Essays and Observations, Physical and Literary*, 2 (1756) 145–56.

Dalton, John *A New System of Chemical Philosophy*. 2 vols. Manchester 1808–27.

Davie, G. E. 'Berkeley's Impact on Scottish Philosophy', *Philosophy*, 40 (1965) 222–34.

— *The Democratic Intellect: Scotland & Her Universities in the Nineteenth Century*. Edinburgh 1961.

Debus, Allen G. *The Chemical Dream of the Renaissance*. Cambridge 1968.

Delaroche, F. and Bérard, J. E. 'Mémoire sur la détermination de la chaleur spécifique des différens gaz', *Annales de Chimie et de Physique*, 85 (1813) 72–110, 113–82.

De Luc, J. A. *Idées sur la météorologie*. 2 vols. London 1786–7.

— *Recherches sur les modifications de l'atmosphère*. 2 vols. Geneva 1772.

— 'To the conductors of the Edinburgh Review', a letter dated 18 April 1805, *Edinburgh Review*, 6 (1805) 502–15.

Derry, T. K. and Williams, T. I. *A Short History of Technology*. Oxford 1961.

Desaguilers, John 'An Account of a book entitul'd *Vegetable Staticks*', *Philosophical Transactions of the Royal Society of London*, 35 (1728, nos. 398 & 399) 264–91.

— 'An Attempt to solve the phenomena of the Rise of Vapours Formation of Clouds, and Descent of Rain', *Philosophical Transactions of the Royal Society of London*, 36 (1729–30, no. 407) 6–22.

— 'Some Thoughts and Conjectures Concerning the Cause of Elasticity', *Philosophical Transactions of the Royal Society of London*, 41 (1739–41, no. 454) 175–85.

— 'Some Thoughts and Experiments Concerning Electricity', *Philosophical Transactions of the Royal Society of London*, 41 (1739–41, no. 454) 186–93.

Dickinson, H. W. *The Garret Workshop of James Watt*. London 1929. (Science Museum Technical Pamphlet No. 1.)

— *James Watt, Craftsman & Engineer*. Cambridge 1936.

Dickinson, H. W. and Jenkins, Rhys *James Watt and the Steam Engine*. Oxford 1927.

Dobbin, Leonard 'A Cullen Chemical Manuscript of 1753', *Annals of Science*, 1 (1936) 138–56.

Du Fay, Charles 'A Letter from Mons. Du Fay, F.R.S. and Royal Academy of Sciences at Paris . . . Concerning Electricity', *Philosophical Transactions of the Royal Society of London*, 38 (1733–4, no. 431) 258–66.

— 'Sur le sel du Chaux', *Mémoires de l'Académie royale des Sciences*, 1724, 88–93.

Duhamel du Monceau, H. L. 'Diverses Experiences sur la Chaux', *Mémoires de l'Académie royale des Sciences*, 1747 (published 1752) 59–81.

— 'Sur la Base du Sel Marin', *Mémoires de l'Académie royale des Sciences*, 1736, 215–32.

Dunbar, James *Smegmatologia, or the Art of Making Potashes, Soap, and Bleaching of Linen*. Edinburgh 1736.

Encyclopedia Britannica; or a Dictionary of Arts and Sciences, Compiled upon a New Plan . . . by a Society of Gentlemen in Scotland. 1st ed. 3 vols. Edinburgh 1771.

— 2nd ed. 10 vols. Edinburgh 1778–83.

Ferguson, Adam 'Minutes of the Life and Character of Joseph Black, M.D.', *Transactions of the*

Royal Society of Edinburgh, 5 (1805) 101–17.

Fichman, Martin 'French Stahlism and Chemical Studies of Air, 1750–1770', *Ambix*, 18 (1971) 94–122.

Fleming, Donald 'Latent Heat and the Invention of the Watt Engine', *Isis*, 43 (1952) 3–5.

Fortuna Domus, A Series of Lectures delivered in the University of Glasgow in Commemoration of the fifth centenary of its Foundation. Glasgow 1952.

Fox, Robert *The Caloric Theory of Gases from Lavoisier to Regnault.* Oxford 1971.

— 'Dalton's Caloric Theory', in *John Dalton & the Progress of Science*, ed. D. S. L. Cardwell. Manchester 1968.

Fussell, G. E. *Crop Nutrition: Science and Practice before Leibig.* Lawrence, Kansas 1971.

Geoffroy, E. F. 'Eclaircissement sur la Table inserée dans les Mémoires de 1718. Concernant les Rapports observés entre différentes Substances', *Mémoires de l'Académie royale des Sciences*, 1720, 20–34.

— 'Table des différents rapports observé en chimie entre différentes substances', *Mémoires de l'Académie royale des Sciences*, 1718, 202–12.

Gibbs, F. W. 'Peter Shaw and the Revival of Chemistry', *Annals of Science*, 7 (1951) 211–37.

Gillispie, Charles, 'The Natural History of Industry', *Isis*, 48 (1957) 398–407.

Gough, J. B. 'Lavoisier's Early Career in Science: an Examination of some New Evidence', *British Journal for the History of Science*, 4 (1968) 52–7.

— 'Nouvelle contribution à l'étude de l'évolution des idées de Lavoisier sur la nature de l'air et sur la calcination des métaux', *Archives internationales d'histoire des sciences*, Number 88–89 (July–December, 1969) 267–75.

Graham, Henry Grey *The Social Life of Scotland in the Eighteenth Century.* 2nd ed. London 1928.

Grant, Sir Alexander *The Story of the University of Edinburgh During its First Three Hundred Years.* 2 vols. London 1884.

Gray, J. *History of the Royal Medical Society, 1737–1937.* Edinburgh 1952.

Grew, Nehemiah 'Experiments in Consort upon the Solution of Salts in Water. Read before the Royal Society, January 18, 1676/7', in Nehemiah Grew *Anatomy of Plants* London 1682.

Guerlac, Henry 'The Background to Dalton's Atomic Theory', in *John Dalton & the Progress of Science*, ed. D S. L. Cardwell. Manchester 1968.

— 'Joseph Black', in *Dictionary of Scientific Biography.* Vol. 2. New York 1970.

— 'Joseph Black and Fixed Air', *Isis*, 48 (1957) 124–51, 433–56.

—*Lavoisier—The Crucial Year; The background and origins of his first experiments on combustion in 1772.* Ithaca, New York 1961.

— 'A Lost Memoir of Lavoisier', *Isis*, 50 (1959) 125–9.

— 'Some French Antecedents of the Chemical Revolution', *Chymia*, 5 (1959) 73–112.

Hales, Stephen *Vegetable Staticks: Or, An Account of some Statical Experiments on the Sap in Vegetables: Being an Essay towards a Natural History of Vegetation. Also, a Specimen of An Attempt to Analyze the Air, By a great Variety of Chymio-Statical Experiments.* London 1727.

Hall, Marie Boas 'The History of the Concept of Element', in *John Dalton & the Progress of Science*, ed. D. S. L. Cardwell. Manchester 1968.

Hamilton, Henry *An Economic History of Scotland in the Eighteenth Century*. Oxford 1963.

— *The Industrial Revolution in Scotland*. Oxford 1932.

Heimann, P. M. and McGuire, J. E. 'Newtonian Forces and Lockean Powers: Concepts of Matter in Eighteenth-Century Thought', in *Historical Studies in the Physical Sciences*, ed. Russell McCormmach. Vol. 3. Philadelphia 1971.

Hoffmann, Frederick 'Animadversiones & experimenta circa magnesiam albam, tutem & gratium insipidum pulverem laxantem', in Frederick Hoffmann *Opera Omnia Physico-Medica*. 6 vols. Geneva 1740. 4. 497–8.

—'De lixivio a coctione salis communis relictio & ex eo prodeunte terra laxante sive magnesia, & sale sic dicto Ebsoniensi', in Frederick Hoffmann *Opera Omnia Physico-Medica*. 6 vols. Geneva 1740. 4. 500–1.

Holmes, J. D. 'Early Years of the Medical Society of Edinburgh', *University of Edinburgh Journal* (Autumn 1968) 333–40.

Holmyard, E. J. *Makers of Chemistry*. Oxford 1931.

Homberg, William 'Observation sur la quantité d'acides absorbées par les alcalis terreux', *Mémoires de l'Académie royale des Sciences*, 1700, 64–71.

Home, Francis *An Essay on the Contents and Virtues of Dunse-Spaw*. Edinburgh 1751.

— *Experiments on Bleaching*. Edinburgh 1756.

— *Experiments on Bleaching*. 2nd ed. Dublin 1771.

Home, Henry *The Gentleman Farmer; being an attempt to improve agriculture by subjecting it to the test of rational principles*. Edinburgh 1776.

[Home, Henry], Lord Kames. 'On Evaporation', *Essays and Observations, Physical and Literary*, 3 (1771) 80–99.

Horn, D. B. *A Short History of the University of Edinburgh*. Edinburgh 1967.

Howell, William S. *Eighteenth-Century British Logic and Rhetoric*. Princeton 1971.

[Hume, David] 'Preface', *Essays and Observations, Physical and Literary*, 1 (1754) iii–viii.

Hutcheson, Francis *A System of Moral Philosophy*. 2 vols. Glasgow 1755.

Irvine, William *Essays Chiefly on Chemical Subjects*. London 1805.

Irvine, William Jr. 'A Letter from Mr. Irvine Concerning the Late Dr. Irvine, of Glasgow, His Doctrine, which Ascribes the Disappearance of Heat, Without Increase of Temperature, to a Change of Capacity in Bodies, and that of Dr. Black, which Supposes Caloric to Become Latent by Chemical Combination with Bodies; with Particular Remarks on the Mistakes of Dr. Thompson, in His Accounts of these Doctrines', (Nicholson's) *Journal of Natural Philosophy, Chemistry, and the Arts*, 6 (1803) 25–31.

Kent, Andrew, ed. *An Eighteenth Century Lectureship in Chemistry*. Glasgow 1950.

— 'Gas-producing Solids, part one', *M & B Laboratory Bulletin*, 6 (no. 1, 1964) 8–10.

King, Lester S. 'Medical Theory and Practice at the Beginning of the 18th Century', *Bulletin of the History of Medicine*, 46 (1972) 1–15.

— *The Medical World of the Eighteenth Century*. Chicago 1958.

Kolakowski, Leszek *The Alienation of Reason, a History of Positivist*

Thought. Garden City, New York 1968.

Koyré, Alexander 'Les *Queries* de l'*Optique*', *Archives internationales d'histoire des Sciences,* 13 (1960) 15–29

Kuhn, Thomas 'The Caloric Theory of Adiabatic Compression', *Isis*, 49 (1958) 132–40

Larder, David F. 'An Unpublished Chemical Essay of James Watt', *Notes and Records of the Royal Society of London*, 25 (1970) 193–210.

Lauden, L. L. 'Thomas Reid and the Newtonian Turn of British Methodological Thought', in *The Methodological Heritage of Newton*, ed. R. E. Butts and J. W. Davis. Toronto 1970.

Lavoisier, A. 'De la combinaison de la matière du feu avec les fluids évaporables, et de la formation des fluides élastiques aëriformes', dated 5 September 1777, *Mémoires de l'Académie royal des Sciences,* 1777 (published, 1780) 420–32.

— 'Expériences sur la Respiration des Animaux, et sur les Changements qui Arrivent à l'Air en Passant par leur Poumon', dated 3 May 1777, *Mémoires de l'Académie royale des Sciences,* 1777 (published 1780) 185–94.

— 'Mémoire sur la combustion en général', dated 5 September 1777, *Mémoires de l'Académie royal des Sciences*, 1777 (published 1780) 592–600.

Lavoisier, A. and Laplace, P. S. 'Mémoire sur la Chaleur', dated 18 June 1783, *Mémoires de l'Académie royale des Sciences*, 1780 (published, 1784) 355–408.

Law, R. J. *James Watt and the Separate Condenser, An Account of the Invention*. London 1969 (A Science Museum Monograph.)

Leechman, William 'Account of the Life, Writings, and Character of the Author', in Hutcheson, Francis, *A System of Moral Philosophy*. 2 vols. Glasgow 1755.

Leicester, Henry M. *Development of Biomedical Concepts from Ancient to Modern Times*. Cambridge, Mass. 1974.

[Leslie], Patrick Dugud. *De Caloris Animalium causa*. Edinburgh 1775.

Leslie, Patrick Dugud *A Philosophical Inquiry into the Causes of Animal Heat: With Incidental Observations on Several Phisiological and Chymical Questions, Connected with the Subject*. London 1778.

Locke, John *An Essay Concerning Human Understanding,* ed. John Yolton. 2 vols. London 1965.

McGuinness, Arthur E. *Henry Home, Lord Kames*. New York 1970.

Mach, Ernst *Die Principien der Wärmelehre*. Leipzig 1896.

McKie, Douglas 'The "Observations" of the Abbé François Rozier (1734–93)—I', *Annals of Science*, 13 (1957, published 1958) 73–89.

McKie, Douglas and Heathcote, Niels H. de V. *The Discovery of Specific and Latent Heats*. London 1935.

Mackie, J.D. 'Glasgow University in the Eighteenth Century', in *An Eighteenth Century Lectureship in Chemistry*, ed. Andrew Kent (Glasgow 1950).

— *A History of Scotland*. Baltimore 1964.

— *The University of Glasgow*. Glasgow 1954.

Maclaurin, Colin *An Account of Sir Isaac Newton's Philosophical Discoveries*. London 1748.

Macquer, Pierre 'Observations sur la chaux et sur le plastre', *Mémoires de l'Académie royale des Sciences*, 1747 (published 1752) 678–96.

Magellan, J.H. *Essai sur la nouvelle théorie du feu élémentaire, et de la chaleur des corps*. London 1780. Published as the fourth *traité* in his *Collection de différens traités sur des instruments d'astronomie, physique*, etc. and republished in Rozier's *Observations sur la Physique*, 17 (1781) 375–86, 411–22.
— 'Extrait d'une Lettre de M. Magellan, de la Société Royale de Londres, sur les Montres nouvelles qui n'ont pas besoin d'être montées, sur celle de M. Mudge & sur l'Ouvrage de M. Crawford', *Observations sur la Physique*, 16 (1780) 60–3.
— 'Lettre de M. Magellan, à l'Auteur de ce Journal, sur le Mémoire suivant [i.e., his *Essai sur la nouvelle théorie du feu élémentaire*]', *Observations sur la Physique*, 17 (1781) 369–75.
Malouin, Paul 'Sur le Sel de la Chaux', *Mémoires de l'Académie royale des Sciences*, 1745, 93–106.
Manuel, Frank E. *A Portrait of Isaac Newton*. Cambridge, Mass. 1968.
Martine, George *Essays Medical and Philosophical*. London 1740.
Mayow, John. *Medico-Physical Works*, trans. A. Crum Brown and Leonard Dobbin. Edinburgh 1907.
Mendelsohn, Everett *Heat and Life; the Development of the Theory of Animal Heat*. Cambridge, Mass. 1964.
Metzger, Hélène *Les Doctrines Chimiques en France du début du XVIIe à la fin du XVIIIe Siècle*. Vol. I. Paris 1923.
— *Newton, Stahl, Boerhaave et la Doctrine Chimique*. Paris 1930.
Mossner, E. C. *The Forgotten Hume: Le bon David*. New York 1943.
— *The Life of David Hume*. Austin, Texas 1954.
Muirhead, James Patrick *The Life of James Watt, with selections from his correspondence*. 2nd ed. revised. London 1859.
— *The Origin and Progress of the Mechanical Inventions of James Watt—Illustrated by his Correspondence with his Friends and the Specifications of his Patents*. 3 vols. London 1854
Multhauf, Robert P. *The Origins of Chemistry*. New York 1967.
Munimenta Alma Universitatis Glasguensis, Records of the University of Glasgow from its Foundation till 1727. 3 vols. Glasgow 1854.
Murray, David *Memoires of the Old College of Glasgow*. Glasgow 1927.
Musschenbroek, Pieter van *Essai de Physique*, trans. P. Massuet. 2 vols. Leyden 1739.
Musson, A. E. and Robinson, Eric *Science and Technology in the Industrial Revolution*. Manchester 1969.
Newton, Sir Isaac. *Opticks, or a Treatise of Reflections, Refractions, Inflections & Colours of Light*. 4th ed. London 1730. Reprinted, New York 1952.
Olson, Richard 'Count Rumford, Sir John Leslie, and the Study of the Nature and Propagation of Heat at the Beginning of the Nineteenth Century', *Annals of Science*, 26 (1970) 273–304.
— 'The Reception of Boscovich's Ideas in Scotland', *Isis*, 60 (1969) 91–103.
Partington, J.R. *A History of Chemistry*. 4 vols. London 1961–70.
Partington, J.R. and McKie, Douglas 'Historical Studies on the Phlogiston Theory', Part I, *Annals of Science*, 2 (1937) 361–404; Part II, *Annals of Science*, 3 (1938) 1–58; Part III, Light and Heat in Combustion', *Annals of Science*, 3 (1938) 337–71.
Passmore, John *Hume's Intentions*. Revised ed. New York 1968.
Phillipson, Nicholas 'Culture and Society in the 18th Century Pro-

vince: The Case of Edinburgh and the Scottish Enlightenment', in *The University in Society*, ed. Lawrence Stone. 2 vols. Princeton 1974.

Phillipson, Nicholas and Mitchison, Rosalind, eds. *Scotland in the Age of Improvement*. Edinburgh 1970.

Playfair, Lyon. *A Century of Chemistry in the University of Edinburgh, being the Introductory lecture to the course of chemistry in 1858*. Edinburgh 1858.

Plummer, Andrew 'Experiments on Neutral Salts, compounded of different acid Liquors, and alcaline Salts, fixt and volatile, read 7 June 1739', *Essays and Observations, Physical and Literary*, 1 (1754) 315–40

Ramsay, Sir William *The Life and Letters of Joseph Black, M.D.* London 1918.

Rappaport, Rhoda 'G.F. Rouelle: An Eighteenth-Century Chemist and Teacher', *Chymia*, 6 (1960) 68–101.

Reid, Thomas *The Works of Thomas Reid*, ed. Sir William Hamilton. Edinburgh 1846.

Riddell, Henry 'The Great Chemist, Joseph Black, his Belfast Friends and Family Connections', *Proceedings of the Belfast Natural History and Philosophical Society*, Session 1919–20, 3 (1920) 49–88.

Ritchies, R. P. *The Early Days of the Royal College of Phistians, Edinburgh*. Edinburgh 1899.

Robinson, Bryan *Dissertation on the Aether of Sir Isaac Newton*. Dublin 1743 and London 1747.

Robinson, Eric and McKie, Douglas, eds. *Partners in Science: Letters of James Watt and Joseph Black*. London 1970.

Robinson, Eric and Musson, A. E. *James Watt and the Steam Revolution; a documentary history*. New York 1969.

Robison, John *A System of Mechanical Philosophy*, ed. David Brewster. 4 vols. Edinburgh 1822.

Rouelle, G.F. 'Sur les sels neutres', *Mémoires de l'Académie royale des Sciences*, 1744, 353–64.

Royal Medical Society 'History of the Medical Society of Edinburgh', in *List of Members, Laws, and Library-Catalogue, of the Medical Society of Edinburgh*. Edinburgh 1820.

Rush, Benjamin 'Eulogium upon Dr. William Cullen', in Rush, Benjamin, *Essays, Literary, Moral & Philosophical*. Philadelphia 1798.

— *Letters of Benjamin Rush; Vol. I: 1761–1792*, ed. L.H. Butterfield. Princeton 1951.

Russell-Wood, J. 'The Scientific Work of William Brownrigg, M.D., F.R.S. (1711–1800)—II', *Annals of Science*, 7 (1951) 77–94.

Schofield, Robert *The Lunar Society of Birmingham*. Oxford 1963.

— *Mechanism and Materialism: British Natural Philosophy in an Age of Reason*. Princeton 1970.

Scott, W.R. *Adam Smith as Student and Professor*. Glasgow 1937.

— *Francis Hutcheson, His Life, Teaching and Position in the History of Philosophy*. Cambridge 1900.

[Senac, J. B.] *Nouveau Cours de Chymie suivant les principes de Newton et de Stahll*. Published anonymously. Paris 1723.

Shaw, Peter *Chemical Lectures, Publickly read at London, In the Years 1731, and 1732; And at Scarborough, in 1733; For the Improvement of Arts, Trades, and Natural Philosophy*. London 1734.

— *Scheme for a Course of Philosophical Chemistry: Or the Application of Chemical Experiments to the Improvement of Natural Knowledge*. Published in *An Essay for Introducing a Portable Laboratory . . . by Peter*

Shaw, M.D. and Francis Hauksbee. London 1731.

Siegfried, Robert 'Lavoisier's View of the Gaseous State and Its Early Application to Pneumatic Chemistry', *Isis*, 63 (1972) 59–78.

Smeaton, W. A. 'Some comments on James Watt's published account of his work on steam and steam engines', *Notes and Records of the Royal Society of London*, 26 (1971) 35–42.

Smiles, Samuel *Lives of Boulton and Watt; principally from the original Soho MSS.* London 1865.

Smith, Norman Kemp *The Philosophy of David Hume.* London 1941.

Smout, T. C. *A History of the Scottish People,* 1560–1830. London 1969.

Stahl, George Ernest *Philosophical Principles of Universal Chemistry*, trans. Peter Shaw. London 1730.

Stone, Lawrence, ed. *The University in Society.* 2 vols. Princeton 1974.

Taylor, Brook 'An Experiment Proving the Expansion of the Liquid in the Thermometer to be as the Degree of Heat', *Philosophical Transactions of the Royal Society of London*, 32 (1723, no. 376) 291.

Thackray, Arnold *Atoms and Powers; An Essay on Newtonian Matter-Theory and the Development of Chemistry.* Cambridge, Mass. 1970.

Thompson, D'Arcy W. 'George Martine, an 18th Century Physician', *The Scotsman*, 27 December 1941, p.7.

Thomson, John. *An Account of the Life, Letters, and Writings of William Cullen, M.D.* 2 vols. Vol. 1 published 1832, republished with vol. 2. Edinburgh 1859.

Trevor-Roper, Hugh 'The Scottish Enlightenment', *Studies on Voltaire and the Eighteenth Century*, 58 (1967) 1635–58.

Tunbridge, Paul A. 'Jean André De Luc, F.R.S.', *Notes and Records of the Royal Society of London*, 26 (1971) 15–33.

Tytler, Alexandre F. *Memoirs of the Life and Writings of the Honourable Henry Home of Kames.* 2 vols. Edinburgh 1807.

Van Helmont, J. B. *Van Helmont's Works.* London 1664. A reissue of the *Oriatrike* (London 1662).

Watt, James *Correspondence of the Late James Watt on his Discovery of the Theory of the Composition of Water,* ed. J. P. Muirhead. London 1846.

Whytt, Robert *An Essay on the Virtues of Lime-Water in the Cure of the Stone.* Edinburgh 1752.

— 'Essay towards the Discovery of a safe Medicine for dissolving the Stone', *Medical Essays and Observations*, 5, part 2 (1744) 667–750.

Wightman, W. P. D. 'Wars of Ideas in Neurological Science—From Willis to Bichet, and From Locke to Condillac', in *History and Philosophy of Knowledge of the Brain and its Functions*, ed F.Poynter, London 1958.

— 'William Cullen and the Teaching of Chemistry', Part 1, *Annals of Science*, 11 (1955) 154–65; Part 11, *Annals of Science*, 12 (1956) 192–205.

Index